Infrared Spectroscopy of Symmetric and Spherical Top Molecules for Space Observation 2

Infrared Spectroscopy Set
coordinated by
Pierre Richard Dahoo and Azzedine Lakhlifi

Volume 4

Infrared Spectroscopy of Symmetric and Spherical Top Molecules for Space Observation 2

Pierre Richard Dahoo
Azzedine Lakhlifi

ISTE

WILEY

First published 2021 in Great Britain and the United States by ISTE Ltd and John Wiley & Sons, Inc.

ISTE Ltd
27-37 St George's Road
London SW19 4EU
UK

www.iste.co.uk

John Wiley & Sons, Inc.
111 River Street
Hoboken, NJ 07030
USA

www.wiley.com

Library of Congress Control Number: 2021938325

British Library Cataloguing-in-Publication Data
A CIP record for this book is available from the British Library
ISBN 978-1-78630-652-4

MIX
Paper from responsible sources
FSC® C013604

Contents

Foreword

Spectroscopy is the high road to physico-chemical measurements of astrophysical objects, such as the interstellar medium, stars or planets and exoplanets. In recent decades, the understanding of the increasingly finer details of the interaction between matter and radiation, as reflected in the spectral range by effects, sometimes unexpected, on the observed spectra, has led to outstanding advances in spectroscopy, from ground- or space-based observations. For this purpose, increasingly sophisticated instruments were developed for space probes, initially inspired by laboratory instruments, in order to cope with the severe constraints of space observation and exploration.

The challenge of laboratory spectroscopy presented in this book, the fourth volume in this series, resides not only in registering increasingly complex spectra, even for the simplest molecules observed at high temperature or excited to high vibrational states, but also in understanding specific molecular mechanisms, such as the spectroscopy of molecules in rare gas matrices, clathrates or physico-chemical mechanisms related to the adsorption on graphite substrates. These mechanisms can be used to extrapolate to the very specific conditions of the interstellar medium, where the very rich chemistry discovered by the observations in the millimetric range in the last 50 years is highly dependent on ion–molecule–substrate mechanisms.

Spatial instrumentation in remote spectroscopy offers the possibility of exploring planets with atmospheres, particularly by means of instruments with medium to high resolution. However, Earth observations should be treated separately, as they have other objectives, particularly operational ones, and a different approach and constraints compared to the various distant exploration missions, even though they involve similar categories of instruments. Fourier transform spectrometers have measured, in particular, the atmospheres of Mars, Venus and giant planets and their satellites, thanks to their performances in combining large spectral extent and (relatively) high resolution. The pioneering missions were Mariner, Venera and Voyager. Grating or prism spectrometers, particularly those using adjustable prisms (AOTF – Acoustic Optic Tunable Filter), have also provided remarkable results on Mars and Venus. After the first observations using variable circular filter spectrophotometry (IKS on Vega probe), comets, with their highly specific atmosphere, were observed with grating spectrometers (VIRTIS on Rosetta), which led to a better understanding of their composition and revealed surprising isotopic ratios.

Substrate spectroscopy issues are connected with planetary studies in order to understand the adsorption chemistry at play in the interstellar medium, as well as in the very high atmosphere of planets: one of the key results of the Cassini mission showed that the complex hydrocarbons observed on Titan were generated not only from the photochemistry of nitrogen or methane atmosphere at the level of high clouds, but also in the ionic chemistry of the very high atmosphere, measured directly using mass spectrometry from the instrument aboard the Cassini during its flight over Titan.

There is no doubt that there is still room for embedded spectroscopy to significantly evolve into future space missions: Raman spectroscopy methods, still fresh on planetary probes (Mars Curiosity and Perseverance missions), LIDAR observations or the CRDS technique will lead the way to new advances, which laboratory techniques and theoretical models will help to interpret. The joint progress of various disciplines of chemistry, spectroscopy and of Earth and universe sciences is a remarkable example of a coordinated multidisciplinary approach that involves the joint research efforts of laboratories of various specializations. This book, the fourth volume in the series "Infrared Spectroscopy of Molecules for Space Observation", focuses on the context and results of this research.

Pierre DROSSART
Paris Institute of Astrophysics, CNRS, Sorbonne University
June 2021

Preface

The observable universe is visible thanks to the light that reaches us from the point being observed. These extreme points form the contour of a sphere, whose limit lies at the cosmological horizon, with the Earth at its center. Other observers, located elsewhere in the universe, see a different observable sphere of the same radius. It is a relative notion. Let us recall that in cosmology, the unit of measurement for distances is the light-year, which is the distance that light travels in one year, which corresponds to about $9.5\ 10^{12}$ km. The megaparsec, which amounts to 3.26 million ($3.26\ 10^{6}$) light-years, is another unit of distance used particularly in extragalactic astrophysics. The Standard Model of Cosmology elaborated at the beginning of this century, by 2000, is perhaps the most successful model presently that offers a description of the evolution of the universe, in terms of the major stages in the history of the observable universe, as well as its current composition resulting from astronomers' observations. In order to explain the myriad of observable galaxies and planetary systems, suns and black holes, cosmologists proposed an inflationary model.

As mentioned in the prefaces of the previous volumes, the universe is a preferred spatial environment for astrophysicists, cosmologists, astronomers and physicists, both for observations and for testing the theory of general relativity of Albert Einstein by means of predicted and observed phenomena, elaboration of new theories and their expected confirmation by observations. This is the context in which the Nobel Prize for Physics was awarded to Sir Roger Penrose in 2020, for his work on the existence of black holes based on the theory of general relativity, and to astronomers Andrea Ghez and Reinhard Genzel, for their research on the phenomena observed at the center of the Milky Way galaxy in the Sagittarius A region, where the existence of a black hole was predicted, and who showed that, indeed, a supermassive black hole was present there.

This book, the fourth volume of this series, offers an overview of several experimental instruments used for the observation and analysis of a physico-chemical system, in order to determine relevant characteristics such as the position, intensity and width of IR absorption lines. Some instruments rely on the principle of light interferences, such as the Fourier transform infrared (FTIR) spectroscopy, or on the operating principle of a resonant laser cavity, such as the cavity ring-down spectroscopy (CRDS), or the frequency comb techniques. The species present in a given atmosphere, such as the tholins on Titan (satellite of Saturn), can be prepared *in situ* in the laboratory and studied by spectroscopic ellipsometry, in order to determine their optical properties. This data can be used to calculate the albedo, for example. A spectrometer employing an acoustic optic tunable filter (AOTF) was used to obtain IR data on Mars and Venus from Mars Express and Venus Express missions. Data analysis made it possible to identify the nature of certain observations in the IR spectrum and to eliminate possible ambiguities. Hence, the absorption by the 628 CO_2 isotopic species could be confirmed at the expense of a possible absorption by a C–H bond in the same IR spectrum. Finally, LIDAR is the preferred instrument for observing distant objects and species present in the atmosphere of planets, in a given environment.

This volume, however, focuses on the applications of the theoretical models described for diatomic and triatomic molecules to symmetric tops such as NH_3 and/or spherical tops such as CH_4, when they are in an environment that may be present in the universe, i.e. when they are trapped in a nanocage of rare gas matrix, clathrate, fullerene or adsorbed on a graphite substrate. All of these models are theoretical instruments for the simulation-based observation of a molecule. The results of these theoretical models should be compared with those of experimental observations, using instruments that are specifically developed for various observation situations. The latter may be related to the observation site, such as a laboratory, an observatory or a space probe, or a telescope in space. They may also relate to the type of molecular system observed, molecules or chemical species (ions, radicals, macromolecules, nanocages, etc.) that compose, on the one hand, the atmospheres of planets, the Earth or the planets of the solar system and possibly their satellites, and on the other hand, the interstellar media, the comets or the exoplanets throughout the spatial extent of the universe, estimated by the cosmologists as 10^{23} km.

As underlined in Volumes 1, 2 and 3, theoretical and experimental approaches to spectroscopies led to the development of methods and instruments for the observation and analysis of the spectral characteristics of chemical species, molecules, radicals and ions in specific environments. Volumes 1 and 2 of this series [DAH 17] describe the theoretical models developed in order to determine the absorption spectra of diatomic molecules (homonuclear or heteronuclear), triatomic molecules (linear or nonlinear, symmetric or non-symmetric) in the gas phase and particularly when they are trapped in the nanocages of inert matrices of rare gases, or hydrate clathrates at a very low temperature. Volume 3 deals with the spectroscopy of symmetric and spherical tops in the gas phase.

This book (Volume 4) is a continuation of Volume 3, focusing on the application of theoretical models for the simulation of IR spectra of symmetric or spherical top species evolving in various media. The method of an extended substitution model is explained in terms of the determination of the type of symmetry of the environment in the immediate vicinity of the trapped molecule. The objective is to propose theoretical spectra that match with experimental IR data and hence identify molecules based on transitions and profiles, not only in the gas phase, but also when they are constrained to evolve in an environment, such as a nanocage or a surface.

Chapter 1 provides a brief overview of the instruments developed and used in the laboratory for the study or observation of molecules, FTIR spectroscopy or laser cavity spectroscopy. An example of embedded instruments such as SPICAM, SPICAV or SOIR aboard an orbiter or a space probe serves to illustrate the instrumental context of space observation and of the international collaboration required to develop measurement instruments, as well as the analysis methods for the identification according to the scientific norms of the molecules that may be present in the probed atmosphere, such as on Mars or Venus during Mars Express and Venus Express missions. Aerosol characterization by spectroscopic ellipsometry is also discussed as a non-standard method. The LIDAR technique, which is used for observing the terrestrial atmosphere, is also described.

Chapter 2 presents various contributions to the interaction potential energy between the studied molecule and its solid environment, considering the hypothesis of binary interactions: “studied molecule – molecule or atom of the environment”. The quantum “dispersion–repulsion” contribution is modeled by an atom–atom Lennard-Jones potential energy type, while the electrical contribution is modeled by a charge–charge potential energy in the case of clathrate nanocages or multipole–multipole in the case of nanocages of rare gas, fullerene or graphite substrate matrix.

Chapter 3 provides a description of the substitution model applied to the study of NH_3 in rare gas matrices. An atom–atom potential is used to calculate the interaction between the trapped molecule and its environment. A numerical method is applied to determine the perturbed motions of the molecule. The IR spectral profiles are determined and compared to the experimental spectra. The influence of lattice phonon modes on the shift and width of spectral lines is also discussed.

Chapters 4 and 5 provide a description of the extended substitution model, based on the effect of the local symmetry around the equilibrium position of the molecule in the cage for calculating IR spectra in natural nanocages, such as clathrates and fullerene. This model relies on the substitution model used for building theoretical models in IR spectroscopy in rare gas matrices. The model is applied to CH_4 and NH_3 molecules included in clathrate nanocages (Chapter 4) and NH_3 in a fullerene nanocage.

In Chapter 6, the theoretical models developed in Volumes 1 and 2 to deal with the adsorption of diatomic and triatomic molecules on a graphite substrate, are applied to the NH_3 molecule in order to determine the IR absorption spectra at a very low temperature. This substrate is often used to model the surface of interstellar dust grains.

Pierre Richard DAHOO
Azzedine LAKHLIFI
June 2021

1

IR Spectra in Space Observation

The IR (infrared) range corresponds to the frequency band between 10^4 GHz and 5×10^5 GHz, therefore to wavelengths ranging between 0.78 μm and 1,000 μm. According to the ISO system (ISO 20473, 2007), this range is divided into three regions corresponding to near IR from 780 nm to 3,000 nm (sunlight), mid-IR from 3 μm to 50 μm (thermal source) and far IR from 50 μm to 1,000 μm (source in astronomy). For a given electronic state, the interaction between an electromagnetic wave in this range and a physical system in gas, liquid, solid or possibly plasma state is reflected by a transition between two vibrational and/or rotational energy states of the system, which are discrete and/or in the continuum. The possible physico-chemical processes, such as the absorption and emission of photons, elastic (Rayleigh) scattering and inelastic (Raman, Compton) scattering, dissociation, ionization and various possible combinations and interferences of these phenomena can be observed using optical instruments, such as spectrometers or telescopes. These instruments can operate on site in a laboratory or from an observatory or in the space (aboard a space probe). Depending on the latter sites and the system under observation, various options should be considered for the successful completion of an observation process.

1.1. Introduction

In a laboratory, experimental IR spectroscopy of atmospheric molecules is mainly studied in the gas phase.

Taking into account the shape of the absorption line (Gaussian by Doppler effect, Lorentzian by pressure-induced broadening) [DAH 88, DAH 97, DAH 16a,

DAH 17], the intensity of a vibration–rotation absorption line in the gas phase is given by the expression:

$$S_{if}^{g} = \int_{\sigma_{\min}}^{\sigma_{\max}} \frac{8\pi^3}{3hc} \sigma_{if} f(\sigma - \sigma_{if}) N \left[1 - \exp\left(-\frac{hc}{k_B T} \sigma_{if} \right) \right] g_i \frac{\exp(-hc\sigma_i / k_B T)}{Q(T)} R_{if}^2 \, d\sigma \quad [1.1]$$

where σ_{if} is the wave number (in cm^{-1}) of the considered rovibrational transition, N is the number of molecules per unit volume, $hc\sigma_i$ is the energy of the initial level, $Q(T)$ is the total partition function of the molecule at temperature T, R_{if} is the moment of rovibrational transition, and finally g_i is the degeneracy due to the nuclear spin of the initial level. The shape of the line is a Voigt function, a convolution product of a Gaussian and a Lorentzian, as the Doppler effect and the broadening by pressure are concomitant, and its integral over the absorption domain is equal to 1 ($1 = \int_{\sigma_{\min}}^{\sigma_{\max}} f(\sigma - \sigma_{if}) \, d\sigma$).

In condensed media, when the observed molecule is trapped in a rare gas matrix or in a nanocage of a solid medium, a corrective term reflecting the effect of the refractive index n of the solid medium must be taken into account. If the rotation–vibration interaction is ignored, and the fundamental state is not degenerate ($g_0 = 1$), the intensity of a vibrational line at low temperature, in relation with a molecule is written as:

$$S_{if}^{m} = \int_{\sigma_{\min}}^{\sigma_{\max}} \frac{1}{n} \left(\frac{n^2+1}{3} \right)^2 \frac{8\pi^3}{3hc} \sigma_{if} f(\sigma - \sigma_{if}) \left| R_{if} \right|^2 d\sigma \quad [1.2]$$

where R_{if} is the vibrational transition moment. Written in this form, relation 1.2 provides essential information about the dipole transition moments of the vibrational modes in a solid matrix. These quantities can be extracted experimentally from the integrated intensity of the absorption spectrum:

$$S_{if}^{m} = \frac{1}{2lN} \int Ln \frac{I_0(\sigma - \sigma_{if})}{I_t(\sigma - \sigma_{if})} d\sigma = \frac{1}{2lN} I_{if} \quad [1.3]$$

where N is the number of molecules per cm^3, $2l$ is the path length in the sample, and the term under the integral is the integrated absorption measured over the IR absorption spectrum domain. The calculation of physical quantities, such as the lifetime of a level based on the integrated absorption, is explained in chapters 6 and 7 of [DAH 16a] and in the Phd dissertations [ABO 73a, DUB 75, GAU 80, BOI 85] for molecules inserted in a solid rare gas matrix. For dipole moments, the method is given, for example, in [DAH 88] for CO_2 and N_2O molecules in the gas phase, or [GHE 79] for the CH_4 molecule.

Chapter 3 of Volume 1 of this series provides a description of the method for the measurement and analysis of an absorption line based on the Bouguer–Lambert law, applied to determine the infinitesimal amount of energy absorbed by an elementary thickness of matter. For an intensity $I(\sigma)$ of a radiation of wave number σ traveling through an elementary thickness dl of a gas under pressure P, the resulting variation $dI(\sigma)$ is proportional to $I(\sigma)$ and dl. These hypotheses lead to the following Bouguer–Lambert law:

$$dI(\sigma) = -k(\sigma, P)\, I(\sigma)\, dl \qquad [1.4]$$

where $k(\sigma, P)$ is the absorption coefficient at wave number σ and pressure P. The intensity of the radiation transmitted at wave number σ results from the integration of the previous equation:

$$I_t(\sigma) = I_0(\sigma)\, e^{-k(\sigma,P)l} \qquad [1.5]$$

where $I_0(\sigma)$ is the intensity of the radiation incident on the gas and l is the gas thickness through which the radiation travels.

The actual shape of an absorption line is defined by the following function:

$$A(\sigma) = \frac{I_a(\sigma)}{I_0(\sigma)} = 1 - \frac{I_t(\sigma)}{I_0(\sigma)} \qquad [1.6]$$

which does not take into account the instrumental line shape (infinite resolution hypothesis), with $I_a(\sigma) = I_0(\sigma)\text{-}I_t(\sigma)$, where $I_a(\sigma)$ is the intensity of the absorbed radiation.

The previous two equations lead to:

$$A(\sigma) = 1 - e^{-k(\sigma,P)l} \qquad [1.7]$$

Moreover, the profile of the spectral line is defined by the variation of the absorption coefficient $k(\sigma, P)$, which depends on the wave number σ and is given by:

$$k(\sigma, P) = \frac{1}{l} \ln\left(\frac{I_0(\sigma)}{I_t(\sigma)}\right) \qquad [1.8]$$

For a number of active molecules absorbing or emitting weak radiation, leading to the property $k(\sigma, P)\, l \ll 1$, the actual shape of the line coincides with the profile and is expressed as:

$$A(\sigma) \approx k(\sigma, P)\, l \qquad [1.9]$$

This expression is valid in the real space of the spectrum. In the measurement space, the effect of the instrumental line shape leads to an apparent shape, which is the convolution product of the instrumental line shape and the actual shape. A deconvolution process is required to retrieve the actual spectrum.

The integrated absorption coefficient of the line is defined from its actual profile, by integration over the absorption domain, which is as follows:

$$S = \int_0^{\infty} k(\sigma, P)\, d\sigma \qquad [1.10]$$

If the active molecules absorb independently of one another (Beer's hypothesis), this coefficient is proportional to the pressure and is written as:

$$S = S^0 P \qquad [1.11]$$

where S^0 is known as the "line intensity".

It is important to note that Beer's hypothesis is not generally verified for each wave number, and as a result, the integrated absorption coefficient is not proportional to the pressure for each wave number. This experimentally determined coefficient is related to the theoretical coefficients expressed by equations [1.1] and [1.2].

The absorption lines corresponding to transitions between energy levels can be determined experimentally, either by frequency-resolved spectroscopy using a spectrometer or an interferometer, or by laser absorption over the absorption spectral range of the line. The recorded spectra are the convolution product of the instrumental line shape and the actual shape of the absorption line. A deconvolution of these spectra is required before the analysis, in terms of the intensity, width and center of the line. The procedure is described in Appendix 1 (section 1.6.1) and relies on the application of the sampling theorem.

Building a setup for experimental observation requires three elements.

The first element is a set containing a light source and a dispersive element for the frequency scanning of the spectral range, if the source is polychromatic. There are classical sources (spectral lamp, black body, etc.) or quantum sources (lasers). The classical light source can be a natural heat source, such as the light from a star or an artificial one, or light emitted by a gas lamp or a heated filament. These are wide spectral band polychromatic sources, whose maximum intensity corresponds to a given wavelength λ_m, which depends on the temperature T according to the law $\lambda_m T$ = constant. For a black body, the constant in Wien's displacement law is equal to 2.897 773 10^{-3} mK. The laser quantum sources are also used in IR spectroscopy in experiments studying system dynamics under radiation, such as laser-induced fluorescence, double resonance (DR), Rayleigh or Raman scattering or the dynamics of radiative or non-radiative relaxation of the populations of energy levels in probe–pump experiments. These are monochromatic or quasi-monochromatic sources of narrow spectral band. A tunable laser can be used to determine the absorption spectrum around a frequency defined by the transition between two energy levels. Resonant cavity spectroscopy, developed as a result of efforts to improve laser sources, involves the insertion of an absorbing molecule in the laser cavity, and then the identification of the absorption due to losses in the laser gain curve.

The second element is the system to be analyzed can be a fluid (gas or liquid), a solid or a plasma medium in a suitable laboratory chamber, or in a natural medium in space. Laboratory settings sometimes use absorption cells filled with fluid (gas or liquid) or a sample holder's surface on which the solid medium is deposited. A multiple-reflection cell is used to increase the absorption path when the transition is weak, in order to reach a convenient signal-to-noise ratio (SNR) for the observation of the absorption line.

Finally, the third element is a data detection and acquisition system that translates the transition induced at each frequency of the incident light on the studied systems into an electrical signal, which after analysis allows the characterization of the system based on its spectral signature.

In formal terms, the experimental setup includes an object space, which is the illuminated system, or an image space, which contains the devices for detection and those of the optical system, which is the set of optical devices in the measurement instrument. IR vibrational and rotational spectroscopy focuses on the processes of photon absorption or emission in a spectral range of low energy molecular spectroscopy. This domain is broader than that corresponding to the IR range, since it spans from microwaves to ultraviolet radiation.

The instruments used in IR spectroscopy essentially rely on electromagnetic wave interferences, in order to separate the spectral components of the radiation from a classical light source and determine the effect of the observed system on each spectral element used as a probe. These are generally prism spectrographs, grating spectrometers, Michelson interferometers or frequency- and time-resolved laser spectroscopy, since laser development [DEM 96].

This chapter focuses on observation and measuring techniques. The techniques that are essentially used in laboratories for measuring campaigns involve interferential spectroscopy, the use of gratings, laser spectroscopy or polarized light spectroscopy. The chapter ends with an example of the device used on the Mars Express and Venus Express orbiters.

Section 1.2 is dedicated to Fourier transform infrared spectroscopy (FTIR). Given the rich bibliographical references on the subject, only the first works in this field are referenced [JAC 60, BRA 95, CON 66, LOE 66, HUN 68, BEL 72a, CHA 73, GRI 75, GRI 86, DEM 96, JAM 02]. This section includes a description of the operating elements related to technical aspects in the setup for observation or data analysis. Two types of interferometers are described, a laboratory interferometer and a commercial interferometer.

First, a Connes-type Fourier transform interferometer (FTIR) with a very large path difference is described. This apparatus was built at the Molecular Spectronomy Laboratory at the Sorbonne University in Paris, Pierre and Marie Curie campus (formerly Pierre and Marie Curie University). This instrument was operational for the study of gas phase molecules in the atmosphere until the beginning of the 21st century, under the leadership of L. Henry, according to a concept proposed by P. Connes and developed by Valentin and Michel. A model of this type of interferometer is still operational in the Molecular and Atmospheric Spectrometry Group at Reims University. It is important to note that P. Connes, who worked at the Aimé-Cotton Laboratory at Paris Saclay University (formerly Paris South University, of Orsay) and the Director of LAC of that time, Jacquinot have made significant contributions to the development of Fourier transform spectroscopy [CON 92, CON 95].

A commercial type (Brucker IFS 113v) FTIR used in research work on the spectroscopy of molecular species isolated in an inert matrix is also described. The experimental device was used to study a certain number of atmospheric molecules using the isolated matrix technique, by the group led by L. Abouaf-Marguin and L. Schriver-Mazuoli at the Laboratory of Molecular Physics and Applications of Sorbonne University in Paris (formerly Pierre and Marie Curie University in Paris) at the beginning of the 1990s [BRO 93a, ABD 93, LUG 94, DES 94, JAS 95, RAD 95, DAH 97, BAH 97, HAL 98, CHA 00a, CHA 01, COA 03].

The spectroscopy of laser absorption in a resonant cavity is another example of spectroscopy that differs from the classical frequency-resolved and dispersion spectroscopy, using prisms or interference with echelette grating. Further examples are the resonant cavity laser absorption spectroscopy of the intracavity laser absorption spectroscopy (ICLAS) type, or the cavity ring-down spectroscopy (CRDS). These techniques can be used to measure weak absorptions. In France, these techniques were mainly developed by the Interdisciplinary Laboratory of Physics at Joseph Fourier University of Grenoble (LIPhy, former Spectrometry Laboratory) [STO 52, CHE 83, CAM 86, CAM 90, CAM 91, ROM 93, ROM 97a]. It is worthwhile to highlight the pioneering work of the LIPhy researchers in the development of cutting-edge laser techniques and instruments for molecular absorption spectroscopy, such as the continuous wave cavity in ring-down spectroscopy (CW-CRDS) [MOR 05], optical-feedback cavity-enhanced absorption spectroscopy (OFCEAS) [ROM 97a] and mode-locked cavity-enhanced absorption spectroscopy (ML-CEAS) [GHE 02], which were invented or developed for the first time at LIPhy, or the use of a quantum cascade diode [GOR 13] or frequency combs to improve the limits of detection [MON 15]. One section focuses on frequency comb spectroscopy (FCS), which is increasingly used to study molecules [HÄN 06, HAL 06, CUN 01, FOR 19, DID 20].

Section 1.4 provides examples of techniques that were applied for the study of molecules in the atmospheres of planets other than the Earth. It includes a brief description of the spectroscopic ellipsometry technique [AZZ 77], which is used for the study of materials deposited as thin layers on substrates. This technique was used to determine the optical indices of aerosols present in the atmospheres of planets or satellites such as Titan ("tholins") [SCI 10, CAR 12, MAH 11, SCI 12, MAH 12a, MAH 12b, MAH 12c, MAH 14]. The final part of the section briefly describes the SOIR device used on the Venus Express for the study of the atmosphere of Venus. For the observations conducted from a space probe, an example is the observation device referred to as SPICAM ("Spectroscopy for Investigation of Characteristics of the Atmosphere of Mars") or SPICAV/SOIR (Solar Occultation in the Infra-Red) aboard the orbiters (Mars Express/Venus Express) and developed (formerly Aeronomy Service SA and former Center for Earth and Planetary studies) in collaboration with Belgian and Russian partners, under the main coordination of Bertaux [NEV 06, BER 07, MAH 08, VIL 08a].

1.2. Fourier transform spectroscopy

Since the development of the first interferometer by Michelson at the end of the 19th century [MIC 81, MIC 82], interferometric spectroscopy or Fourier transform spectroscopy using interferometers has gradually replaced classical spectroscopy, by grating spectrometers for the measurement and analysis of absorption spectra in the

IR range. The technique of interferometric spectroscopy, later referred to as Fourier transform spectroscopy, was proposed by Jacquinot [JAC 60, CON 92] in the 1960s. The state-of-the-art for the optical field in France in the 1960s can be found in [AMA 62]. Jacquinot [JAC 60] describes the methods for selecting wavelengths by the frequency of interferometric modulation. The spectrum is not directly recorded in its common form, but in the form of an interferogram that is the Fourier transform of the spectrum. Similar to the photographic methods, the main advantage of this method is that the time required for recording the interferogram does not depend on the width of the spectral range. The resolution and the luminosity of the measuring process are considered, as well as the various methods for spectrum reconstruction from the interferogram.

The evolution of grating spectrometers or spectrographs towards interferometers was made possible by the works of the following scientists:

– Fellgett [FEL 49, FEL 58] on the advantages of the "multiplex" method or Fellgett's advantage; at each time t, an FTIR spectrometer simultaneously records the spectral data at all of the wavelengths at high resolution, over an extended spectral range. A grating spectrometer or a spectrograph measures the intensity of an absorption or emission line over a very narrow spectral range around a given wavelength, during the measurement of a signal whose noise is mainly that of the detector. A multiplex measurement like that registered by a Fourier transform spectrometer can improve the SNR, compared to that of a scanning grating spectrometer, of the order of the square root of m, where m is the number of points in the spectrum.

– Jacquinot [JAC 48, JAC 54a, JAC 54b, JAC 60] on the optical throughput advantage or the Jacquinot advantage, which is linked to the intrinsic brightness of interferometers. An interferometer transmits the whole IR radiation that travels through the sample to the detection system. The circular apertures of the interferometer do not limit the illumination of the sample, unlike the slits in a grating spectrometer, which are required for a good resolution. Indeed, the finer the slit, the better the resolution, but in return, IR light is attenuated. The advantage of a Fourier transform spectrometer differs depending on wave number and resolution.

– J. Connes and P. Connes [CON 61a, CON 61b, CON 61c, CON 61d, CON 66] on the advantage of precise sampling through the interference fringes of a laser source, allowing the continuous recording of the measured spectrum; this technique can be used to precisely determine the optical path difference and also the data digitization during the interferogram acquisition. It is important to note that prior to the availability of the first gas lasers, P. Connes used a monochromatic cadmium lamp beam propagating through the interferometer, along the same path as the primary sample beam.

– Finally, Cooley–Tuckey [COO 65] on how Cooley–Tuckey fast Fourier transform (FFT) algorithm can be used to numerically obtain the inverse Fourier transform of the recorded interferogram, taking into account the constraints related to the instrumental line shape, apodization and optimum sampling interval, thanks to the works of J. Connes [CON 61]. The term Fourier transform infrared spectroscopy originated in this mathematical processing required for the conversion of interferogram data into the line spectrum.

FTIR spectrometer is a commonly used term nowadays. This technique is sometimes divided into two phases: the first one involves the recording of an IR spectrum of a sample as an interferogram, and the second one, the IR spectrum calculation by Fourier transform of the interferogram. The commercial brands of instruments such as Brucker, Bomem or Nicolet record the interferogram in numerical form in a file and then the corresponding spectra can be displayed on a screen after numerical Fourier transform by FFT.

1.2.1. *Principle of IR spectrum acquisition by interferometry*

[DAH 21] gives a description of a homodyne monochromatic interferometer using a laser as the source of monochromatic light of wavelength λ. It is made up of a fixed beamsplitter and a compensator, parallel to each other (Figure 1.1). After traveling through the beamsplitter–compensator set, the light beam splits into two paths (r_1 and r_2) and is then reflected by two parallel mirrors, a fixed one denoted by M_1 and a mobile one, denoted by M_2 (Figure 1.1). It can be noted that the rays reflected by M_2 travel through the beamsplitter twice, while those reflected by M_1 go through the beamsplitter once, leading to a phase shift. The compensator thus compensates and symmetrizes the optical paths in the two Michelson arms.

The beam recombined in the beamsplitter–compensator set generates interference fringes on the detector. These are constructive or destructive interference patterns, depending on the path difference Δx between the two rays, depending on the relative position of the two mirrors M_1 and M_2. The intensity can be written as a function of the electrical field E_0 of the radiation, which is as follows:

$$\frac{I(\Delta x)}{\varepsilon_0 c} = \frac{1}{2}\left|\vec{E}_0\right|^2 + \left|\vec{E}_0\right|^2 \cos\left(\frac{2\pi}{\lambda}r_2 - \frac{2\pi}{\lambda}r_1\right) = \frac{1}{2}I_0(\sigma)(1 + \cos(2\pi\sigma n(\Delta x)) \qquad [1.12]$$

where c is the speed of light and ε_0 is the vacuum permittivity.

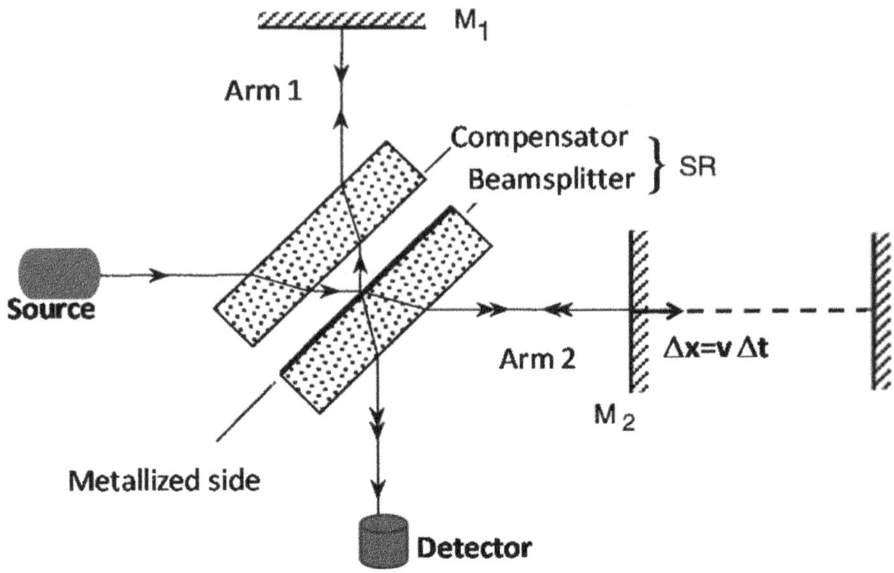

Figure 1.1. *Interferometer diagram. For a color version of this figure, see www.iste.co.uk/dahoo/infrared2.zip*

Interference fringes are formed in the observation plane, coinciding with the detector surface. When the mobile mirror moves, they are transformed into a modulated electric signal that follows the light intensity variations. This signal is directed towards a counter that counts the number of displayed fringes. A fringe is a complete cycle in the variation of light intensity, which passes periodically from a maximum value to zero and then to a maximum value. Each cycle corresponds to a displacement such that $\sigma n \Delta x = 2$. If the path difference Δx corresponding to a cycle and the index n of the surrounding medium are known, the wave number σ or the wavelength λ can be deduced. Conversely, if n and λ are known, Δx can be deduced.

In an interferometer used for studying the spectrum of molecules absorbing light in the IR range, the source used emits a complex polychromatic radiation. If the incident monochromatic flux of wave number σ is denoted by $\phi(\sigma)$, then for a path difference $\Delta = n\Delta x$, the expression of the emergent flux is written as:

$$\phi(\Delta) = \frac{1}{2}\int_0^{\infty} \phi(\sigma)(1 + \cos(2\pi\sigma\Delta))d\sigma \qquad [1.13]$$

The variable part $\phi(\Delta) = \frac{1}{2}\int_0^{\infty} \phi(\sigma)(\cos(2\pi\sigma\Delta)d\sigma$ is referred to as the interferogram. In practice, it is the intensity $I(\Delta)$ that is recorded as a function of the path difference Δ, and whose expression is given by:

$$I(\Delta) = \int_0^{\infty} B(\sigma)\cos(2\pi\sigma\Delta)d\sigma \qquad [1.14]$$

where $B(\sigma)$ is the recorded spectrum.

For a given path difference, the detector receives a signal that contains all of the information about the spectrum, and the energy corresponding to each spectral element is integrated over the measurement time T. In the classical spectrographs or spectrometers, the energy corresponding to the same spectral element is integrated over the time interval T/N of the measurement, where N represents the number of spectral elements received by the detector.

Then, the spectrum is reconstructed by inversion using a Fourier transform such that:

$$B(\sigma) = \int_0^{\infty} I(\Delta)\cos(2\pi\sigma\Delta)d\Delta \qquad [1.15]$$

The calculation of $B(\sigma)$ with formula [1.15] implies that $I(\Delta)$ is known over all of the values ranging between 0 and infinity. In practice, the maximum path difference is fixed at a value Δ_M that is the limit of the displacement of the mobile mirror. Under these conditions, the product of $I(\Delta)$ by the rectangular function $H(\Delta) = \text{rect}(\Delta/2\Delta_M)$ is retrieved instead. Consequently, the calculation of $B(\sigma)$ actually corresponds to that of the Fourier transform of the product $I(\Delta)\text{x}H(\Delta)$, or the convolution product $B(\sigma)*A(\sigma)$, where $A(\sigma)$ is the apparatus function of the interferometer, which depends on the maximum possible path difference. It is important to note that the higher this value, the narrower the apparatus function.

1.2.2. *Design and operation of a long path difference interferometer*

This section mainly describes the optical and electronic techniques developed for the operation of an interferometer over a long path difference. In order to guarantee measurement stability and increase the SNR over a very long distance corresponding to a path difference of 14 m, the step-by-step interferometer, operating in continuous regime, built by the molecular spectronomy laboratory, is based on the principle of Connes interferometer [CON 66, PIN 67, CON 75, PLA 86].

It features a device composed of a concave mirror and a plane mirror placed in its focus (cat's eye) [CUI 67], so that the ray reflected by the mobile mirror remains parallel to the ray incident on the fixed mirror throughout the displacement path, and can also be modulated in position by a piezoelectric ceramics support.

The path difference is measured using a He–Ne laser stabilized in the central dip of its gain profile ("Lamb dip" or saturated absorption). The monochromatic laser beam of wavelength λ_0 is concentric to the beam of the polychromatic IR source illuminating the sample being studied.

The two beams at the output of the beamsplitter–compensator device each travel through a quarter-wave plate oriented so that each beam is circularly (one right-handed and the other left-handed) polarized [DAH 21]. Then, when these beams recombine, a linearly polarized wave is obtained, whose azimuth depends on the path difference. The linear wave is then split into two beams and sent to two photomultipliers through $\pi/4$ polarizers. The signals at the output of the photomultipliers measure the azimuth of the linearly polarized wave, hence the rotation position of this polarization, which depends algebraically on the path difference.

To achieve the step-by-step mode, the path difference is modulated at a frequency $f_0 = 100$ kHz by applying a sinusoidal voltage to piezo-electric ceramics glued on the mirror of one of the cat's eyes. Thanks to this modulation, a detection synchronous with the position of this mirror is achieved. The signals at the output of the photomultipliers are directed towards a broad band filter centered on the frequency f_0 and with a very steep slope to dampen the harmonics in the output signals. These signals are processed and put in phase quadrature at the frequency f_0 and are then added to obtain a signal of constant amplitude, whose phase depends on the path difference Δ. Since the carriage of the mobile part of the interferometer moves at constant speed, the frequency of the detected signal varies depending on the direction of motion. This property is used to precisely localize the position of the carriage. The signal of constant amplitude and variable phase is applied to a logical circuit that sends a pulse at each zero crossing. These pulses are then subtracted from pulses synchronous with the pulses of the signal at f_0. The adder output represents the error signal in digital form. The response range is $\pm$ 20 µm. This error signal is sent through a control loop after passage through a DAC (digital to analogue converter). The path difference variations are obtained by sending pulses to the adder. Each pulse creates a path difference of $\lambda_0/8$, where λ_0 is the wavelength of the He–Ne laser serving as reference.

An optical system composed of mirrors allowed the use of absorption cells of various lengths and designs (multiple reflection cells of paths ranging between 5 m and 40 m, or cells of lengths below 1 m). The adjustment of the optical throughput was defined by the entrance aperture of the interferometer, among which are the entrance aperture of diameter 10 mm, the image of the beamsplitter on the collimator of diameter 31 mm and the distance between the entrance aperture and the beamsplitter image of 754 mm.

The internal modulation of the path difference eliminates the continuous background accompanying the interferogram. For a given value Δ_0 of the path difference, the value measured on the interferogram $I(\Delta_0)$ is obtained by the synchronous detection of the signal detected in the observation plane. In the case of square wave modulation, $I(\Delta_0)$ is written as:

$$I(\Delta_0) = \int_0^{\infty} B(\sigma) sin\pi\sigma\Delta_M \sin(2\pi\sigma\Delta_0) d\sigma \quad [1.16]$$

where Δ_M is the amplitude of internal modulation.

By choosing $\sigma\Delta_M = 1/2$, the reception system records the derivative of the interferogram, which is more interesting for the search of the zero path difference from which each recording starts. Moreover, with the spectrum being bounded, the application of Shannon's theorem makes it possible to determine the optimum sampling interval that depends on the studied spectral width. The Cooley–Tuckey FFT algorithm is then applied to numerical data in order to calculate the sample spectrum from the interferogram.

The width at mid-height of the apparatus function is imposed by the maximal path difference of the interferometer, $\Delta_{Max} = 14$ m, so that $\Delta\sigma \cong 1/(\beta\Delta_{Max})$, where the value of coefficient β depends on the apodization functions and varies between 1.2 and 1.8. The maximum resolution power is given by $R \cong \sigma / \Delta\sigma \cong \Delta_{Max} / \lambda$. For $\lambda = 14$ µm, $R = 10^6$.

Data acquisition was performed using the microcomputer MICRAL from the R2E company, according to a procedure described in [DAH 89], which was developed during supervised training in a research laboratory and proposed to engineering schools. Anecdotally, it is important to note that MICRAL is the first microcomputer whose invention was attributed to F. Grenelle in 1998, who thus promoted the use of the INTEL 8008 microprocessor. The development of MICRAL originated in a call for projects launched by INRA (French National Institute for Agricultural Research) in 1972, the purpose of which was to control the watering of fields, depending on ambient temperature and humidity conditions. The paradigm of Artificial Intelligence had not yet been formulated in those times.

The program written in low level assembly language was used to manage the interruptions between the acquisition device of the interferometer and MICRAL, data acquisition (16 bits per point of the interferogram), the formatting of data coded in hexadecimal ASCII, its visualization on the screen and its writing on a floppy disk by a block of 256 measurements, the commands for the management of the dialogue between the interferometer and MICRAL and the management of floppy disk writing.

The communication between the interferometer and MICRAL allows the management of a step-by-step operation mode, by controlling the acquisition of data from the MICRAL using a read command by the micro-processor. It also permits a continuous operation mode by which data acquisition was automatically achieved at the scanning rate imposed by the mobile carriage of the interferometer, moving from the zeroth path difference to the desired maximal distance from a read command, satisfying Shannon's criterion of non-overlapping spectra. The acquisition speed was generally 100 measurements per second, and an interruption made it possible to write all 256 measurements on a data block. It is important to note that the correction of the wave number at the top of the lines, which takes into account the variation of the air index between the visible and the 10 μm region, was determined by the formula:

$$\sigma_0 = (1 + 0.00034)\sigma + 0.0014 \quad [1.17]$$

The recording characteristics depend on the spectral range of the measurements. In the spectral range from 930 to 1,170 cm^{-1}, a HgCdTe photovoltaic detector was used, cooled with liquid nitrogen. The sampling interval was fixed at $206\lambda_0/8$, the internal modulation amplitude at $62\lambda_0/8$, and the frequency of synchronous detection at 377 Hz. In the spectral range from 718 to 1,077 cm^{-1}, a GeHg detector was used, cooled with liquid helium, featuring a wide-band cold filter (77 K) cutting at 1,300 cm^{-1}. The sampling interval was fixed at $176\lambda_0/8$, the amplitude of the internal modulation at $60\lambda_0/8$ and the frequency of synchronous detection at 328 Hz. For a cold filter cutting at 970 cm^{-1}, the sampling interval was fixed at $184\lambda_0/8$, the amplitude of internal modulation at $62\lambda_0/8$ and the frequency of synchronous detection at 332 Hz.

1.2.3. *FTIR absorption spectroscopy in matrices*

The matrix technique consists of the isolation of neutral species, pairs of ions or radicals in inert solids at low temperatures, for their study. It can be used in particular, for the study of unstable molecular species. Applied for the first time by Lewis [LEW 41] who studied the phosphorescence of aromatic molecules suspended in a rigid vitreous medium at low temperature, it was revisited a dozen years later by

Porter [POR 50, POR 68] and Pimentel [DOW 54, PIM 60, PIM 62, PIM 63]. Argon and nitrogen are used as matrices to isolate molecules in order to obtain free radicals by photolysis, which can be studied using IR spectroscopy. Pimentel and his collaborators had a huge contribution to the development of this technique afterwards. The matrix technique involves the isolation of neutral species, pairs of ions or radicals in inert solids at low temperatures, and their subsequent study by frequency- and time-resolved spectroscopy. The research on the absorption spectroscopy of molecular chemical species isolated in an inert matrix requires a spectrometer. At the beginning of the 1990s, in the group led by L. Abouaf-Marguin and L. Schriver-Mazuoli at the Laboratory of Molecular Physics and Applications of Sorbonne University in Paris (formerly Pierre and Marie Curie University) [BRO 93, ABD 93, LUG 94, DES 94, JAS 95, RAD 95, DAH 97, BAH 97, HAL 98, CHA 00a, CHA 01, COA 03], a Fourier transform spectrometer (Bruker IFS 113V) was used to record the IR absorption spectra for the spectroscopic study of molecules trapped in an inert matrix. The recorded spectra can also be used to control the preparation of the crystalline sample on the sample holder surface inside a cryostat.

The experimental setup was used for frequency-resolved and time-resolved spectroscopic studies [ABO 73, DUB 75, GAU 80, BOI 85, ZON 86]. Figure 1.2 presents an example of the experimental setup. The apparatus used for preparing the gas mixture and then the crystalline sample during deposition is independent from the excitation and analysis system. It includes a cryogenic set for the fabrication of the crystalline sample (Figure 1.3) from a mixture of the matrix and molecular species to be studied in the gas phase, prepared in a chemical apparatus under secondary vacuum.

A commercial cryostat equipped with a reflective sample holder can be used to condense the gas mixture on a metallic support cooled with liquid helium. The sample holder consists of the polished and gilded side of a copper block covered with a gold layer, which also serves as a mirror. The spectroscopic study is conducted by reflection on this mirror, with the beam traveling through the sample twice. A control system composed of a diode and a logical set is used to reach the desired temperature. Probes embedded in the sample holder are used to measure its temperature: a platinum resistance for temperatures above 30 K and a carbon resistance for temperatures ranging between 4.2 and 30 K. In the setup used, the platinum is located above the sample holder block, and the carbon is lower, in the block. A diode is located at the level of the vaporization chamber and of a heating resistor. The base of the cryostat has three windows in KBr, allowing the passage, on the one hand, of the IR beam, and, on the other hand, of the pump pulse and the observation of fluorescence for the vibrational relaxation experiments. The set is maintained under secondary vacuum (10^{-5} Torr).

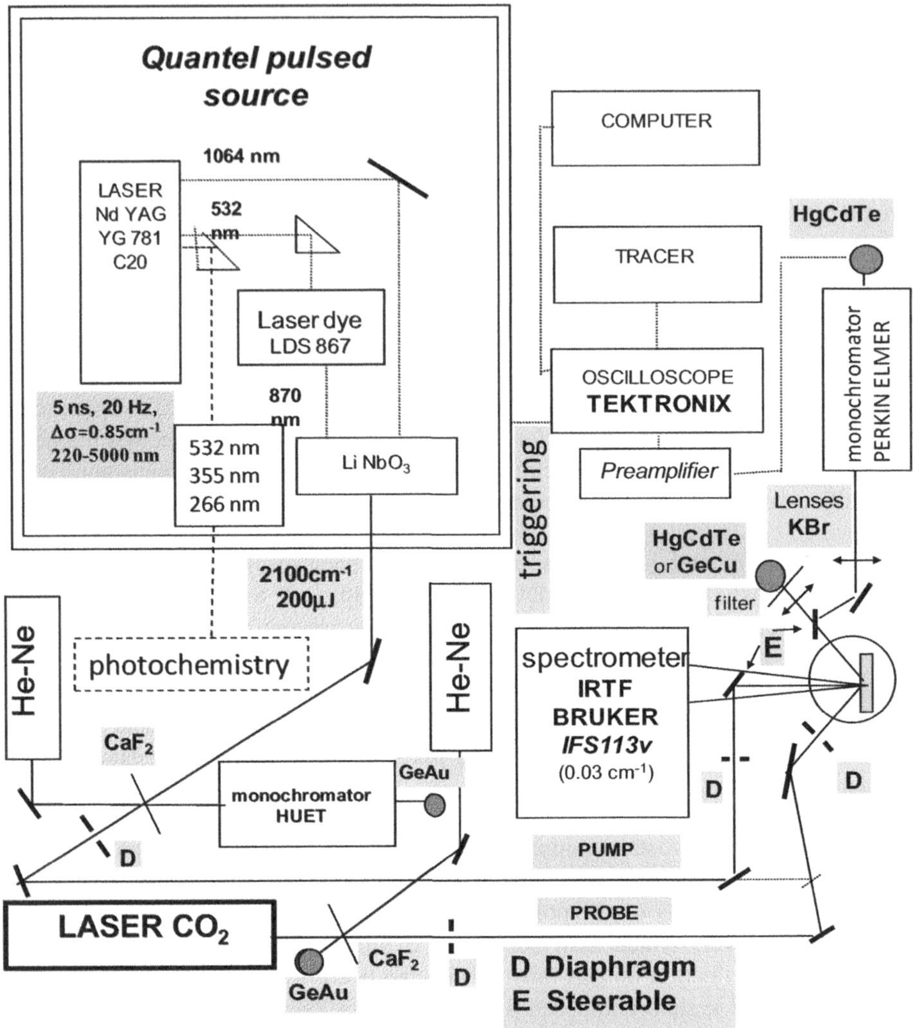

Figure 1.2. *Experimental setup for laser spectroscopy: induced fluorescence and IR–IR double resonance*

The spectroscopic study is conducted using a Fourier transform IR spectrometer made by Bruker (IFS 113V), operating under primary vacuum. The set composed of a Michelson interferometer and a He–Ne laser embedded in the interferometric block allows the sampling of the IR interferogram. The spectra are recorded between 500 and 4,000 cm^{-1} with a resolution of 0.5 cm^{-1} or of 0.15 cm^{-1}, depending on the sample, using a DTGS detector operating at ambient temperature. The interferometer embedded software is operated to control the linearity of the sample deposit and the analysis of the IR absorption spectra (Figure 1.4).

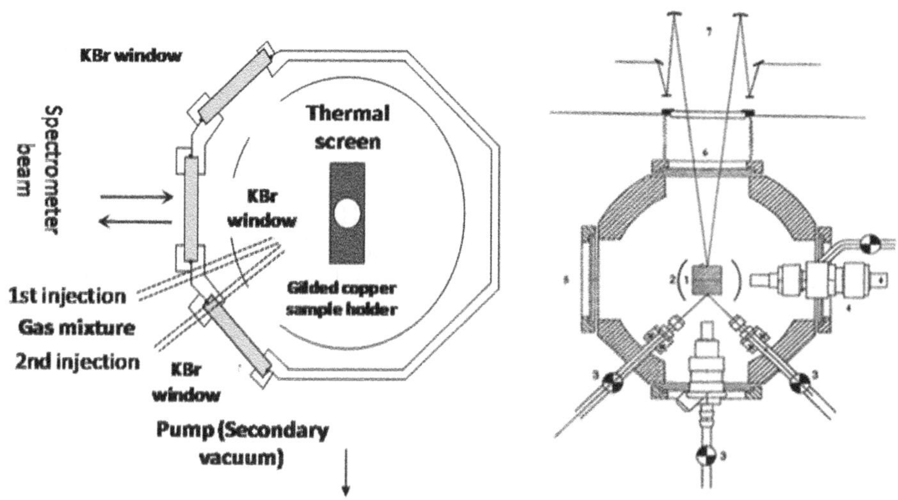

Figure 1.3. *Experimental setup of the cryostat head*

A characteristic of this device is that the beamsplitter–compensator support continuously moves on an air cushion. A maximum resolution without an apodization of 0.03 cm^{-1} is reached for a path that is half of the one traveled by the mobile mirror of a classical interferometer. The duration of spectrum acquisition is thus reduced. The maximum path difference is 32 cm. Conducting fine spectroscopic studies sometimes requires a reduction of the spectral region to be investigated, by means of an interchangeable beam splitter. In medium infrared (MIR), a potassium bromide (KBr) plate containing a deposit of germanium is used as the beamsplitter. For far IR, a mylar sheet is used for the spectroscopic study in the region of vibrational mode ν_2 of molecules N_2O and CO_2.

The beam entrance flux is controlled by an adjustable aperture located at the exit of the compartment containing the IR source (Globar: silicon spiral heated at 1,200 K). The light is sent on the sample holder mirror of the cryostat and is then reflected towards an MCT detector cooled with liquid nitrogen for rapid acquisitions (100 scans in 10 minutes). The quantitative measurements are conducted with a DTGS detector operating at ambient temperature, with a higher linearity than MCT, but at a slower rate (100 scans in 60 minutes). The diameter of this aperture at the source exit must be compatible with the desired resolution. Therefore, a resolution of 0.5 cm^{-1} corresponds to a wide open aperture (10 mm). On the other hand, for a resolution of 0.15 cm^{-1}, the diameter must be reduced to 5 mm. In this case, the energy received by the detector is lower and may prove insufficient for analyzing the spectra of certain samples. The lack of transmitted energy can be critical, as with the xenon and krypton matrices. These two matrices are very scattered and reduce the transmitted intensity of the IR beam. For thick deposits, the SNR is then too low, and the resolution must be reduced to 0.5 cm^{-1}.

The more rapid MCT detector, cooled with liquid nitrogen, can be used for spectra acquisition with a higher scanning speed. However, at the same resolution, the SNR is not better, since in order to avoid saturation of MCT, which is more sensitive than DTGS, the globar source flux on the detector must be limited. On the other hand, resolution can be increased for the studies of thick deposits.

The spectrometer is controlled by a computer that exclusively uses Bruker software. Depending on the studies to be conducted, the operating parameters such as scanning speed, number of scans, resolution, diameter of the source diaphragm aperture, beamsplitter, detector, compartment (backside for our experiments) and electronic filters to reduce the number of points to be taken during the recording of the interferogram, are input as data and are then validated by the program prior to spectra acquisition. The Fourier transform of the interferogram is then calculated using an apodization function proposed by Bruker software. It is important to note that, depending on the desired resolution, a set of graphs can be used to make the best choice of electronic filters, scanning speed and source diaphragm aperture.

The Bruker interferometer is placed in front of the central window of the cryostat (see Figure 1.4). It is made up of a beam extractor, whose design allows the recording of spectra by reflection at 30 cm from the interferometer exit, thanks to an optimization of the position of its mirrors. The IR beam travels in the air over a distance of several centimeters between the two devices. In order to avoid absorption by the carbon dioxide or water vapor molecules in the air, this volume is flushed with nitrogen gas. Before deposit, two spectra without samples are recorded at 5 K with resolutions of 0.5 cm^{-1} and 0.15 cm^{-1}. They serve as references to calculate absorbance spectra, which are obtained by taking the logarithm of the ratio of the reference spectrum to the sample spectrum. When a sample is deposited, a preliminary spectroscopic study is simultaneously conducted to make sure that the optical density in the absorption spectra increases linearly, depending on the number of torrs deposited.

The space available between the window of the cryostat and the spectrometer allows the positioning of optical devices for time-resolved spectroscopy studies. All of the optical paths inside the spectrometer are pumped under permanent vacuum. Before the deposition, a reference spectrum without a sample is acquired at a resolution of 0.15 cm^{-1}, at a temperature of 5 K. The spectrum of the sample is calculated in absorbance form by taking the logarithm of the ratio of reference spectrum and the spectrum of the sample.

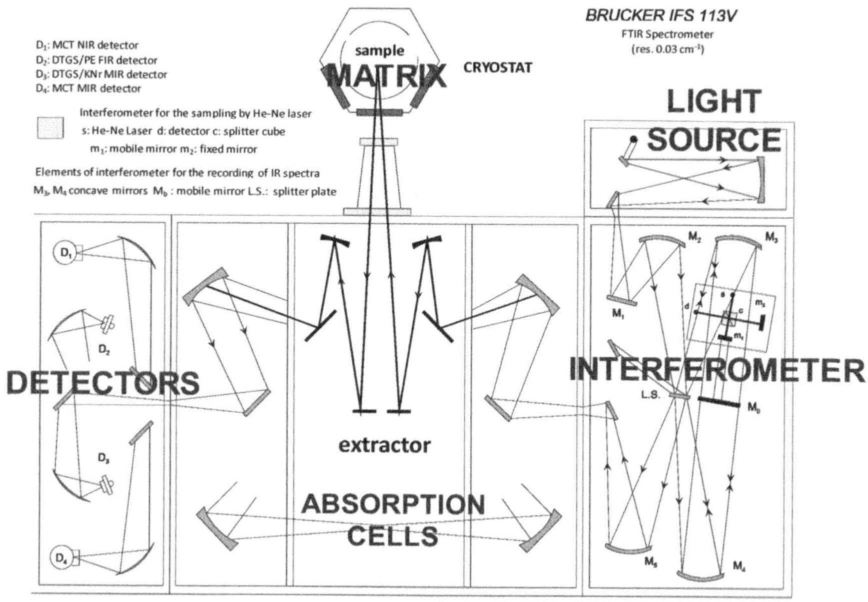

Figure 1.4. *FTIR spectrometer and experimental setup. For a color version of this figure, see www.iste.co.uk/dahoo/infrared2.zip*

Among the results obtained with this experimental setup, it is important to highlight the observation of simple and double substitution sites for the trapping of the O_3 molecule in rare gas matrices (Figure 1.5) [BRO 93], and also the presence of two CO_2 trapping sites and the degeneracy lifting of mode ν_2 of CO_2 in a double substitution site, termed as an unstable site (Figure 1.6) [DAH 99, CHA 00c], and of the presence of only one trapping site for the N_2O molecule, a double substitution site that lifts the degeneracy of mode ν_2 (Figure 1.7) [CHA 00d].

The spectroscopic studies conducted with this experimental device are explained in the following references: [BRO 93, ABD 93, LUG 94, DES 94, JAS 95, RAD 95, DAH 96, BAH 97, HAL 98, CHA 00a, CHA 01, COA 03].

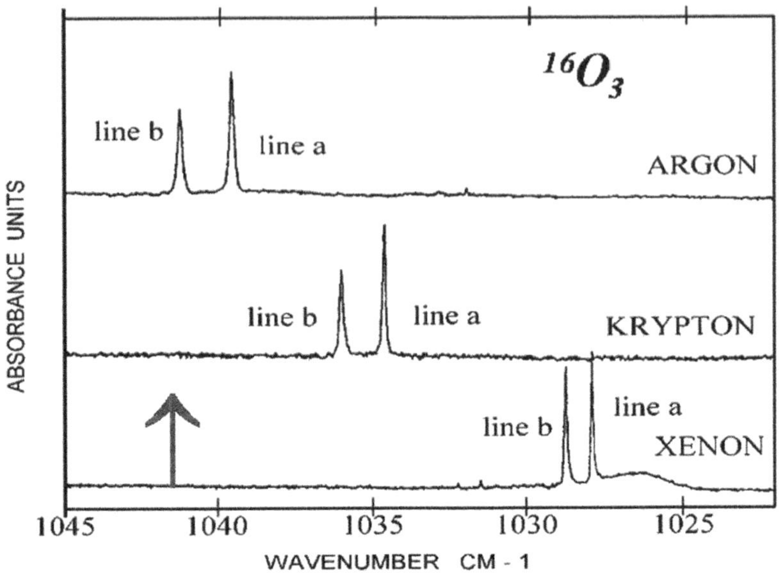

Figure 1.5. *O_3 region ν_3: (a) HF double site and (b) BF simple site (gas: 1042.08 cm^{-1}). For a color version of this figure, see www.iste.co.uk/dahoo/infrared2.zip*

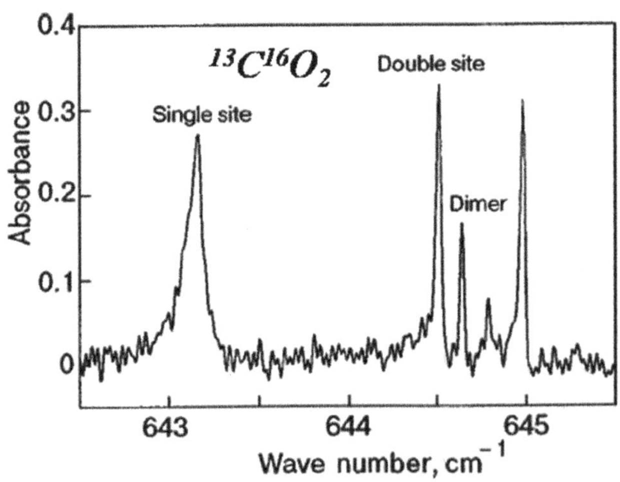

Figure 1.6. *CO_2 region ν_2: (a) BF stable site and (b) HF unstable site (gas: 648.5 cm^{-1})*

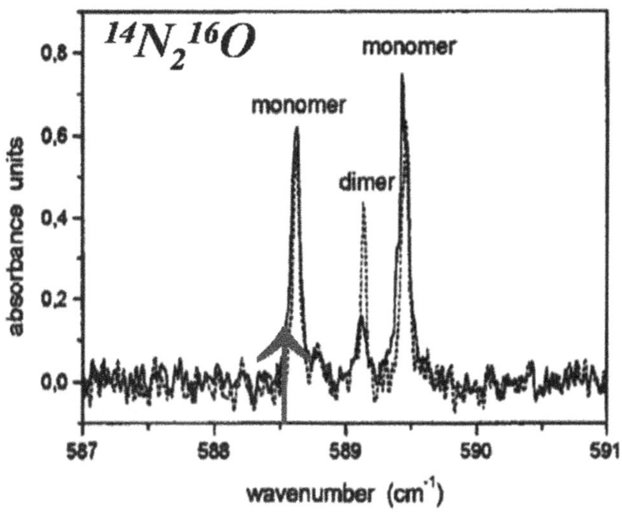

Figure 1.7. *N_2O region ν_2: double site (gas: 588.77 cm^{-1})*

1.2.4. *LIF and DR IR–IR spectroscopies in matrices*

The time-resolved laser spectroscopy conducted with this experimental device was described in Chapter 7 of [DAH 16]. The experimental techniques of laser-induced fluorescence and double resonance (LIF and DR) originate in the works of Kastler and Brossel [BRO 49] for DR, and Kastler [KAS 50] for optical pumping. They were applied within the framework of optical detection methods of magnetic resonance that can be used for the study of atomic level structures. The progress of time-resolved laser spectroscopy techniques, LIF and DR (LIF for Laser-Induced Fluorescence and DR for Double Resonance), unfolded in parallel with the development of lasers. Based on the theoretical model of Schawlow and Townes [SCH 58] showing the feasibility of IR and optical lasers, the development of the first lasers, the pulsed ruby laser, at 3.39 μm, invented by Maiman in 1960 [MAI 60] and the continuous He–Ne gas laser, at 1.15 μm, invented by Javan *et al.* [JAV 61] in 1961, revolutionized LIF spectroscopy. Indeed, the coherent light of lasers is better adapted to the LIF technique than the light emitted by classical sources.

The first works in this field were conducted by the inventors of laser sources themselves. Javan *et al.* studied the vibrational relaxation of CO_2 using the LIF technique in 1966 [HOC 66] and Schawlow *et al.* studied the laser-induced selective photo-catalysis of bromine [TIF 67] using a laser source to excite the bromine molecules in the gas phase up to the first quantum bound states, close to the continuum of dissociation in 1967. The paths opened up by the intense monochromatic laser sources (high power per unit surface) for the studies of photochemistry, molecular spectroscopy or energy transfers by lasers were underlined in the article [TAN 68]. A review of the LIF technique is given in [ZAR 12]. In France, time-resolved laser spectroscopy on molecular systems trapped in matrices was developed by Abouaf-Marguin, after post-doctoral studies conducted in the United States, under the coordination of Jacox [MIL 67, ABO 77] at the Laboratory for Molecular Photo-Physics at Paris Saclay University (formerly Paris Sud University) [ABO 73, DUB 75, GAU 80, BOI 85, ZON 86].

Laser-induced fluorescence (LIF) spectroscopy is the process by which a molecule trapped in an inert matrix is excited by a laser beam, and the light subsequently emitted by the excited molecule is detected. With respect to absorption spectroscopy, LIF detection features a better detection sensitivity, as the signal is observed against a dark background, which limits the ambient noise. Since fluorescence is emitted in all directions, it can be collected under various angles with respect to the incident laser beam, which makes it possible to obtain two- and three-dimensional images. Moreover, by using a monochromator for frequency-resolved fluorescence, it is also possible to determine the transitions from the excited state to various lower levels of the excited molecule. The time delay between the events related to excitation and detection makes it possible to study the processes through which the excited molecule returns to its initial state in the time interval between the onset of the laser excitation and that of the fluorescence detection.

The experimental setup presented in Figure 1.2 was used to determine the effect of an inert matrix on the atmospheric molecules trapped in substitution sites by time-resolved spectroscopy. The laser-induced fluorescence and IR–IR double resonance experiments conducted on O_3, CO_2 and N_2O involved a set of lasers (pump laser) (Figure 1.8) (YAG, laser dye, doubling (visible), tripling, quadrupling, quintupling (UV) and mixer (IR) crystals), completed by a continuous CO_2 laser (probe laser) and a system composed of a cryostat and Brucker IFS 133v interferometer, to prepare and characterize the samples.

A flash lamp is used for optical pumping and achieving the energy level population inversion in the YAG rod. The rod, which is the active medium, is a yttrium–aluminum–garnet (YAG) crystal weakly doped (0.1%) with neodymium (Nd^{3+}) ions. The active medium emits a fluorescence on the $4F_{3/2}$ -$4I_{11/2}$ transition line at 1,064 nm of Nd^{3+} ions. The high power flash lamp is composed of a cylinder located at one of the foci of a cylindro-elliptical mirror, while the YAG rod is located in the other focus. The rod and the lamp are thermostatically regulated by controlled water circulation. A dichroic crystal is used for the second harmonic generation at 532 nm, a tripling crystal gives the 355 nm line and finally a quadrupling crystal gives the 266 nm line. In order to increase population inversion, the laser operates under switched mode by a delay in the input of laser oscillations, with respect to the lamp flash. The switched mode of the laser is controlled by a Pockels cell and by a polarizer, making it possible to reach opening times of nearly 5 ns with a frequency of 20 Hz.

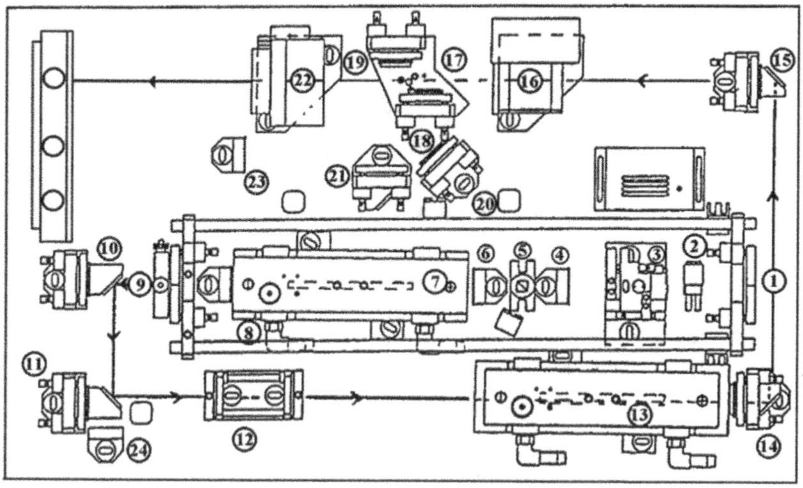

Figure 1.8. *Nd-Yag laser: (1) cavity bottom mirror (Rmax = 1,064 nm), (2) shutter, (3) Pockels cell, (4), (6), (8), (23) quarter wave plate, (5) horizontal polarizer (GLAN), (7) laser head, (9) output mirror (cavity bottom) of variable reflectivity, (10), (11), (14), (15), (19), (21) reflective mirrors Rmax = 1,064 nm, (12) telescope, (13) amplifier, (16) doubling crystal, (17) stage with two oscillating mirrors, of which (18) one mirror with Rmax = 532 nm, Tmax = 1,064 nm and (19) one mirror with Rmax = 532 nm, (20) reflective mirrors with Rmax = 532 nm and (22) tripling or quadrupling crystal (24) beam blocker*

The IR module of the dye laser set has three core elements: the frequency doubled YAG laser, a dye laser and a mixer crystal to produce mid-IR (Figure 1.9).

Its operation relies on these three devices. Laser pulses are generated by a flash pumped YAG rod, whose frequency is then doubled through a doubling crystal. The dye laser involves an oscillating cavity and two amplifying cells containing a dye solution (LDS 867) pumped by the 532 nm radiation. The set emits IR pulses (35 mJ/pulse) at about 870 nm. The frequency calibration is determined by the position of a toothed wheel that rotates a grating (2,400 traits/mm) in the dye oscillator cavity. Wheel rotation can be manually controlled by an operating wheel, or automatically controlled by a control unit or a computer. This control system can be used to precisely adjust the frequency (0.02 cm^{-1}). The spectral width of the pulse can be reduced from 2.5 cm^{-1} to 0.8 cm^{-1} by means of a beam expander located in the cavity of the oscillator. Finally, a $LiNbO_3$ crystal placed in a last module generates the IR radiation by parametric conversion. The beam issued by the dye laser (35 mJ/pulse) and that of the residual YAG laser (40 mJ/pulse) cross the crystal and intersect in its center. A third wave resulting from the subtraction of frequencies of the two incident beams can be available at the exit. The efficiency of this nonlinear effect depends on the geometrical phase matching of the wave vectors of the YAG and dye beams through the crystal. Each frequency of the dye is related to a specific inclination of the crystal. The orientation of the latter, placed on a holder that can be adjusted by an operating wheel, must be modified at each frequency change, in order to maintain the IR laser intensity. In an optimized configuration, the crystal provides pulses of about 200 μJ with the same spectral and temporal qualities as the pulse of the dye ($\Delta\nu = 0.8$ cm^{-1} and $\tau = 5$ ns). The frequency of the emitted radiation ranges between 2 μm and 4.8 μm, depending on the frequency of the dye. It is used around 2,100 cm^{-1} in resonance with the transition $\nu_1+\nu_3$ of the ozone in matrices for laser spectroscopy.

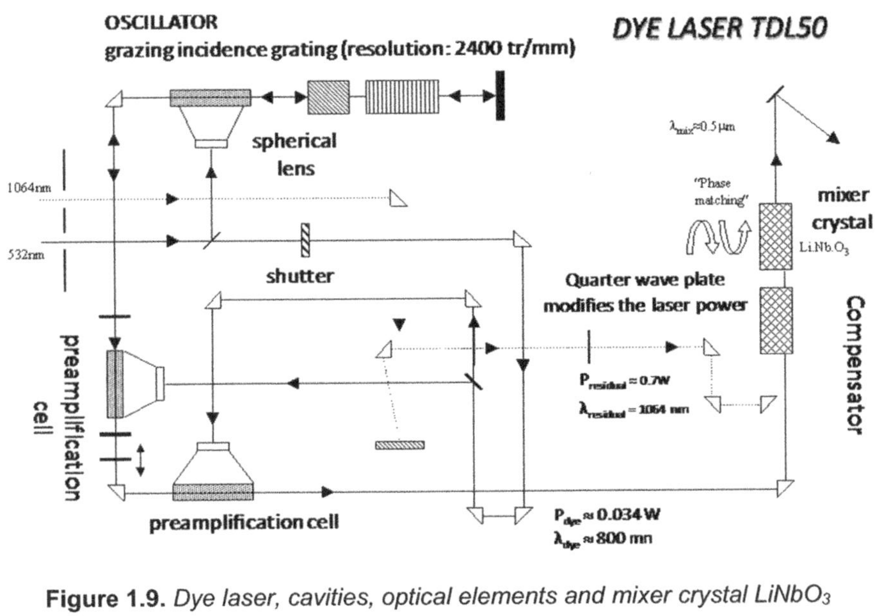

Figure 1.9. *Dye laser, cavities, optical elements and mixer crystal $LiNbO_3$*

The results are published in the following references: [JAS 94, JAS 95, BER 96, DAH 97, DAH 98, JAS 98, CHA 00a, CHA 00b, CHA 00c, CHA 02, VAS 03]. For example, the global fluorescence observed for two trapping sites of CO_2 in the argon matrix (Figure 1.10(a) and (b)) is more chaotic for the unstable site, with signals constituted of short (one at 10 μm and two at 16 μm) and long (one at 16 μm) component superpositions in the time domain.

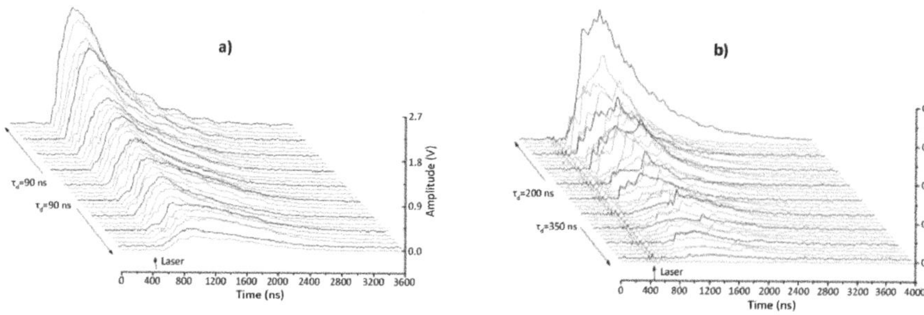

Figure 1.10. *Global fluorescence: (a) stable site and (b) unstable site, Ar/$^{13}CO_2$ = 2 000, T = 5 K [CHA 00a, DAH 16]*

Chapter 7 of [DAH 16] presents a summary of the time-resolved spectra that were observed. Fluorescence was observed after excitation in the IR range in the region around 4.5 μm, of the vibrational mode $\nu_1+\nu_3$ of the fundamental electronic state of $^{16}O_3$ [JAS 94, JAS 95, DAH 97, JAS 98, CHA 00b] and of the vibrational mode ν_3 of $^{13}CO_2$ [BER 96, DAH 97, CHA 98, CHA 00a, CHA 00c, CHA 02, VAS 03] and $^{14}N_2O$ [BER 96, DAH 97, CHA 00b, CHA 02]. Fluorescence differs from the pump pulse by a temporal delay and a weaker intensity. Correlated with the absorption zone, it fully disappears if laser frequency shifts outside the range. Its intensity is optimized by shifting the laser impact on the sample.

The results of the application of the IR–IR double resonance technique are essentially related to the studies conducted with the CO_2 probe laser on ozone in GR and nitrogen matrices [JAS 95, DAH 97, DAH 98]. A double resonance (DR) signal is characterized by three temporal parameters, τ_M, τ_R and τ. The time τ_M is the time for the signal to reach the maximum and corresponds to the mean duration of depopulation or repopulation of the initial level in the probed transition. The time τ_R is the characteristic rise time of the signal or decay time of the signal, and corresponds to the dynamics of interaction between the probe and the initial level. The decay time τ is usually determined by adjusting an exponential $s(t) = Aexp(-t/\tau)$ to a properly chosen portion of the signal. It is sometimes necessary to use two exponentials for the adjustment, by taking $s(t) = A_1\ exp(-t/\tau_1)+A_2\ exp(-t/\tau_2)$ when two independent phenomena are juxtaposed. These times are associated with the return of the initial level to its initial population state.

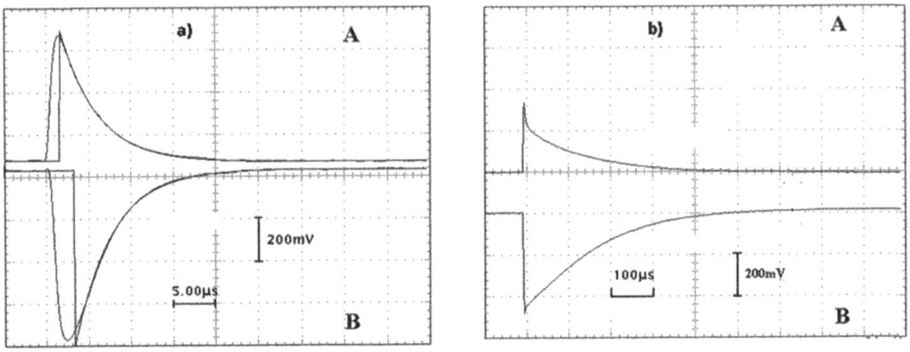

Figure 1.11. *Double resonance signals: (a) carried by the line P_{26} (A: 4.2 µs) and the line P_{42} (B: 4.1 µs) of the CO_2 laser (O_3/Ar = 1/2 000, thickness: 110 mm); (b) superimposed probe (P_{26}) and pump (A) and probe (P_{26}) and pump laterally shifted (B) [JAS 95, DAH 97, DAH 16]*

A signal, due to a thermal effect (Figure 1.11(b)) that occurs as soon as an excited molecule releases its energy to the environment, is generally superimposed to the DR signal, on a time scale that is above the measured times (τ_1 and τ_2). In order to analyze the DR signal on its time scale, the amplitude of the thermal signal must be negligible compared to that of the DR signal. A second type of thermal effect, (Figure 1.11(b)) known in laser theory as the lens effect, must also be minimized to correctly analyze the DR signal. This effect is manifested in the superposition of a second positive or negative signal, depending on the relative positions of the pump and probe beams. Its decay (or rise) occurs over the same time scale as the thermal signal of the first type. This effect can be minimized by adjusting the relative positions of the beams, so that the positive signal compensates the negative one.

1.3. Resonant cavity laser absorption spectroscopy

Resonant cavity laser absorption spectroscopy involves the use of a laser source and a resonant cavity composed of highly reflective mirrors, of reflectivity nearly equal to 1, for the study of molecules in the form of trace. It is important to note that as early as 1962, Kastler [KAS 62] suggested the use of a Fabry–Pérot cavity for measuring atomic absorption and emission.

It is a technique through which the absorption of a medium is measured after multiple passages in a laser cavity, which operates as a multiple reflection cell. It can be used for the study of weak absorptions that are due to chemical species in the form of traces, or to weak transitions. In practice, the laser cavity operates as a filter that fixes the (longitudinal or transverse) modes (see Chapter 6 of [DAH 16] or the references [SIE 86], [DEM 96] and [GRY 97]) of the laser and on which laser amplification can occur (the cavity length is assumed to be constant by a feedback control). The introduction of a spatial filter in the form of a variable aperture diaphragm can be used to select only the transverse mode TEM_{00}. In this case, if the broadening of the laser gain profile is inhomogeneous, there is a gain increase after each passage of the laser beam in the cavity, and for the longitudinal modes that are within the gain profile and coincide with the absorption lines of the absorbing medium, lines appear in the laser profile of saturated gain (Figure 1.12). Saturation occurs from a threshold defined as the equality between the pumping gains and the back-and-forth beam path in the cavity and the intracavity losses [SIE 86, DEM 96, GRY 97].

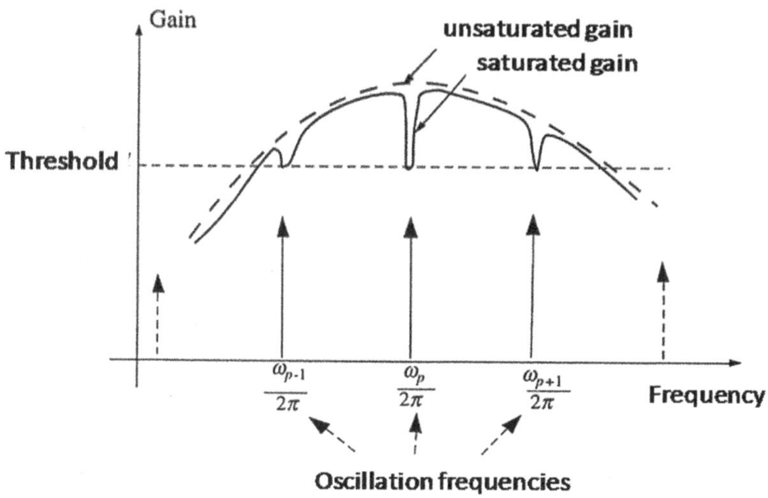

Figure 1.12. *Effect of gain saturation for an inhomogeneous widening: hole burning in the gain profile [DEM 96, GRY 97]*

With this technique, the absorbing medium is thus integrated in the source device in the laser cavity. The saturation of amplification or the decrease in gain when light intensity increases only affects the molecules of the amplifying medium, whose frequency is tuned on the active frequency. The simultaneous oscillation of all of the modes allowed by the threshold condition is observed. This phenomenon is possible because the broadening is inhomogeneous, which means that gain saturation only affects the molecules of the amplifier medium, whose frequency is tuned on the active frequency. At a given frequency, the laser effect does not affect the gain at another frequency. On the other hand, if the broadening is homogeneous, the response of each molecule is similarly affected by the gain saturation. The gain curve as a whole undergoes the effect of saturation.

Two techniques, namely the intracavity laser absorption spectroscopy (ICLAS) and the cavity ring-down spectroscopy (CRDS), are based on the multipassage intracavity absorption. The principle relies on the increase of the optical path, following multiple reflections of the laser beam on the mirrors in the cavity, and consequently, the absorption of these molecules. The finesse and the cavity gain are also parameters to be considered among the conditions allowing for the observation of the absorption of molecules inserted in the cavity. This technique follows the research works conducted on laser cavities, whose objective was to limit cavity losses in order to improve the laser output power and the gain profile. In this context, the objective was to improve the reflectivity of the cavity mirrors and to reduce the other sources of loss potentially due to impurities in the cavity.

In the ICLAS method, absorbing species are placed inside the cavity of a broad band laser. Because of the positive feedback mechanism of lasers, even tiny quantities of an absorber produce holes in the spectrum, where the laser exit is partially turned off. ICLAS was proposed in the early 1970s [PAK 70, PET 71, KEL 72, BEL 72]. It relies on the high sensitivity of a broad band multimode laser, compared to the selective losses inside the laser cavity.

For example, Pakhomycheva *et al.* [PAK 70] proposed that the line structure of the spectrum generated by a laser from an inhomogeneously broadened line is due to the high sensitivity of the laser spectrum to the presence of losses, depending on the frequency in the resonator. If the losses described by the following formula:

$$\frac{1}{T(\omega)} = \frac{1}{T_0} - \Delta\left(\frac{1}{T}\right)\cos\left(\frac{\pi r \omega}{\Delta\omega}\right) \quad [1.18]$$

are introduced into the laser cavity, the gain profile presents dips in the spectrum, with a relative intensity given by:

$$\frac{\Delta I}{I_0} = \frac{\Delta(1/T)}{1/T_0}\left(\frac{ST_0}{I_0} + \frac{I_0}{PT_0}e^{-\left(\frac{2\pi\gamma}{\Delta\omega}\right)}\right) \quad [1.19]$$

where ST_0, I_0 and P are, respectively, the spontaneous noise power, the generated power averaged over the frequency and the frequency density of the pumping power (power per unit frequency interval). The hypotheses leading to the expression of formula 1.19 are an infinite inhomogeneous width and a homogeneous broadening of the line profile characterized by a width γ. The proof of this effect and the validation of the equations used were studied with benzene and chlorobenzene, and also with the atmospheric molecules NH_3 and CH_4 inside a laser cavity.

Belikova *et al.* [BEL 72] described a method for the detection of weak absorption lines of gas, whose absorption coefficients are as low as 10^{-8} cm^{-1}. This method relies on the multiple passages of light through an absorbing substance during laser action. They show that the gain of a laser with an inhomogeneously broadened gain profile remains constant over the whole range of frequencies of coherent emission. This guarantees a high stability of the method. The theoretically predicted changes in the coherent emission spectra resulting from the presence of losses, depending on the frequency in the resonator, are confirmed by experimental results. The proposed method was used for the study of absorption spectra of ammonia, acetylene and water vapor in the spectral range of 1.06 μm.

Petersen *et al.* [PET 71] described a new technique developed to improve the detection of absorption spectra of weak absorbents. A laser with organic dye normally emits a continuous spectrum with a bandwidth from 2 nm to 10 nm. The effect of placing a sample of absorbing gas inside a dye laser cavity leads to an extinction of the laser at wavelengths where the absorption of the sample exceeds a minimal absorption threshold. The threshold of deactivation by absorption seems to be very low; an absorption of about 0.5% drives an alteration of the laser action at absorbing wavelengths and a shift of the laser emission towards wavelengths where the absorption is below the level of critical commutation. It is possible that the limits of detection of atoms and molecules by means of ICLAS, are by more than two orders of magnitude lower than those that can be reached by conventional absorption spectroscopy.

Keller *et al.* [KEL 72], in the same group as Petersen *et al.*, indicated that the insertion of samples in a broad band dye laser cavity improves the photographic detectability of absorption traces by about three orders of magnitude. Optical densities as low as 0.0004 were easily detected in a diluted solution of $Eu(NO_3)_3$ species. The technique by [PET 71] was modified, the bandwidth of the dye laser being reduced by the use of prisms. The wavelength of the narrow band is then used for a scanning through absorption. The modified technique allowed the detection of optical densities as low as 0.01. The solutions to the theoretical equations describing the two processes are in agreement with the experimental results and give an overview of the mechanism of the effect. The conclusion is that the measurements with broad band laser could be conducted in 100 ns or less, and that consequently, this technique should provide a powerful tool for the detection of short duration and very low concentration transients.

CRDS is a highly sensitive absorption method, based on the measurement of the decay profile of a laser pulse injected in an optical cavity. The length of the absorption path depends on the number of reflections required for the intensity to decrease by $1/e$. In the visible range, using the best available mirrors, the total length of the path can be about several dozen kilometers. Unlike the standard multiple passage absorption methods, as in the multiple reflection cells, the probed volume is limited to the cavity modes. Though described by Anderson *et al.* [AND 84], the proof of CRDS is attributed to O'Keefe and Deacon [OKE 88], who explicitly use the term CRDS to describe their invention. CRDS has since been used in many applications.

Anderson *et al.* [AND 84] described their invention as a reflectometer that can be used to measure the reflectivity of low-loss highly reflective mirror coatings and the transmission of low-loss antireflective coatings. The technique directly measures the intensity decay time of an optical cavity composed of low loss elements. They developed the theoretical framework of the instrument and discussed the conditions under which and extent to which the decay time represents a genuine measure of the mirror reflectivity. The instrument described provides a resolution of the decay time of 10 and shows a resolution of total loss of the cavity of 5 ppm.

O'Keefe and Deacon [OKE 84] developed the CRDS technique, which can be used to conduct measurements of optical absorption using a pulsed light source, which offers a far higher sensitivity than that obtained with continuous light sources. The technique relies on the measurement of the absorption rate, rather than of the absorption amplitude of a light pulse confined in a closed optical cavity. The decay in light intensity in the cavity is a simple exponential, with losses due to loss on the mirrors, broad band scattering (Rayleigh, Mie) and molecular absorption. The narrow band absorption spectra are recorded by scanning the output of a pulsed laser (that is injected in the optical cavity) by means of a resonant absorption. [OKE 84] prove the sensitivity of the technique by the measurement of several bands in the forbidden transition $b\Sigma_g^- - X\Sigma_g^-$ of very low intensity of the oxygen molecule in the gas phase. Absorption signals below one part per million are also detected.

1.3.1. *Intracavity laser absorption spectroscopy (ICLAS)*

The study of weak molecular transitions at high resolution can be conducted using the intracavity laser absorption spectroscopy (ICLAS) [STO 66] technique. This method relies on the high sensitivity of a broad spectral band multimode laser, in relation to the low intracavity laser absorptions, which are selective losses. The minimum detectable absorption coefficient is about 10^{-9} cm^{-1} for measurement times of several seconds. Typically, with the ICLAS technique, a substance is inserted in the laser cavity oscillating over a range of wavelengths. If the sample absorbs within the range of wavelengths emitted by the laser, the absorption characteristics affect the laser spectrum by an amount that is measurable. ICLAS is very sensitive, as it allows extremely long effective path lengths and high spectral resolution. The technique relies on the temporal dependence and the high sensitivity of a multimode laser, whose losses depend on the intracavity frequency when the spectral width of the absorbing species is greater than the spacing between modes and narrower than the gain profile.

Intracavity absorption in the spectrum of a laser with a homogeneously broadened gain profile is represented in a logarithmic scale by $N\sigma(\nu)\ ct_g = N\sigma(\nu)$ Leq, where N is the density of absorbing species, $\sigma(\nu)$ is the effective absorption cross section, c is the speed of light and t_g is the generation time (the time between the onset of the laser emission and the observation time). The product ct_g can be defined as an equivalent absorption path Leq. The linear growth of the equivalent absorption path is limited by the spontaneous emission at a characteristic time $\tau \cong \langle M \rangle/\gamma$, where $\langle M \rangle$ is the average number of photons in the mode and γ is the intracavity loss rate. The characteristic time τ_{sp} expressed through the parameters of the cavity and of the active medium provides an estimate of the fundamental sensitivity limit for an intracavity spectrometer:

$$\tau_{sp} \cong \frac{(\eta-1)^2 w_0^4 L}{32\sigma_0^2\Gamma^2\tau^2c^3 ln(R^{-1}(1-T_i)^{-2})} \qquad [1.20]$$

where L is the cavity length, w_0 is the laser beam radius inside the active medium, R and T_i are the output mirror reflectance and the intracavity losses, respectively, η is a pump power excess above the threshold, and σ_0, Γ and τ are the emission cross section, the gain linewidth and the fluorescence lifetime of the active medium, respectively. For a typical Ti:sapphire laser, this yields $\tau_{sp} \cong 0.7$ s, and $Leq \cong 210{,}000$ km! [KAC 94].

The very weak absorption lines of an absorber inside the cavity of a multimode laser can modify the laser spectrum and appear superimposed on the broader spectrum of the laser. This phenomenon is universal, meaning that any laser with a homogeneous gain profile (such as dye lasers, color center or Ti:sapphire) can be easily transformed into an ICLAS spectrometer [CAM 90]. This technique is very sensitive, as an equivalent absorption path up to a 100 kilometers can be reached. Consequently, a very low absorption coefficient can be detected, and it is proportional to the absorber density and to the line intensity. ICLAS was used to observe a weak absorption of NO_2 and CH_4, cooled in a supersonic jet through a slit. By using this method, the thermally congested spectrum of molecules is considerably simplified and the analysis becomes possible [CAM 91]. The major application domains of the ICLAS technique are the detection of low-density absorbers, the observation of very weak absorption transitions (forbidden transitions) and the study of harmonic transitions in the visible range [CAM 90].

The dynamics of a multimode laser (dye laser or Ti:sapphire) is described by a system of coupled balance equations between the population inversion of the amplifying medium and the photon density in each cavity mode (as, for example, in

Chapter 6, [DAH 16]), so that the temporal evolution, after gain saturation, of the averaged laser spectrum can be expressed by:

$$I(\sigma, t_g) \cong \frac{I_0}{\Delta\sigma}\sqrt{\frac{\Gamma t_g}{\pi}}\exp(-(\frac{\sigma - \sigma_0}{\Delta\sigma})^2 \Gamma t_g) \quad [1.21]$$

where $\Gamma = 1/T$ is the inverse of the lifetime of a photon in the cavity, σ_0 is the center of the gain profile, $\Delta\sigma$ is the laser gain linewidth, and t_g is the laser generation or operation time.

The exponential function describes the temporal evolution of the Gaussian base line of the laser spectrum. Its amplitude increases as $\sqrt{t_g}$, and its width decreases in time as $1/\sqrt{t_g}$. As already described, by introducing a gas in the cavity, the modes of the laser in coincidence with a transition, grow at a slower rate than that of the neighboring modes, the absorption in this case being considered as selective losses [PAK 70, PET 71, BEL 72, KEL 72]. The lines appear in the laser spectrum and grow with laser generation time. The observed spectrum can be considered as a superposition of the laser spectrum and that of the absorption spectrum of the gas inserted in the cavity.

The intensity transmitted by the cavity $I_t(\sigma)$ [PER 96] can thus be calculated using equation [1.5], according to the Beer–Lambert law:

$$I_t(\sigma, t_g) \cong I(\sigma, t_g)\exp(-\alpha(\sigma) l_{eq}) \quad [1.22]$$

where $I(\sigma)$ is the intensity of the intracavity laser radiation and l_{eq} is the thickness of the gas through which it travels, or the equivalent absorption length following multiple reflections in the cavity such that:

$$l_{eq} = \frac{l_{gas}}{l_{cavity}} c t_g \quad [1.23]$$

where l_{gas} is the optical path of the beam in the gas, l_{cavity} is the optical length of the cavity, and c is the speed of light.

For a time of 100 μs, an occupancy rate of 33% leads to an equivalent length traveled in the gas of about 10 km.

The quantitative measurements require the control of the temporal evolution of the laser and the capability to record the laser spectrum at a given instant equal to t_g. When a laser is initiated by the optical noise generated by the spontaneous emission resulting from vacuum fluctuations, the process is stochastic and the spontaneously emitted photons are distributed over all of the longitudinal modes (it is admitted that a diaphragm placed in the waist can limit the modes transversal to TEM_{00}) of the cavity. The average of N spectra makes it possible to reduce the noise according to $\sqrt{N}$. Losses due to the scattering centers should be avoided, although they cannot always be eliminated despite the use of thick mirrors. One method involves the use of a ring laser, which also has the advantage of increasing the equivalent path length. Laser generation times that are too long should also be avoided, as the Beer–Lambert law is then no longer applicable. The absorption spectral width must remain below the gain spectral width of the laser. The laser setup also comprises a high resolution monochromator, coupled to an array of photodiodes for the detection of the light issued from the laser cavity. Two synchronized acousto-optical modulators can be used, on the one hand, to interrupt the laser operation, and, on the other hand, to inject the laser output in a spectrograph after a time interval t_g. A synthesis on laser intracavity absorption spectroscopy can be found in [BAE 99] and [RAH 05].

1.3.2. *Cavity ring-down spectroscopy (CRDS)*

While Fourier transform spectroscopy originated in the 1960s, CRDS or similar techniques date back to the end of the 1980s. As previously described, there is an earlier work [PAK 70], which shows that the structure of the line spectrum of a laser, with inhomogeneous broadening of the amplification line, was due to losses depending on the intracavity laser frequency. The first CRDS study is usually cited in reference to the work of O'Keefe and Deacon [OKE 88] because they describe this spectroscopic technique by using the term cavity ring-down (CRD). It relies on the measurement of the absorption rate, in opposition to the measurement of the absorption amplitude of a light pulse confined in a closed optical cavity, with a high quality factor Q.

In a patent, Lehmann proposed the use of CRDS with a continuous monomode laser involving a monomode excitation of the cavity and an improvement of sensitivity [LEH 96]. The first CW-CRDS (Continuous Wave-Cavity Ring-Down Spectroscopy) measurements were conducted by Romanini *et al.* [ROM 97a, ROM 97b]. Subsequently, various CRDS variants are then applied with continuous lasers, in order to optimize cavity injection, such as cavity-locked ring-down spectroscopy [PAL 98] and OF-CRDS (Optical-Feedback CRDS) [MOR 04b]. CRDS is applied to an extended spectral range (spanning from UV to IR) for various applications: the fundamental spectroscopic characterization of weak transitions [MEJ 08, PEA 96, ROM 93, ROM 95], trace detection [MEI 94, SCH 94] or the detection of radicals in flames [MEI 94]. A synthesis of CRDS can be found in [SCH 97].

A typical CRDS device in pulsed mode (Figure 1.13) is composed of a pulsed laser source (dye lasers in the visible range), a high finesse optical cavity (of length L) and a detection system. It is important to recall that the finesse increases with the reflective power of the cavity mirrors. If the duration of the laser pulse t_L is small compared to the back and forth time in the cavity ($t_{ar} = 2L/c = 10^{-20}$ ns for $L = 1.5$ 10^{-12} m), light does not interfere with itself in the cavity. A stable fraction of the laser pulse (of about 10^{-5}) is injected by transmission in the cavity by the input mirror. For a typical energy pulse of 1 mJ, this fraction corresponds to 3×10^{10} photons.

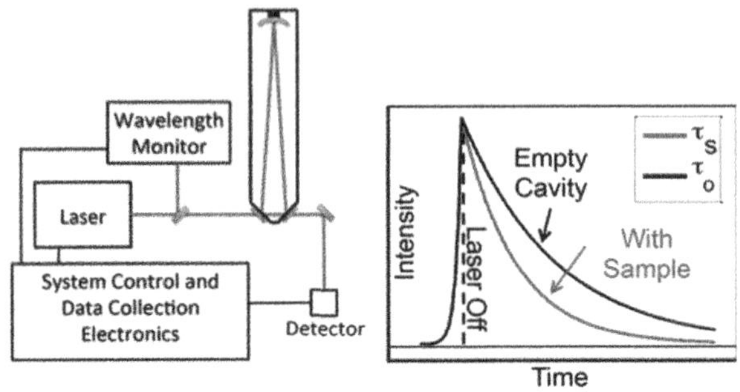

Figure 1.13. *Diagram of a CRDS setup and temporal profile of the optical signal in the absence and in the presence of an absorbent. For a color version of this figure, see www.iste.co.uk/dahoo/infrared2.zip*

The laser pulse trapped inside the cavity undergoes multiple reflections, and at each reflection, it loses a part of its energy, which is released after transmission on the mirror in front of the cavity. The temporal profile of the observed transmitted signal is then formed by a succession of pulses, whose intensity gradually decreases (Figures 1.13 and 1.14) $I(t) = I_0 \exp(-\tau/t)$, where the time constant τ depends on the presence or absence of an absorbent in the cavity.

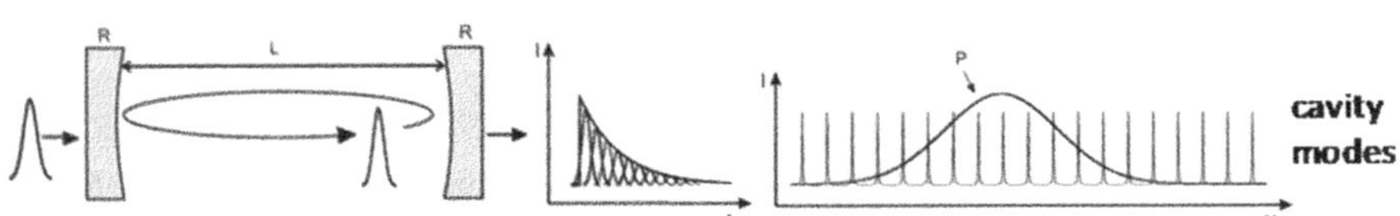

Figure 1.14. *Temporal profile of the transmission of a laser pulse and spectral profile P of the laser pulse on the cavity mode comb*

It is important to note that the laser pulse spectrum splits over several longitudinal resonance modes of the cavity ($\Delta\nu = c/2L$ and the transverse modes are not considered). Laser resolution (dye laser ≅ 0.07 cm^{-1}, Ti-sapphire ≅ 0.17 cm^{-1}, $\Delta\nu = c/2L = 0.0115$ cm^{-1} for $L = 43.5$ cm) must be higher than the gap between modes, so that the absorption lines appear in the laser gain profile.

Assuming that the Beer–Lambert law can be applied, the detected intensity is given by $I_t(\sigma) = I_0 \exp(-\sigma Cl)$, where $I_t(\sigma)$ is the intensity transmitted at the wave number σ, C is the concentration of the absorbent and l is the length of the sample traveled by the wave, following multiple back and forths of the beam in the cavity. The absorption coefficient α is defined by the product Cl so that the transmitted intensity can be written according to equation [1.22]: $I_t(\sigma, t_g) \cong I(\sigma, t_g)\exp(-\alpha(\sigma)l_{eq})$.

The calculation of the intensity after n passes in the cavity leads to the expression:

$$I_n = (Rexp(-\alpha L))^{2n} I_1 = I_1 \exp(-2n(-lnR + \alpha L)) \qquad [1.24]$$

which expresses the laser intensity detected by the photoconductor, after n back and forths in the cavity and where $I_1 = T^2 I_0 \exp(-\alpha L)$, with R being the reflection coefficient of the mirrors, T the transmission coefficient of the mirrors and I_0 the intensity of the laser injected in the cavity. With the reflection coefficient being close to unity, lnR is approximated by $-(1 - R)$. By putting $t = 2nL/c$ for the time duration of the path on n back and forths in the cavity, an expression depending on time t is obtained:

$$I(t) = I_1 \exp\left(-\frac{t}{\tau_S}\right) \qquad [1.25]$$

with:

$$\tau_S = \frac{L}{c((1-R)+\alpha L)} \qquad [1.26]$$

The CRDS decay time in the absence of an absorbent has the following expression:

$$\tau_0 = \frac{L}{c(1-R)} \qquad [1.27]$$

where L is the distance between the mirrors, c is the speed of light, and R is the reflectivity of the mirrors. τ_0 is obtained by putting $\alpha = 0$.

Equation [1.26] directly gives the absorption coefficient α in the form:

$$\alpha = \frac{1}{c\,\tau_S} - \frac{1-R}{L} = \frac{1}{c\,\tau_S} - \frac{1}{c\,\tau_0} \qquad [1.28]$$

The time τ_0 or τ_s, referred to as decay time or ring-down time, represents the average duration of the trapping of the light energy in the resonator. The measurement of this time is the fundamental principle of CRDS. The ring-down time is measured for a given number of waves and leads to the absorption coefficient. It is important to note that the effective length L_{eq} traveled by light is equal to $L_{eq} = c\tau$. For a cavity of length $L = 1$ m and a reflectivity coefficient $R = 0.9998$, $\tau = 16.7$ µs. This corresponds to an effective length traveled $L_{eq} = c\tau = 5$ km. For low intensity signals, the formula [1.28] leads to $\delta\tau/\tau = \delta\alpha/\alpha$. The sensitivity of the technique leads to a detection limit imposed by the noise $\delta\tau_N$, which is $\delta\alpha_N = \delta\tau_N/(\tau L_{eq})$. For a molecule such as methane, CH_4, which presents isolated absorption lines, the latter appear on a nearly constant horizontal base line, depending on the wave number related to the term $(1-R)/L$. For triatomic molecules such as CO_2, which present a dense absorption spectrum, it is necessary to have an independent measurement of the base line obtained with an empty cavity, in order to measure the absorption coefficient of isolated lines.

The pulse injection in the optical cavity can be easily achieved with pulsed lasers. The technique is insensitive to the fluctuations in intensity of the laser pulse, which is one of the main noise sources with direct absorption measurement methods. For the dye lasers, these fluctuations are on average about 10%. The dye lasers (Figure 1.9) make it possible to cover a spectral range that extends from the IR to near ultraviolet by using doubling or mixer crystals. Given that the developed systems are robust and shielded from vibrations, the minimal detectable absorption coefficient ranges between 10^{-6} cm^{-1} and 10^{-10} cm^{-1}.

In pulsed CRDS, the width of the laser pulse spectrum and the poor quality of the spatial profile of the laser beam are drawbacks that led to the development of continuous wave CRDS (CW-CRDS). In practice, the spectrum of the laser pulse splits over several longitudinal modes (Figure 1.14) and over several higher-order transverse modes. The result is a profile gap with respect to the exponential function of the intracavity field ring-down. It is important to note that the spectral width of the laser pulse must be below that of the targeted space to be able to detect its absorption. These limitations of pulsed lasers are eliminated when a continuous emission mono-frequency laser is coupled to the resonance modes of the cavity.

CW-CRDS involves tuning a continuous emission monochromatic (mono-frequency) laser on a resonance mode of a high-finesse optical cavity. Compared to the pulsed CRDS, CW-CRDS is more difficult to implement, due to the need to effectively inject over a sufficient time duration, the narrow laser spectrum (of the order of MHz for a telecommunication laser diode, for example) in the narrower mode of the cavity modes ($\simeq 10$ kHz). CW-CRDS was developed by Romanini *et al.* at the end of the 1990s, first with a dye laser [ROM 97a] and then with a laser source with an external cavity [ROM 97b], and in 2004 with a fiber telecommunication laser diode (of distributed feedback (DFB) type) [MOR 04].

The diagram of the setup used by Romanini *et al.* [ROM 97a] is illustrated in Figure 1.15. The cavity consisting of two mirrors of very high reflectivity is linear and the injection in the cavity is achieved by a simple passage through resonance, without frequency control. The cavity length is modulated by applying an electrical voltage on the piezoelectric plate (P) that supports one of the mirrors. The intensity transmitted by the cavity is focused on the photodiode. Thanks to modulation and continuous and single-frequency emission, scrolling of the $TEM_{m,n}$ modes of the cavity can be displayed. This configuration can be used for precise optical alignment in order to reach a spatial agreement between the laser and a TEM_{00} mode of the cavity.

The signal emitted from the length-controlled cavity is directed towards a threshold detector circuit. When the state of the system is such that the fixed frequency of the laser coincides with that of the resonance, and the intracavity field reaches a threshold value, the circuit rapidly (≈ <1 μs) interrupts the acousto-optical modulator that deflects the beam towards the cavity entrance. The modulator plays the role of a switch and sends a significant fraction (80%) of the laser beam towards the cavity in position 2. A ring-down signal can be recorded with a high SNR, in theory, with a transmitted intensity equal to the incident intensity. At the end of the recording, the modulator sends the laser beam to its initial position waiting for a new injection in the cavity. The transmission signal of the cavity is digitized and digitally processed by software.

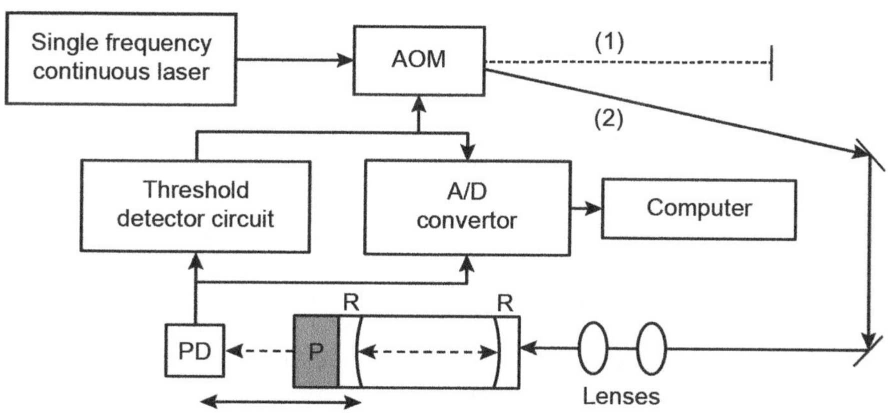

Figure 1.15. *CW-CRDS by injection through resonance [ROM 97a]: P, piezoelectric; AOM, acousto-optical modulator; PD, photodiode; A/D, analog-to-digital converter [DOM 97a, DOM 97b]*

The duration of the measurement can be less than one second for an average over 100 measured decay times. The measurement of the ring-down time is not affected by the amplitude noise from the conversion of the laser phase noise by the cavity, and which results from its greater width than that of the longitudinal mode of the cavity. CW-CRDS typically allows a detection dynamic covering 3 to 4 orders of magnitude (varying from 10^{-5} cm^{-1} to 10^{-9} cm^{-1}).

The repetition rate of the ring-down time is limited by the modulation frequency of the cavity (>100 Hz) in CW-CRDS, while in pulsed CRDS it depends on the laser used (often $\simeq$10 Hz only). Reproducibility of the ring-down time values, from one measurement to another, is of about 0.1% in CW-CRDS and of about 1% with systems employing pulsed CRDS. This difference is directly linked to the stability of the laser source. With CW-CRDS, the laser is spectrally and spatially stable, while a pulsed laser presents spectral distribution fluctuations and spatial fluctuations. Therefore, from one measurement to another, the excited transverse and longitudinal modes are not the same, which leads to different ring-down time values.

Since the spectral width is generally determined by the laser width, CW-CRDS provides a gain in spectral resolution compared to pulsed CRDS. In the first case, it is of the order of MHz (for a DFB diode) and in the second case, it is of the order of GHz. Hence, it is suitable for atmospheric trace-gas detection. Moreover, the use of laser diodes in CW-CRDS allows for a significant reduction in the size of the optical system and its cost. A disadvantage of CW-CRDS is the limited availability of laser sources, in particular of laser diodes in certain spectral regions, which moreover only provide a limited scanning range.

1.3.3. *Frequency comb spectroscopy (FCS)*

At the end of the 1990s, the mode-locking technique led to a breakthrough in the precision of frequency and time measurements thanks to laser frequency combs [HÄN 06, HAL 06]. The steady pulse train of a mode-locked femtosecond laser can give rise to a steady comb spectrum composed of 1 million modes with equal spacing at the pulse repetition frequency. The frequency of the electromagnetic waves in this optical range varies between $4{,}010^{14}$ Hz and $7{,}510^{14}$ Hz. The measurements of absolute phase coherence of these frequencies proved difficult [CUN 01]. Electronic techniques can be used to measure and directly generate radiofrequencies (RF) up to several hundred GHz. Nowadays, the most stable and precise frequency standards are achieved in the RF range on the basis of hyperfine transitions of cesium at 9.2 GHz and of hydrogen at 1.42 GHz. The transition of cesium is used to define the second in the international system of units (SI).

The development of studies related to mode-locking dates back to the beginning of the 1960s [DID 64, HAR 64, YAV 65] with the description of the generation of a continuous train of pulses by the modulation of the internal losses of a He–Ne laser cavity by means of an acousto-optical modulator. Mode-locking involves phase synchronization of the longitudinal modes of a laser, therefore reaching an in-phase state. If at a given instant t, the electrical field of each mode is at a maximum, the laser output is a pulse centered on the value t. The greater the number of synchronized modes, the shorter the duration of this pulse. There are a variety of locking techniques, the most commonly used being dye lasers with saturable absorption, the use of an acousto-optical modulator or the Pockels cell. The temporal width of the pulse depends on the width of the gain envelope, since the number of longitudinal modes on which the laser can operate depends on the losses of the cavity (Figure 1.12). From 1960 to 1970, the mode-locking technique was implemented on various types of lasers: gas lasers [DEM 65, DEC 68, WOO 68], solid lasers [DEU 65, DEM 66, DID 66], liquid dye lasers [GLEN 68, BRA 68, SAM 68, SCH 68] and semi-conductor lasers [MOR 68].

In a multimode laser, there is *a priori* no phase relation between the longitudinal modes oscillating in the cavity, since their phases are not fixed and randomly vary, mainly due to the action of spontaneous emission. Consequently, interference between longitudinal modes are not constructive, and the power fluctuations they generate are on average zero, leading to the emission of constant optical power. If the modes are interconnected by a fixed phase relation, they can constructively interfere. The beats then produce a modulation of the intracavity optical power, potentially leading to the generation of pulses (Figure 1.16), whose period corresponds to the time it takes light to go back and forth in the cavity. In this case, the laser is known as mode-locked laser.

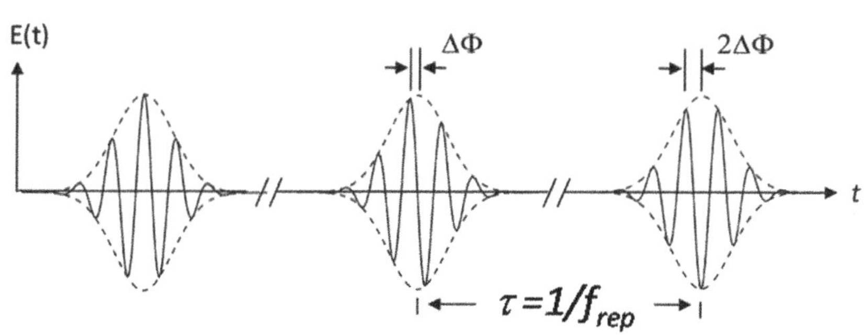

Figure 1.16. *Pulse train emitted by a laser source in the temporal range*

In the frequency range (Fourier transform), the intensity spectrum of a pulse train is a frequency comb (Figure 1.17). The phase difference $\Delta\Phi$ between the envelope and the pulse carrier leads to an offset f_0 in the frequency spectrum. Each peak is separated by the repetition frequency f_{rep} of the pulses. The line frequency f_n is given by the relation $f_n = nf_{rep} + f_0$. It is essential to properly control the values of parameters f_{rep} and f_0. In this case, the device behaves like a set of single-frequency lasers, or thousands of lasers that can be used as an optical rule. In practice, mode locking in a laser cavity is classified into two categories: active and passive.

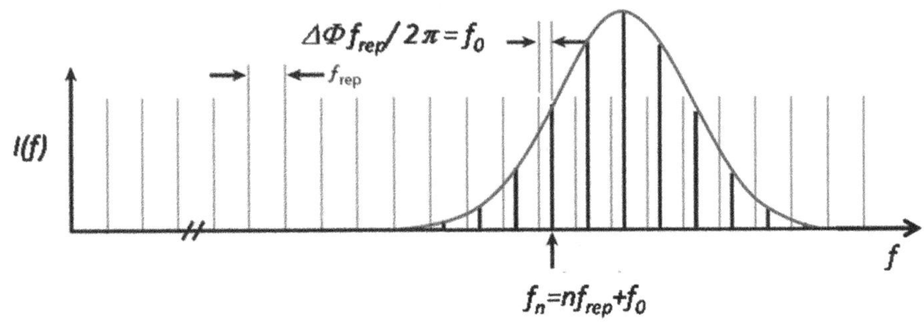

Figure 1.17. *Pulse train emitted by a laser source providing a frequency comb in the time domain. For a color version of this figure, see www.iste.co.uk/dahoo/infrared2.zip*

Active mode locking can be done by an external signal that generates a modulation of the intracavity field. The external modulated signal can be electrical, by modulation of the current injected into the device, or optical, by non-resonant injection into the cavity of a modulated optical signal. This amplitude modulation generates sidebands that are superimposed onto the cavity modes when the modulation frequency is identical to the spectral interval between modes. The fixed phase relation between these modulation sidebands is then transferred to the cavity modes, leading to their phase locking. The active mode locking with loss modulation can be achieved with acousto-optical modulation with an external RF signal. An acousto-optical modulator for mode locking differs from that used for Q-switching: an acousto-optical mode locker is a continuously operating Bragg diffraction unit with stationary wave, while an acousto-optical Q-switch is a Bragg diffraction unit that is switched on and off in order to switch the Q cavity between two different values. Synchronous pumping can also be used to generate ultra-short pulses, considered to be active mode locking by gain modulation. The gain medium is periodically pumped, either optically or electrically, with a train of very short pulses at the same repetition rate as that of the periodic arrival in the gain medium of the pulse circulating inside the laser cavity.

Passive mode locking can be achieved by using a storable absorber located near an edge of a linear cavity. Sometimes, a saturable absorber used for the passive

Q-switch can also be used for passive mode locking, but the requirements for passive mode locking are very different from those for passive Q-switching. Passive mode-locking techniques rely on the intrinsic properties of the elements of the laser cavity to generate self-modulation of the optical power, i.e. without using an external signal. The phase synchronization of modes can be achieved by using a saturable absorber inside the laser cavity [NEW 74, HAU 75], the optical Kerr effect [NEG 91, SPE 91, ASA 93] or nonlinear gain effects [STA 67, NOM 02]. For example, passive mode locking with a saturable absorber can be achieved in a ring configuration. In this configuration, two intracavity laser pulses travel through the saturable absorber in opposite directions, rapidly decreasing the gain at the profile center (Lamb dip).

With a Ti:sapphire laser, which has a broadband gain profile, it is possible to obtain pulses of the order of the femtosecond, typically of 5 fs, since the number of locked modes can be high. Hence, mode locking makes it possible to obtain very short duration optical pulses of the order of the femtosecond. The development of mode-locked lasers and their applications to metrology was thus made possible thanks to the development of femtosecond sapphire titan lasers (Ti^{3+}:Al_2O_3). Moreover, combined with the width of their gain band, the mode locking phenomenon induced by the Kerr effect in this type of laser allows the formation of an optical frequency comb. The very strong phase coherence between the modes of the laser cavity, forming a comb whose spectral interval is constant, is adapted to the precise measurement of optical frequencies. This led to the joint award of the Nobel Prize for Physics in 2005 to J. L. Hall and T. W. Hänsch. Since its discovery, the optical frequency comb has revolutionized the field of metrology. More specifically, it provides spectroscopy with a multipurpose tool for the synthesis and measurement of optical frequencies with extremely low relative uncertainties, from 10^{13} to 10^{15}. The evolution of the frequency comb technique is described in a number of publications, among which are the following: [STO 08], [YE 10], [DID 10], [PIC 19].

The extension of frequency combs based on mode locking lasers in the MIR range is not easy [SCH 12]. Though it is possible to develop an average IR comb by using a nonlinear frequency conversion from a near-IR frequency comb [MAD 06, DEL 07, MAL 08, ADL 09, GUB 09, WON 10, LEI 11, KEI 12, ULV 14, PAP 14], these technologies are still quite complex. Consequently, MIR high-precision spectroscopy is often conducted by referencing the MIR laser source to a NIR optical frequency comb, via a nonlinear frequency conversion. In this IR range, recent works are described in [MEN 20, GAT 20, LIN 20b, DID 20].

The authors of [GAT 20] describe the development of a high power mid-IR frequency comb with a repetition frequency of 100 MHz and a pulse duration of 100 fs. The frequency difference is generated between two branches derived from an Er:fiber comb, amplified separately in Yb:fiber and Er:fiber amplifiers. Average powers of 6.7 W and 14.9 W are, respectively, generated in the forwarding signal of 2.9 µm and 1.6 µm. With a high average power, an excellent quality of the beam and a passive phase stabilization of the carrier envelope, this light source is a platform that allows the generation of broadband frequency combs in far IR, visible and far ultraviolet.

In [LIN 20b], IR frequency combs are developed for the atmospheric window in mid-IR from 3 to 5.5 µm and contain valuable information concerning the molecular composition and the function that is significant in the atmospheric processes for fundamental and applied spectroscopy. Using a fiber laser source with robust mode locking, of pulse < 11 fs in the near IR, the quadratic nonlinear optical processes ($\chi(2)$) were investigated, which led to the generation of frequency combs over any atmospheric window in mid-IR. Experiments and modeling proved the generation of intra-pulse difference-frequency that produces average IR pulses of several cycles in a single passage through lithium niobate of periodic polarization. Moreover, the harmonic and cascade $\chi(2)$ nonlinearities provide direct access to the carrier envelope offset frequency of the driving pulse train in the near IR. The high-frequency stability of the mid-IR frequency comb is used for the spectroscopy of acetone and for carbonyl sulfide with simultaneous bandwidths above 11 THz and with a spectral resolution of 0.003 cm^{-1}. The combination of a low noise and wide spectral coverage allows the detection of gas traces of concentrations of about one part per billion.

The rise of frequency combs pushes the frontiers of very high resolution molecular spectroscopy into the spectral range of mid-IR. The works of Chen, Hänsch and Picqué [CHE 19] show that a multiplexed metrology of line shapes is expected by Fourier transform spectroscopy in the mid-IR using two frequency combs. This method can be used to limit the effect of measurement distortion on the line shape, a high SNR and the possibility of having access to a broad spectral bandwidth for the high-precision spectroscopy of small molecules.

[FOR 19] and [DID 20] describe the development and improvement of optical combs throughout the last 20 years. Optical clocks, based on optical transitions of atoms, operate at a much higher frequency than microwave atomic clocks currently used as synchronization norms. Better stability is reached with the prospect of being able to redefine the second. Development of stable and self-referenced optical frequency combs that span from microwaves to optical wavelengths was the key to research efforts. Many applications go from precision synchronization to spectroscopy, high resolution imaging, range finding and navigation or space exploration.

1.4. Spectroscopy for space observation

This section describes two other examples of devices used for space observation and analysis, one *ex situ* and the other aboard a space probe. Ellipsometry was used to determine the optical constants of tholins, aerosols present in the atmosphere of planets, namely the Titan satellite, which orbits Saturn. A spectrometer (SOIR) based on an AOTF was used to collect data on the molecules present in the atmosphere of Venus.

1.4.1. *Spectroscopic ellipsometry for space observation*

Ellipsometry is a non-destructive optical technique that can be used for the study of materials in the form of thin layers. The technique is sensitive to surfaces and interfaces. Samples can be prepared by deposition on a substrate and then a spectroscopic ellipsometer, in the near IR–mid-IR range or in the visible range, can be used to determine the optical refractive indices. Studies on tholins present in the atmosphere of Titan were conducted in this context. This technique is widely described in the literature [AZZ 77, DRE 82, BOC 93, PIC 95, VED 98, TOM 99, OHL 00, TOM 2005, FUJ 07].

A synthesis of this technique, as well as the inverse method required for the analysis of ellipsometric data, can also be found in [DAH 19b] and [DAH 16]. Ellipsometry involves the study of the polarization state of light after reflection on the surface of a sample in order to determine its physical and optical characteristics, particularly its refraction index, and also its thickness and rugosity. The reflected electric field E_r can generally be decomposed into two components of orthogonal polarizations, E_{rp} and E_{rs}, whose amplitudes determine the ellipticity of the reflected wave. This ellipticity is characterized by the ratio:

$$\frac{E_{rp}}{E_{rs}} = \frac{r_p Ei_p}{r_s Ei_s} = \rho \frac{E_{ip}}{E_{is}} \quad \text{with } \rho = \frac{r_p}{r_s} = (\tan\psi)e^{i\Delta} \qquad [1.29]$$

where E_{ip} and E_{is} correspond to the amplitudes of the electric field of the incident wave. The reflection coefficients r_s and r_p in amplitude of waves p (parallel to the incidence plane) and s (perpendicular to the incidence plane) are expressed in terms of complex indices $\tilde{n}_i = n_i + ik_i$ (n_i is a real index characterizing the polarization of matter by the electric field of the wave and k_i is an imaginary index characterizing the absorption of light energy by matter, Chapter 5, equations [5.11] and [5.12], [DAH 16]) and the reflection and refraction angles (Fresnel formulas, equations [1.33] and [1.34]) of the two media on both sides of the surface or diopter).

Based on the initial polarization state of the incident wave E_i and the measurement of the polarization state after reflection, an ellipsometer provides the ellipsometric parameter ρ (equation [1.29]) that can be expressed as a function of two other ellipsometric parameters Ψ and Δ, which are related to variations of the ellipse shape (ratio of minor and major axes, orientation) and can be used to determine ρ. Irrespective of the technique employed, the instrumentation comprises the optical instruments shown in Figure 1.18, two inclinable arms and a sample support. One of the arms is composed of a source of light and an optical setup yielding an incident wave of known polarization. The other arm is made up of an analyzer and a detector for the measurement of the polarization state of the wave reflected by the sample.

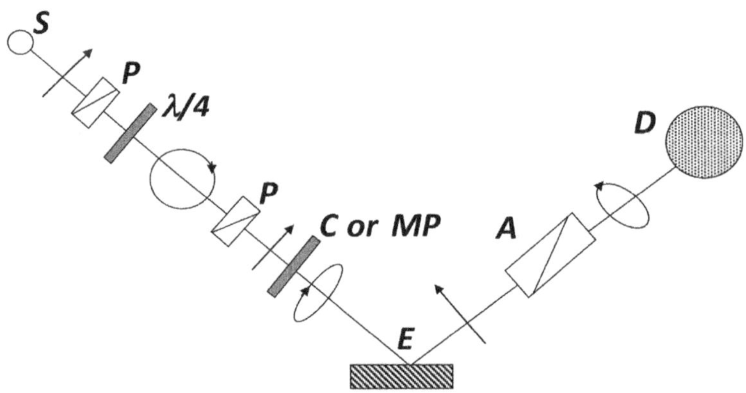

Figure 1.18. *Elements of an ellipsometer: S, source; P, polarizer; λ/4, quarter wave plate; C, compensator; MP, phase modulator; E, sample; A, analyzer; D, detector [DAH 16]*

The measurement and analysis process comprises the following stages:

– preparation of the initial state of polarization of the incident light;

– light–matter (sample) interaction and reflection;

– measurement of the polarization state of the reflected wave;

– calculation of the ellipsometric parameters Ψ and Δ or $tan(\Psi)$ and $cos(\Delta)$);

– evaluation of random and systematic errors on Ψ and Δ;

– determination of the physical properties of the material (n_i, k_i, e_i, etc.), complex indices and thicknesses from Ψ and Δ using a numerical optimization method (inverse problem).

The analysis of ellipsometry data is an inverse problem of deterministic optimization, such as that described in Chapter 2 of [DAH 16]. The optimization is conducted by comparing the experimental values to the theoretical ellipsometric parameters calculated from a model [AZZ 77, DRE 82, BOC 93, PIC 95, VED 98, TOM 99, OHL 00, TOM 2005, FUJ 07, DAH 04a, DAH 04b, FRA 05, NOU 07, LOU 07, LOU 08, DAH 10, SCI 10, MAH 11, CAR 12, SCI 12, MAH 14].

The ellipsometric parameters Ψ and Δ or ($tan(\Psi)$ and $cos(\Delta)$) measured by the ellipsometer are stored for each wavelength λ in digital form in a file. Each wavelength is identified after scattering through a grating on a diode array or a set of photomultiplier tubes. The data are stored in discrete form for each wavelength detected.

The analysis involves the comparison, for each wavelength, of the values of parameters ψ_{th} or $tan(\psi_{th})$ and Δ_{th} or $cos(\Delta t_h)$ calculated from structural models (multilayers) and optical models (dielectric function or complex index) to the measured values of ψ_{exp} or $tan(\psi_{exp})$ and of Δ_{exp} or $cos(\Delta_{exp})$. The structural model corresponds to the type of multilayer considered for the structure of the studied layer. For example, for a system composed of air, layer, substrate, there are three media to be considered for each wavelength. When the surface rugosity is taken into account, the multilayer comprises an additional layer (Figure 1.18).

In order to calculate the various reflection and transmission coefficients at the interface of various layers forming the structure of a sample, the matrix methods of Abeles [ABE 50] or of Hayfield and White [HAY 64] can be applied.

Considering the multilayers in Figure 1.18, the axis Oz being oriented downwards, the thickness of layer i is denoted by e_i and its complex index by $\tilde{n}_i = n_i + k_i$, where n_i and k_i are the real and imaginary parts of the complex index, respectively. θ_i is the angle of incidence of the incident ray in layer i. These parameters can be used to calculate the global Fresnel coefficients for a sample by means of the Abeles matrix formalism [ABE 50] and then deduce by inversion the ellipsometric parameters of the sample.

At the coordinate z, the total electric field is the sum of two components, one that propagates towards ascending z (hence towards the substrate) and the other towards descending z (towards the surface of the sample) denoted, respectively, by $E^+(z)$ and $E^-(z)$. The beam is linearly polarized either p or s, and this property is preserved across the layers. The field at two depths z and z' is connected by the matrix transformation:

$$\begin{pmatrix} E^+(z') \\ E^-(z') \end{pmatrix} = \begin{pmatrix} S_{11} & S_{12} \\ S_{21} & S_{22} \end{pmatrix} \begin{pmatrix} E^+(z) \\ E^-(z) \end{pmatrix} \quad [1.30]$$

Assuming that z and z' are the coordinates on both sides of the interface $i/(i+1)$, the expression of the interface crossing matrix $I_{i,i+1}$ is:

$$I_{i,i+1} = \frac{1}{t_{i,i+1}} \begin{pmatrix} 1 & r_{i,i+1} \\ r_{i,i+1} & 1 \end{pmatrix} \quad [1.31]$$

where r and t are Fresnel coefficients.

Depending on the type of polarization p or s, these coefficients are expressed as:

$$r_{pi,i+1} = \frac{n_{i+1}\cos\theta_i - n_i\cos\theta_{i+1}}{n_{i+1}\cos\theta_i + n_i\cos\theta_{i+1}} \qquad r_{si,i+1} = \frac{n_i\cos\theta_i - n_{i+1}\cos\theta_{i+1}}{n_i\cos\theta_i + n_{i+1}\cos\theta_{i+1}} \quad [1.32]$$

and:

$$t_{pi,i+1} = \frac{2n_i\cos\theta_i}{n_{i+1}\cos\theta_i + n_i\cos\theta_{i+1}} \qquad t_{si,i+1} = \frac{2n_i\cos\theta_i}{n_i\cos\theta_i + n_{i+1}\cos\theta_{i+1}} \quad [1.33]$$

Two matrices $I_{i,i+1}$ are calculated, one for each type of polarization.

Relating z and z' to the layer i, of thickness e_i, the propagation matrix in the layer is expressed as:

$$\mathrm{T}_i = \begin{pmatrix} e^{i\delta_i} & 0 \\ 0 & e^{-i\delta_i} \end{pmatrix} \quad [1.34]$$

where $\delta_i = \frac{2\pi}{\lambda} e_i n_i \cos(\theta_i)$

The expressions of matrices T correspond to an isotropic layer and do not depend on polarization. The response corresponding to the stacking of N layers crossed by the beam is obtained after multiplication of the matrices by placing z' at the level of the substrate and z at the level of the ambient medium. The successive angles in each layer are related by the Snell–Descartes law. The expression of the matrix corresponding to a layer defined by N multilayers is given by:

$$S = I_{0,1}T_1T_{1,2}T_2 \ldots I_{i,i+1}T_{i+1} \ldots T_{N-1}I_{N-1,N} = \prod_{i=0}^{N-1}\left(I_{i,i+1}T_i\right) \quad [1.35]$$

The matrix product for each polarization leads to the expressions of global Fresnel coefficients in the following form:

$$r_p = \frac{S_{21p}}{S_{11p}} \text{ and } r_s = \frac{S_{21s}}{S_{11s}} \qquad [1.36]$$

This leads to:

$$\rho = \frac{r_P}{r_S} = \frac{S_{21p}S_{11s}}{S_{11p}S_{21s}} = (\tan\psi)e^{i\Delta} \qquad [1.37]$$

For example, for an air, layer, substrate system, for each wavelength λ_i, ψ and Δ are functions of five parameters, three indices that lead to three real parts and two imaginary parts and a thickness such that:

$$\Psi = f(n_a, n_s, k_s, n_f, k_f, e_f) \qquad [1.38]$$

and:

$$\Delta = g(n_a, n_s, k_s, n_f, k_f, e_f) \qquad [1.39]$$

where indices *a*, *s* and *f* correspond, respectively, to the ambient medium, the substrate and the film.

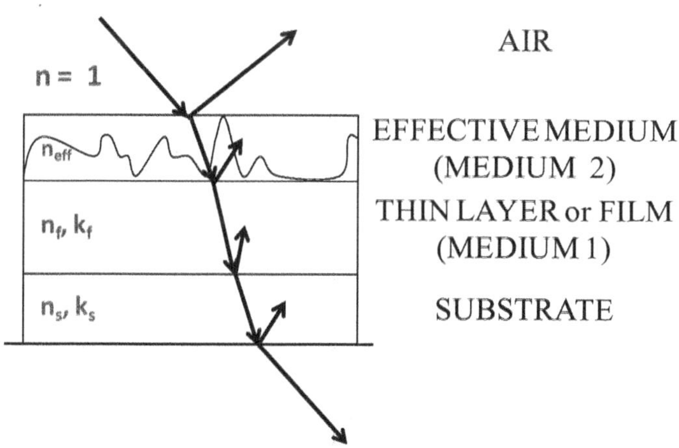

Figure 1.19. *Rough surface modeled by an effective medium. For a color version of this figure, see www.iste.co.uk/dahoo/infrared2.zip*

A cost or objective function (see Chapter 2, section 2.4.1 [DAH 16], [AZZ 77]) is defined based on ellipsometric parameters ρ_{th} and ρ_{exp}, in order to implement the minimization process. An example of cost function σ or MSE (*Mean Square Error*) that is used to estimate the gap between the calculated values and the measured values for a given angle of incidence can be expressed as follows:

$$\sigma = \frac{1}{n-m-1}\sum_{j=1}^{n}\frac{\left|\rho_{exp}(\lambda_j) - \rho_{th}(\lambda_j,\ X_j)\right|^2}{\left|\partial\rho_{exp}(\lambda_j)\right|^2} \qquad [1.40]$$

where n represents the number of experimental points, m is the number of relations between the equations (n-m-1 is the number of independent parameters, the air index is assumed to be known and equal to 1). In this expression, the denominator takes into account the SNR that is different on the measurement spectral range, which allows better adjustment of experimental data.

Depending on the ellipsometric angles, the following expression is used:

$$\sigma = \frac{1}{n-m-1}\sum_{j=1}^{n}\left(\frac{\left|\Psi_{exp}(\lambda_j) - \Psi_{th}(\lambda_j,X_j)\right|^2}{\left|\partial\Psi_{exp}(\lambda_j)\right|^2} + \frac{\left|\Delta_{exp}(\lambda_j) - \Delta_{th}(\lambda_j,X_j)\right|^2}{\left|\partial\Delta_{exp}(\lambda_j)\right|^2}\right) \qquad [1.41]$$

where $\partial\Psi$ or $\partial\Delta$ are the standard deviations of experimental data.

Two values, one for ψ and one for Δ, are measured per wavelength. The contribution of noise-contaminated measurements to the adjustment is minimized by the weighting of data by the inverse of standard deviations. In this expression, $X_j = (x_1, x_2, \ldots x_k)$ is a vector whose components are the various parameters to be adjusted. For example, in the case of a transparent isotropic film on an absorbent substrate, $X_j = (x_1= n_f, x_2= e_f, x_3= n_s, x_4= k_s)$ and therefore $k = 4$, the indices f and s corresponding to the film and substrate, respectively.

In the case of an absorbent film on a substrate, if the ambient medium is air ($n_{air} = 1$) of known index, there are only five unknowns to be determined ($k = 5$) with $X_j = (x_1= n_f, x_2= k_f, x_3= e_f, x_4= n_s, x_5= k_s)$. There is interdependency between certain parameters of X_j, which complicates analysis and optimization. For example, there is a strong correlation between the optical index and the thickness, and in order to reduce the effect of this coupling, a certain number of measures should be implemented, depending on the number of parameters to be adjusted. It is possible to combine the data for several angles of incidence, for example. For three parameters to be adjusted (e_f, n_f and k_f), six series of measurements are required, therefore three angles of incidence (each time Ψ and Δ are measured).

A model of a more complex layer can be used to take into account the presence of roughness at the surface of a thin layer or at the interface between two layers. In this case, the film is decomposed into two other homogeneous layers, consisting of a medium of different indices, with plane surfaces (Figure 1.19). The first layer (medium 1) is composed of a material of unknown index and thickness. Above it, a second layer consists of an effective medium, corresponding to an air–material mixture, with an effective index and a certain amount of air inclusion for modeling the presence of rugosity (medium 2).

Two models are used (Figure 1.20): the Maxwell–Garnett (MG) model, which corresponds to inclusions (proportion $p_m << 1$) in a host matrix, and the Bruggeman (BR) model, which is applied when various materials are present in similar proportions.

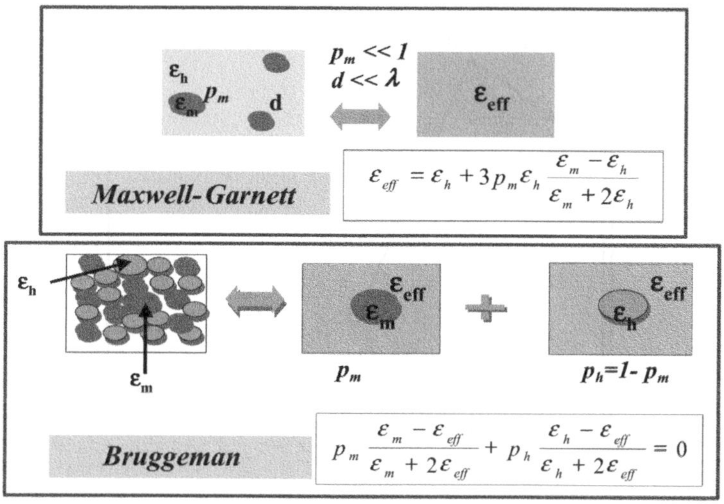

Figure 1.20. *Maxwell–Garnett and Bruggeman mean-field models. For a color version of this figure, see www.iste.co.uk/dahoo/infrared2.zip*

The following formula is used:

$$\frac{n_e^2 - n_h^2}{n_e^2 + 2n_h^2} = \sum_i f_i \frac{n_i^2 - n_h^2}{n_i^2 + 2n_h^2} \qquad [1.42]$$

where n_i is the index of inclusion, n_h is the index of the host medium, n_e is the index of the effective medium, and f_i is the inclusion fraction for the calculations of indices. The indices are related to the electric permittivity by the well-known relation $n_i = \sqrt{\varepsilon_i}$. The MG model considers that one of the two materials is the host material, i.e. $n_h = n_1$

or n_2 if two materials are present in the medium. The BR model considers the effective medium to be the host medium, and the index of the host material is replaced by $n_h = n_e$. For example, with the MG model, a certain thickness whose order of magnitude is close to the physical situation must be fixed, i.e. the thickness of the average roughness, as well as a fraction of the inclusion of air in the layer.

All of these parameters must also be adjusted by the fitting algorithm in order to minimize the gaps between the calculated and experimental values. Optimization is the search for the maximal or minimal value of a function *f(x)* (Chapter 2, section 2.2, [DAH 16]) that can present local minima or maxima (Figure 1.21). This can always be dealt with as a minimization problem (in this case *-f(x)*) is optimized and the algorithm used must be able to reach the minimal value while avoiding local minima.

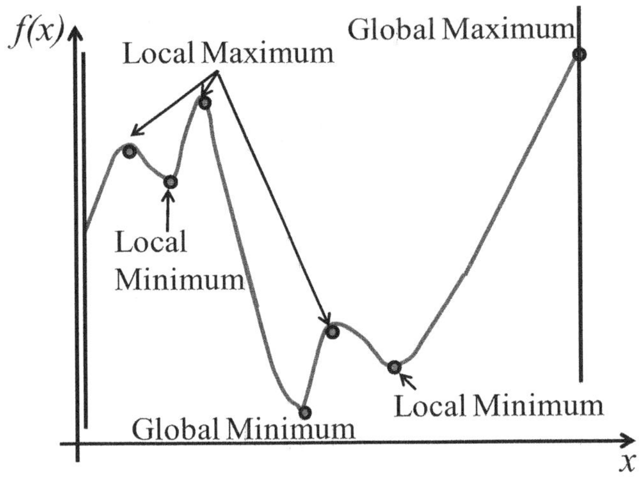

Figure 1.21. *Maxima and minima of a function. For a color version of this figure, see www.iste.co.uk/dahoo/infrared2.zip*

Generally speaking, since the models depend nonlinearly on the variables to be adjusted, the cost function σ is a nonlinear function of the parameters of $X_j = (x_1, x_2, \ldots x_k)$. The minimum can only be obtained by an iterative method: an initial estimation of parameters is refined at each stage until the parameters no longer vary. Inversion algorithms are generally used and they can be classified into two categories: those that require the calculation of gradient, such as gradient descent or conjugate gradient, of Newton, Levenberg–Marquardt (LM) [LEV 44, MAR 63], Davidon–Fletcher–Powell (DFP) [DAV 59, FLE 63] or Broyden–Fletcher–Goldfarb–Shanno (BFGS) [BRO 69, FLE 70, GOL 70, SHA 70]; and those that minimize the function without calculating the gradient, such as the Simplex method

[DAN 90] or closed methods [POW 64, BRE 73]. The commonly used algorithms in ellipsometry are the nonlinear simplex method of Nelder and Mead [NEL 65, PRE 86, pp. 521–528], a zeroth-order method that does not require the calculation of gradient, and of the LM and the BFGS methods, second-order methods that require the calculation of a gradient of the cost function for the evaluation of second-order derivatives. Second-order methods approach the cost function by an expansion limited to the second order.

The ellipsometry technique can be used to determine the optical constants of aerosols present in the atmospheres of planets or satellites of planets. This section presents the studies conducted on tholins, aerosols present in the atmosphere of Titan, satellite of Saturn [SCI 10, CAR 12, MAH 11, SCI 12, MAH 12a, MAH 12b, MAH 12c, MAH 14]. Sunlight scattering and absorption by the aerosols are an integral part of understanding the radiative balance of the planetary atmosphere covered by mist such as Titan's, and possibly the Earth's, early atmosphere. A key optical parameter of an aerosol is its refractive index.

Titan, the largest moon of Saturn, is surrounded by a very dense atmosphere. Since the beginning of the exploration of giant planets in the 1980s, Titan has focused a lot of attention on its unique atmosphere, characterized by thick mist. Based on radial intensity scans performed by the Voyager 2 space probe at high phase angles, Rages and Pollack [RAG 80] were able to determine the vertical distribution of scattered particles and to reveal a significant reduction in the density of aerosol particles under this layer.

Testing the numerical models used for the analysis of Titan observation data (albedo, radiative transfer, vertical profile of the mist, surface contribution, etc.) requires precise knowledge of the optical constants of Titan aerosol analogues or tholins. On Titan, the dissociation of N_2 and CH_4 by solar UV radiation and the electron bombardment of the Saturn magnetosphere induce a complex organic chemistry that leads to the production of solid aerosols in the stratosphere and upper atmosphere. These aerosols explain the orange mist that gives Titan its characteristic color. There is strong astrobiological interest in Titan's aerosols as they play an important role in the climate, composition and properties of Titan's atmosphere and surface.

In [SCI 10] and [SCI 12], the ellipsometry technique is used to determine the optical constants of the tholins produced by an RF plasma discharge in a gas mixture (95% N_2–5% CH_4) simulating the main atmospheric composition of Titan and deposited as a film on an Al–SiO_2 substrate (Figure 1.22).

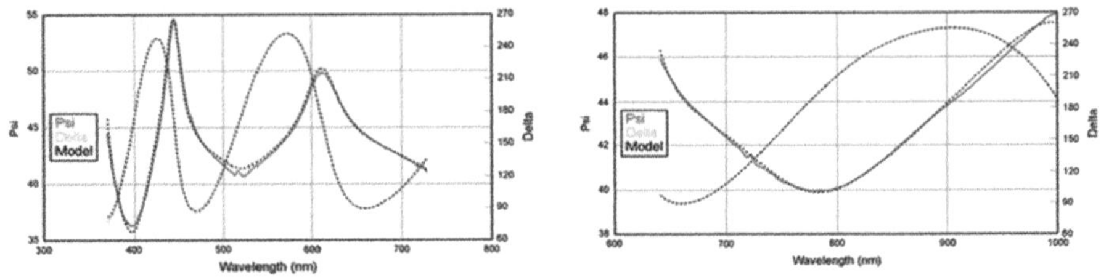

Figure 1.22. *Ellipsometric parameters ψ (Psi) and Δ (Delta) on the spectral range of 370–730 nm and 640–1,000 nm and adjustment in the dotted line. For a color version of this figure, see www.iste.co.uk/dahoo/infrared2.zip*

The real and imaginary parts, *n* and *k*, of the complex refractive index of tholins are determined over the spectral range from 370 nm to 900 nm by spectroscopic ellipsometry. The angle of incidence is fixed at 70°, close to Brewster's angle of SiO_2 substrate deposited on Ag or Au, and whose optical characteristics are preliminarily determined using Sellmeier's formula (equation [1.44]] adapted for transparent substrates. The optical constants of tholins are compared to those determined by other authors, and to the constants of Titan's aerosols, extracted from observation data provided by space probes (Figure 1.22), and are determined by Cauchy formulas for the real part and an exponential form for the imaginary part (equations [1.45] and [1.46]). Surface rugosity is modeled by the MG model

$$(n(\lambda))^2 = 1 + \sum_j {A_j\lambda^2}/{\lambda^2 - \lambda_j^2} \qquad [1.43]$$

$$n(\lambda) = A + {B}/{\lambda^2} + {C}/{\lambda^4} \qquad [1.44]$$

$$k(\lambda) = \alpha e^{\beta hc\left({1}/{\lambda} - {1}/{\gamma}\right)} \qquad [1.45]$$

In formula [1.44], λ_j is the wavelength of the resonant absorption *j* and A_j is its absorption line force. In formulas 1.45 and 1.46, given *n* and *k, A, B, C,* α and β are adjustable parameters, γ is the limit of the forbidden band fixed at 250 nm for the adjustments, λ is the wavelength, *h* is Planck's constant, and *c* is the speed of light. The parameter *A* is connected to the approximate value of the amplitude of the refractive index of the material, and *B* and *C* are the parameters that give the profile of the index variation depending on the wavelength. The parameters α and β determine the profile of the extinction coefficient *k*. In the exponential formula (Urbach tail model) used for modeling the imaginary part of index *k*, the parameter γ is related to the forbidden band of the material, which can vary depending on its structure and composition.

In Figure 1.23, the real part of the refractive index, n, decreases from $n = 1.64$ (at 370 nm) to $n = 1.57$ (at 900 nm). The values n that are given are in agreement with those of tholins resulting from photochemistry [TRA 03] or discharge plasma DC [RAM 02]. The order of magnitude of n does not seem to depend on the production process of tholins. The values of k show two types of variations: an exponential decrease from 370 nm to 500 nm, with $k = 12.4\ e^{-0.018\lambda}$ (where λ is expressed in nm), followed by a plateau, with $k = (1.8 \pm 0.2)10^{-3}$.

This characteristic of a decreasing variation in two regimes is also observed in other studies of tholins though the change of slope does not always occur at the same wavelength. This may be related to the forbidden band in the absorption region and which depends on the structure of the material.

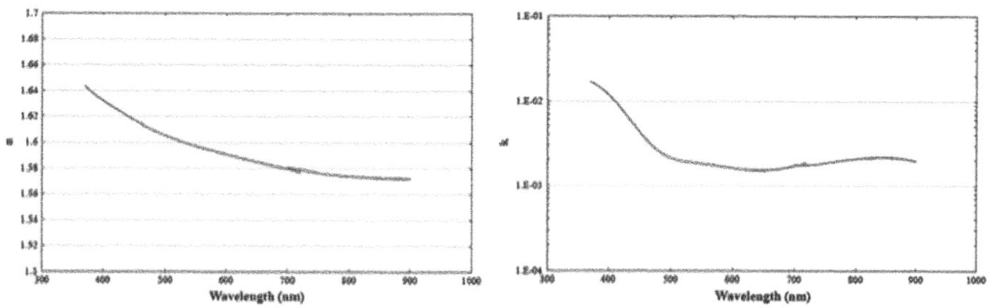

Figure 1.23. *Optical constants n and k in the 370–900 nm range. For a color version of this figure, see www.iste.co.uk/dahoo/infrared2.zip*

[MAH 12a], [MAH 12b] and [CAR 12] present the results of a study on the effect of various initial concentrations of methane during the preparation of tholin aerosols on the optical constants of tholins. Spectroscopic ellipsometry is used in the 370–1,000 nm range of wavelengths. As the thin layers of tholins are amorphous, the Tauc–Lorentz model (equation [1.47]) is used for the optical characterization

$$\varepsilon_2 = AE_0C\left(E - E_g\right)^2/E((E^2 - E_0^2)^2 + C^2E^2) \quad E > E_g$$
$$\varepsilon_2 = 0 \quad E < E_g \qquad [1.46]$$

The expression of ε_2, the imaginary part of dielectric permittivity, comprises four energy parameters: the forbidden band, for example, the peak of joint density of states E_0, the broadening parameter C and the prefactor A, which includes the elements of the optical transition matrix [JEL 96]. The real part of the dielectric permittivity is then calculated using Kramers–Kronig relations (Chapter 5, equations [5.13] and [5.14], [DAH 16]).

Figure 1.24 (TF_i with i%) shows the optical constants of a layer of tholins on Al–SiO_2 determined with the Tauc–Lorentz model for various percentages of methane in the initial gas mixtures of 1%, 2%, 5% and 10%. The initial quantity of CH_4 injected into the reactor is about twice that involved in the production of thin films.

The values are comparable to those obtained with the B-spline model ([MAH 12a, MAH 12b]). A strong dependency of optical constants correlated with the initial concentrations of methane in the gas phase of the mixture used for the fabrication of tholins is observed. The imaginary optical index (k) decreases with the initial concentration of CH_4 from 2.3×10^{-2} down to 2.7×10^{-3} at a wavelength of 1,000 nm, while the real optical index (n) increases from 1.48 to 1.58 at 1,000 nm.

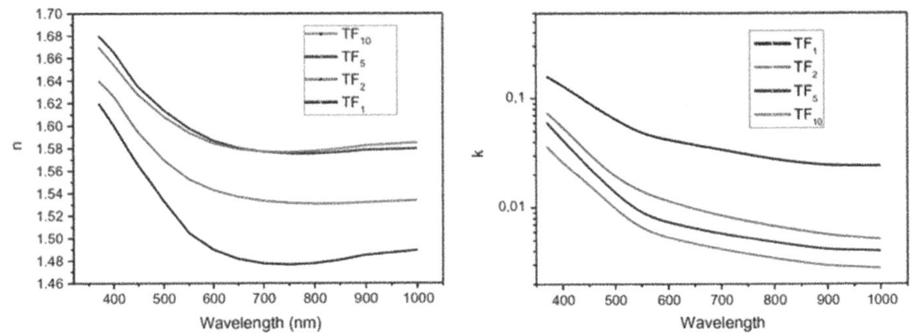

Figure 1.24. *Optical constants n and k of tholins between 400 nm and 1,000 nm for various initial concentrations (TFi: i%) of methane [MAH 12a]. For a color version of this figure, see www.iste.co.uk/dahoo/infrared2.zip*

The mid-IR spectra [SCI 10, CAR 12, MAH 12a, MAH 12b, GAU 12] recorded for thin films of tholins produced at various initial concentrations of methane are shown in Figure 1.25. Analysis of this spectrum shows that the increase in the signature of secondary and primary amines in absorption in mid-IR is correlated with higher absorption of tholins in the visible region (high k) for the thin layers produced when the percentage of methane is low. The signatures in mid-IR of VIMS (Visual and Infrared Mapping Spectrometer) and CIRS (Composite Infrared Spectrometer) are in agreement with an aerosol dominated by aliphatic carbon content.

The values of k of tholin films are compared to those of Titan aerosols deduced from Voyager [RAG 80] and Cassini/VIMS (Visual and Infrared Mapping Spectrometer) [RAN 10] observations (Figure 1.26).

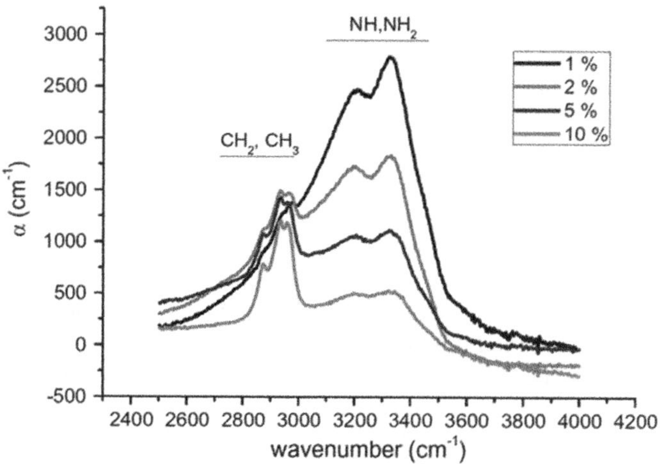

Figure 1.25. *Absorption spectra of tholins in mid-infrared for various initial concentrations of CH_4 [MAH 12a]. For a color version of this figure, see www.iste.co.uk/dahoo/infrared2.zip*

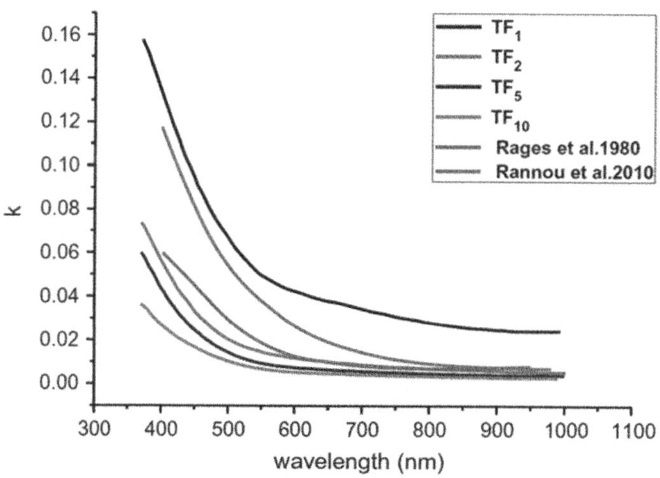

Figure 1.26. *Comparison of optical constants k of tholins for various initial concentrations (TFi: i%) of methane [MAH 12a] and the values deduced from data provided by planetary missions [RAG 80, RAN 10]. For a color version of this figure, see www.iste.co.uk/dahoo/infrared2.zip*

Rannou *et al.* [RAN 10] deduced their values from data provided by the DISR (Descent Imager/Spectral Radiometer) at low altitude (<140 m), using a model of scattering by fractal aggregates. The values of k thus obtained are particularly high and range between those of TF_1 (1%) and TF_2 (2%). These high values of k obtained from Titan observations are in definite disagreement with the aliphatic-rich aerosols. On the other hand, the extinction coefficient k calculated for Titan from Voyager data [RAG 80] is closer to the imaginary indices obtained for TF_2 (2%), TF_5 (5%) and TF_{10} (10%) which are relatively close to one another. A possible explanation for this apparent disagreement is the difference in the chemical composition of Titan aerosols at low and high altitudes. These conclusions on various compositions in the atmospheric layers of Titan seem to be supported by the works of Doose *et al.* [DOS 16] who reexamined the DISR (Descent Imager/Spectral Radiometer) data in the atmosphere of Titan with two constraints issued from measurements conducted outside the atmosphere, relying on the detection of layers of stratospheric clouds by the Cassini composite infrared spectrometer (CIRS). DISR data led to proposing three thin layers and a thicker layer of atmospheric aerosols. The initial aerosol model does not correspond to the geometric albedo or to the size of Titan's disc. The new aerosol model proposed by Doose *et al.* is compatible with DISR observations conducted inside and outside the atmosphere. DISR data correspond to a nearly constant low aerosol extinction coefficient at nearly 55 km altitude. The aerosols grow significantly darker from 200 km to approximately 80 km in altitude.

1.4.2. *Space-borne spectroscopy*

Mars Express (MEX) is the first European planetary mission on the Red Planet. The Mars Express orbiter (1,220 kg) was launched in June 2003 and placed on an elliptical polar orbit around Mars on December 25, 2003, with a pericenter altitude of 259 km, an apocenter altitude of 11,559 km and a period of 7.57 h. The spectroscopic study of the UV and IR signatures, characteristic of Mars' atmosphere, was conducted with a device referred to as the SPICAM (Spectroscopy for the Investigation of Characteristics of the Atmosphere of Mars). It is a spectrometer operating on an ultraviolet channel and a channel for IR radiation mainly dedicated to the study of Mars' atmosphere. It was developed through a collaboration between LATMOS (formerly Aeronomy Service: J.L. Bertaux) in France, BIRA-IASB (Royal Belgian Institute for Space Aeronomy: P. Simon) in Belgium and IKI (Russian Institute for Space Research: O. Korablev).

SPICAM is a light (4.7 kg) UV–IR double spectrometer, the first instrument to make stellar occultations on Mars. The UV imaging spectrometer (118–320 nm, resolution ≅ 1.5 nm, intensified CCD detector) was designed to obtain atmospheric vertical profiles and its fabrication relied on SPICAM plans, which was aboard Mars 96, which could not complete its mission. It is composed of two sets of instruments:

one for solar occultations, and one for stellar occultations and NADIR observation. SPICAM instrumentation is composed of two blocks: one sensor unit with a total mass of 3.8 kg, which comprises the UV and near-IR spectrometers, and a data processing unit (DPU, 0.9 kg) providing the interface of these two channels (Figure 1.27) with the spacecraft.

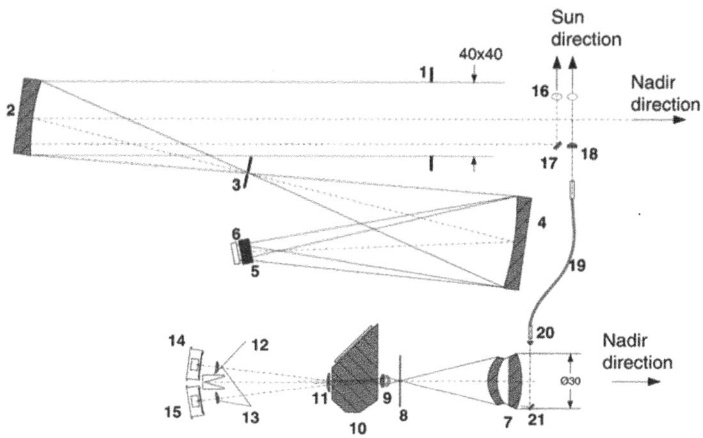

Figure 1.27. *Optical diagram of UV and IR channels of SPICAM light. (1) UV channel aperture; (2) off-axis parabolic mirror; (3) slit (mechanically - controlled aperture of variable width from wide to narrow); (4) concave UV grating; (5) intensifier; (6) CCD; (7) IR channel objective; (8) IR FOV diaphragm; (9) and (11) collimation lenses; (10) AOTF crystal; (12) light trap for non-diffracted light; (13) detector proximity lenses; (14) extraordinary beam detector; (15) ordinary beam detector; (16) solar aperture (closed by a shutter when the Sun is not observed); (17) and (21) plane mirror; (18) IR solar entry; (19) optical fiber; (20) fiber collimator [BER 06]. For a color version of this figure, see www.iste.co.uk/dahoo/infrared2.zip*

The UV imaging spectrometer (118–320 nm, resolution 1 nm, intensified CCD) of SPICAM is mainly dedicated to stellar occultations on Mars [BER 06]. The IR channel of SPICAM is a separate AOTF spectrometer integrated into the optical block of SPICAM with the UV spectrometer [NEV 06]. For the first time in planetary research, this spectrometer used acoustic-optic tunable filter (AOTF) technology, which allowed an unprecedented mass reduction for this type of instrument: 0.7 kg. The electrically controlled acoustic-optic filter sequentially scans the entire spectral range at a desired sampling rate, with random access. The absorption bands of H_2O (1.1 mm and 1.37 mm) are recorded with bands of CO_2 (1.43, 1.58 and 1.65 mm) making it possible to recover relative contents of H_2O. The near-IR albedo of Mars is measured with high spectral resolution, revealing bands characteristic of H_2O and CO_2 ice, though it is not possible to distinguish if the origin of these characteristics is related to the atmosphere (clouds) or to the surface (ice and frost). As expected, the emission of O_2 (a1D g) at 1.27 mm, occurring in the process of photodissociation of ozone, is easily detected by the IR SPICAM, allowing still another way to measure the ozone of Mars besides the UV measurements. The emission of O_2 (a1D g) allows the probing of altitudes mainly above 15–20 km; the reaction producing this emission is severely depleted by non-radiative quenching below this altitude. The two detectors of the AOTF spectrometer are sensitive to two orthogonal polarizations.

The ESA Venus Express orbiter was the first mission to apply solar and stellar occultation techniques to the atmosphere of Venus, with SPICAV/SOIR ((Solar Occultation in the Infra-Red) instruments. SPICAV (V for Venus) on the SPICAM model is composed of a series of three spectrometers with varied scientific objectives [BER 07].

SOIR is one of the three high-resolution spectrometers with a new optical design, which allows it to reach a resolution power of $\lambda / \Delta\lambda$ = 15,000–20,000 in a perfectly reasonable volume and mass (6 kg) balance, operating in the 2.2–4.3 μm range. It is the highest resolution achieved in planet exploration. The optical design comprises a high dispersion grating that operates at high orders and an acoustic-optic tunable filter (AOTF) to sort one order at a time. A detailed description of SOIR can be found in Nevejans *et al.* [NEV 06, VIL 08a], while in-flight performances are described in Mahieux *et al.* [MAH 08].

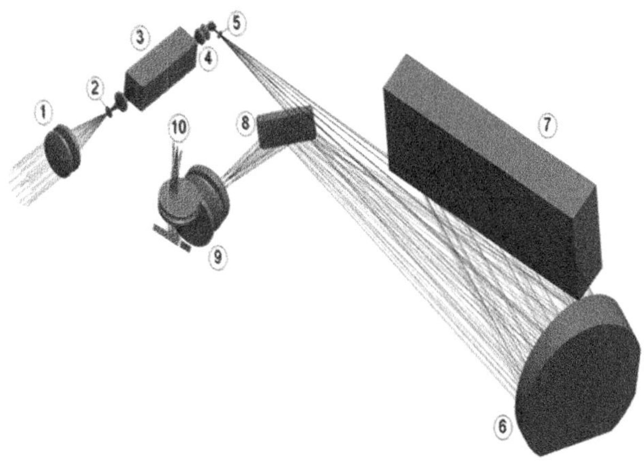

Figure 1.28. *Optical diagram of SOIR: (1) entrance optics, (2) diaphragm, (3) AOTF, (4) exit optics, (5) entrance slit of the spectrometer, (6) off-axis parabolic collimation and camera mirror, (7) echellette grating, (8) plane mirror, (9) detector optics and (10) detector assembly [NEV 06, MAH 08, VIL 08a]*

Venus is a very hot and dry planet, with a dense atmosphere mainly composed of carbon dioxide (CO_2, 96.5%) and nitrogen (N_2, 3.5%). Chemically active species, such as sulfur-based gases (OCS, SO_2) and halogenides (HCl, HF), had already been reported but since measurements had essentially already been conducted in the mesosphere below 100 km and below clouds, information on minor atmospheric components, their concentration, their reactions, their sources and their wells were incomplete. Venus is fully wrapped in clouds of enormous vertical extent of over 50 km. These clouds are mainly found in a permanent cloudy bridge located between 45 and 70 km altitude, with fine mists above and below. They are mainly composed of sulfuric acid (H_2SO_4) aerosol particles. SOIR is the only instrument aboard the Venus Express that can contribute to the study of this region. Above 100 km altitude, between 120 and 300 km, is the ionosphere composed mainly of O_+ and O_2^+ ions in the lower layers. Note the absence of a magnetic field on Venus.

The main objective assigned to SOIR was to determine the vertical distribution of HDO and H_2O, as high as possible in the atmosphere, to see whether a splitting mechanism (such as condensation) existed that would modify the D/H ratio in the upper atmosphere, and perhaps prevent D atoms from escaping Venus. In fact, as previously reported [BER 07], the HDO/H_2O ratio does not vary much with altitude until 90 km. Thanks to a high spectral resolution power of R ~ 15,000–20,000, unprecedented in space exploration using SOIR, a new gas absorption band was rapidly detected in the atmospheric transmission spectra around 2,982 cm^{-1}, showing a structure resembling a non-resolved Q-branch and a number of isolated lines with a regular profile. This absorption was not correlated to any species contained in the HITRAN or GEISA databases at that time but was found to be very similar to an absorption model observed by an American team in the spectrum of solar light reflected by the ground of Mars [VIL 08b]. In the scientific literature, an absorption around 2,982 cm^{-1} is typical for an asymmetric C–H bond, and research in other databases, mainly of organic materials, was conducted. None of the absorption characteristics found near 2,982 cm^{-1} correspond to the observed signature. Finally, this team suggested that the absorption was due to an uncatalogued transition of the molecule $^{16}O^{12}C^{18}O$. The possible existence of this isotopic species of CO_2 was quickly confirmed by theoretical considerations by Perevalov and Tachkun [BER 08].

The discovery and attribution of the observed transitions were thus described by Bertaux *et al.* [BER 08]. "Then on December 22, 2006 one of us (Bertaux) received a message from Mike Mumma, asking what SOIR was seeing on Venus around 2,982 cm^{-1}, and willing to compare to a new absorption feature that his group had detected in the solar spectrum scattered by Mars, as observed from the Earth's ground. When we compared the signatures, they were identical (within error bars). It should have rang a bell in our mind: what is most common to Venus and Mars atmosphere? CO_2! In spite of this hint, we continued to search for organic material signature, until the group of Mumma and colleagues proposed to us in April 2007 that the absorption signature common to Mars and Venus was due to an uncatalogued band of 628 CO_2 isotope (628 standing by convention for $^{16}O^{12}C^{18}O$, dropping the 1 in front of 16, 12, 18). This identification was soon after confirmed by Perevalov and Tashkun (personal communication) on theoretical grounds."

It is the $\nu_2 + \nu_3$ vibration–rotation transition. Its existence in this region of the spectrum could have been predicted from the study of transitions of 628 CO_2 molecules observed, either in the gas phase or in argon and krypton matrices, giving the energy levels of ν_2 (656.09 cm^{-1}) and ν_3 (2,323.91 cm^{-1}) transitions (Dahoo *et al.* [DAH 06]), with $\nu_2 + \nu_3 = 2{,}980.0$ cm^{-1}. This transition does not exist for the main isotopologue 626 of CO_2, as it is forbidden by the symmetry of the molecule, which being linear, has an inversion center. The replacement of one ^{16}O atom by a ^{18}O atom breaks the symmetry, making the transition possible for the 628 CO_2 variety. The selection rules also allow transitions with $\Delta l = 1$, where l is the quantum number of internal rotation for the degenerate mode ν_2. A central line can then be observed at 2,982.0 cm^{-1} since 2,980.0 cm^{-1} corresponds to $\Delta l = 0$. It is important to note that for a symmetric molecule (626 or 636), besides $\Delta l = 1$, the transitions are not authorized unless at the same time $\Delta V_2 + \Delta V_3$ is odd, where V_2 and V_3 are vibrational quantum numbers of modes ν_2 and ν_3, respectively. Hence, SOIR led to the discovery of this absorption of the isotopic variety of CO_2 in the atmosphere of Venus, making it possible to exclude any confusion with an absorption by C–H bonds. This discovery highlights the role of CO_2 isotopologues (as well as that of H_2O and HDO) as important greenhouse gases in the atmosphere of Venus.

1.4.3. *LIDAR spectroscopy for space observation*

The measurement of the profile of the atmospheric concentrations of particles (aerosols), water vapor and ozone depending on the altitude H, requires either measurements from planes or balloons, or techniques that rely on the use of a high power pulsed LASER (Figure 1.29). The first LIDAR (LIght Detection And Ranging) measurements were conducted in 1968 in Sierra Nevada and the first publications were authored by Collis *et al.* [COL 68, VIE 73]. According to the studies, pulsed lasers could be used for locating clouds and aerosols and studying the atmospheric dynamics. The LIDAR equipment, which consists of a laser scanner, a global positioning system (GPS) and an inertial navigation system (INS), can be airborne on a small plane. The laser sends short pulses to the ground surface, which reflects them or sends them back to the laser emission device. Detecting the return pulses, the equipment records the time needed for the pulse to travel from the laser scanner to the ground and back. The distance between the laser and the ground is then calculated according to the speed of light.

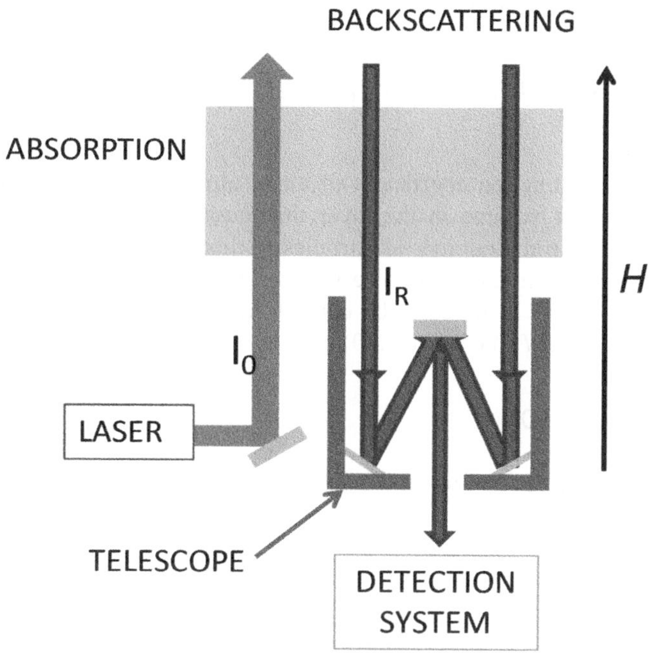

Figure 1.29. *Diagram of the LIDAR operating principle. For a color version of this figure, see www.iste.co.uk/dahoo/infrared2.zip*

The operating principle of a LIDAR relies on the backscattering of an electromagnetic wave in the visible range, similar to RADARs, which operate in the microwave range. A short and intense laser pulse is sent vertically into the atmosphere. The molecules and particles of the atmosphere absorb, transmit and scatter the laser light. A small part of the energy emitted by the laser is backscattered by aerosols in the atmosphere in the direction of a telescope and of a detection chain measuring the backscattered intensity IR(t) as a function of time. The backscattered wave detected by the LIDAR carries information on the properties of the atmosphere. The wave collected by the LIDAR system after propagation through the atmosphere and backscattering by an aerosol is obviously of very low intensity and its detection is not easy. A direct detection system or heterodyne detection can be used. The analysis of parameters such as intensity, backscattered wave frequency delay or offset leads to data that carries some amount of information and knowledge. Intensity measurement provides quantitative information on aerosol concentration. The delay gives the distance traveled by the laser and therefore the position of aerosols. The frequency offset gives the speed by Doppler effect.

The LIDAR equation is expressed as:

$$n(l) = n^0\Gamma(l)\frac{S}{l^2}\beta(l)\frac{c\tau_D}{2}\exp\left[-2\int_0^l \alpha(z)dz\right] \qquad [1.47]$$

where $\alpha(z)$ is the extinction coefficient of air at altitude z, β is the backscattering coefficient of the air volume in the layer thickness, n is the number of detected photons depending on the number of particles in this layer, and n^0 is the number of initial photons in the initial pulse. The solid angle under which the backscattered photons are detected is assumed to be equal to S/l^2 and the detection efficiency is given by the parameter $\Gamma(l)$ (Figure 1.30).

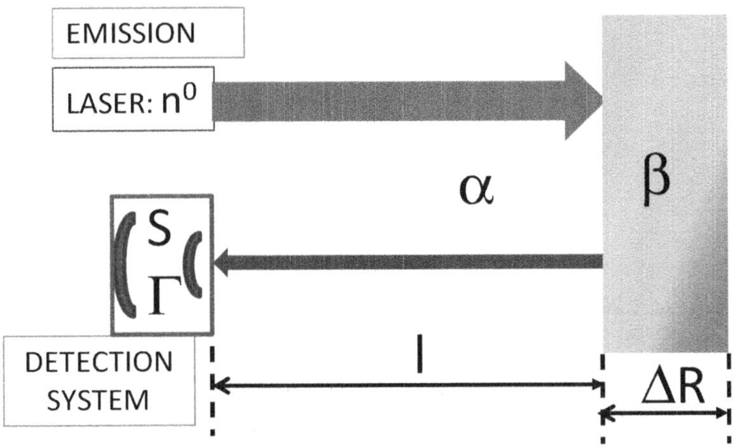

Figure 1.30. *Diagram of LIDAR measurement parameters. For a color version of this figure, see www.iste.co.uk/dahoo/infrared2.zip*

The number of backscattered photons is proportional to the number of initial photons n^0 in the initial pulse; $\Gamma(l)$, the detection efficiency depends on path l, as the number of backscattered photons depends on l and Γ depends on the number of photons collected by the detector. The number of collected photons is limited by the solid angle S/l^2 of the telescope aperture. β, the backscattering coefficient of the air volume in the layer thickness, also depends on path l. In the integral, factor 2 corresponds to the return path and the integral corresponds to the continuous sum of various factors involved in the extinction of light (absorptions, scattering, reflection coefficient, etc.) depending on the dimensions and properties of the particles encountered on the laser path.

The backscattering and extinction coefficients can be calculated from the interaction between light and the air components. If the laser wavelength is within the absorption band of a species, the extinction coefficient splits into two parts: $\alpha_{tot}(z) = \alpha(z) + \sigma C(z)$, where σ is the absorption cross section for the species, and C(z) its concentration at altitude z. In this case, in order to take into account the species absorption, the LIDAR equation is written as:

$$n(l) = n^0 \Gamma(l) \, {}^{S}\!/_{l^2} \, \beta(l) \, {}^{c\tau_D}\!/_{2} \exp(-2 \int_0^l (\alpha(z) + \sigma C(z)))dz \qquad [1.48]$$

In terms of emitted power P^0 and detected power P, the detector response τ_D can be replaced by the laser pulse duration τ, $n(l)$ and n^0 can be replaced, respectively, by $P(l)$ and P^0 to obtain:

$$P(l) = P^0 \Gamma(l) \, {}^{S}\!/_{l^2} \, \beta(l) \, {}^{c\tau}\!/_{2} \exp(-2 \int_0^l (\alpha(z) + \sigma C(z)))dz \qquad [1.49]$$

The detection chain measures an intensity $I_R(t)$ that depends on time. Assuming that the laser emits a pulse between the very close instants $t = 0$ and $t = \tau$, there is a simple relation between the distance l of particles that backscatter light and the instant of the pulse return. Indeed, the scattering corresponding to the start of the pulse emitted at t=0 returns at instant $t_1 = 2l/c$ and that corresponding to the end emitted at τ returns at instant $t_2 = \tau + 2l/c$, where $2l$ corresponds to the back and forth path. The maximum duration of the laser pulse so that the spatial resolution of the LIDAR Δl in altitude is of about 20 m, for example, is in this case given by $2\Delta l/c$; thus, 133 ns. If τ_D is the response time of photons backscattered by the detector towards the active surface of the telescope S, then the thickness ΔR of the observable layer is given by $\Delta R = 2c\,\tau_D$

Elastic scattering (without frequency change) depends on the size of the particles: Mie scattering ($\lambda << d$) and Rayleigh scattering ($d << \lambda$). Consequently, Rayleigh scattering is due to molecules, while Mie scattering is due to aerosols, since the maximum size of molecules is of the order of nm and that of aerosols is of the order of μm.

A particularly important application of LIDARs is the Doppler wind LIDAR, used in meteorology, which uses the backscattering of erbium laser light (wavelength of 1.56 μm, harmless to the eyes) on the aerosols in the atmosphere (Figure 1.31).

For example, LIDAR measurements contribute to the reliability of meteorological predictions. A spatial resolution of the order of the meter corresponds to a temporal resolution Δt of the detection system of about 6 ns, which is feasible. The minimal bandwidth Δf of detection electronics is in this case equal to 30 MHz and is determined by Fourier transform.

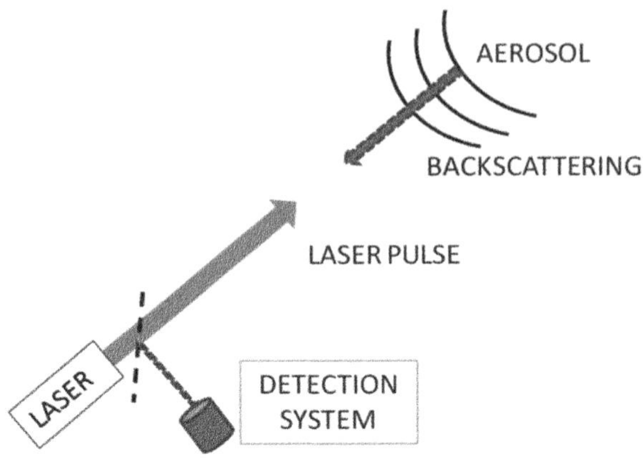

Figure 1.31. *Diagram of the Doppler wind LIDAR observation principle. For a color version of this figure, see www.iste.co.uk/dahoo/infrared2.zip*

1.5. Conclusion

This chapter presented several measuring instruments that are used in the laboratory to observe the IR spectra of molecules. These techniques are based on wave interference by Fourier transform or the various measurement methods based on laser spectroscopy aimed at achieving better precision in absorption measurements, at being able to measure weak transitions or finally at identifying species present in negligible quantities. An example of a technique using polarized light through spectroscopic ellipsometry was also provided for the characterization of aerosols in the solid phase, such as the tholins present in the atmosphere of Titan orbiting Saturn. This chapter also presented an example of observation instruments aboard orbiters (Mars Express and Venus Express) allowing better identification of the nature of the observed species (628 CO_2 variety or other isotopic species). Other instruments such as the LIDAR are used for the study of molecules and complete the panorama of different devices, which are part of a larger set of instruments used for the spectroscopic study of molecules.

1.6. Appendices

1.6.1. *Appendix 1: Measurement distortion and data processing*

The observation or detection of an absorption line requires further processing to determine its intensity. It is the result of measurement distortion inherent in any measurement by an instrument and which requires the processing of collected data.

In general, the recorded spectrum $A_O(\sigma)$ of a line is the convolution product of the apparatus function $f(\sigma)$ and of the real shape $A_v(\sigma)$ of the line. Parameters such as intensity, width and center of an absorption line are obtained after line deconvolution. The relation between $A_O(\sigma)$ and $A_v(\sigma)$ is expressed as an integral:

$$A_O(\sigma_0) = \int_{\sigma_{min}}^{\sigma_{max}} f(\sigma - \sigma_0) A_V(\sigma) d\sigma$$

where the integral is calculated over the spectral range of the absorption line.

In general, the method used for line deconvolution relies on the criteria of the sampling theorem, which is applied to determine the adequate sampling interval for digitizing the continuous spectrum. The deconvolution of the recorded spectrum is then conducted digitally by algorithms based on the sampling theorem.

Consider $h(x) = f(x)*g(x)$, the convolution product of two real functions. $F(\nu) = TF\ f(x)$ and $G(\nu) = TF\ g(x)$ are the Fourier transforms of $f(x)$ and of $g(x)$ of finite domain $[-1/2h_0, +1/2h_0]$. Using the sampling theorem with an interval h_0, this product can be written as an infinite series. Applying the sampling theorem to the function $f(x)$ leads to expressing $f(x)$ in the following form:

$$f(x) = P_h\, f(x) * sinc\ ({}^{x}/_{h})$$

where sinc $(x/h) = sin(\pi x)/\pi x$ and $P_h\, f(x)$ is the sampled function of $f(x)$ of interval h such that:

$$P_h\, f(x) = \sum_{-\infty}^{+\infty} f(ph)\,.\,\delta(x - ph)$$

where $f(ph)$ are the values of function $f(x)$ at points ph.

Considering the above equations, the convolution product $h(x)$ is expressed in the following form:

$$h(x) = \sum_{-\infty}^{+\infty} f(ph)\,.\,\delta(x - ph) * sinc\ ({}^{x}/_{h}) * g(x)$$

In the Fourier space, the expression of this equation is given by:

$$H(v) = TF\left(\sum_{-\infty}^{+\infty} f(ph)\,.\,\delta(x - ph)\right).\,h\, rect\ (hv).\,G(v)$$

where $h\ rect\ (h\nu)$ is the Fourier transform of sinc x/h, such that:

$$rect\ (h\nu) \begin{cases} = 1 & |\nu| < {}^{1}/_{2h} \\ = 0 & |\nu| > {}^{1}/_{2h} \end{cases}$$

Since $h < h_0$, $[-1/2h_0, +1/2h_0]$, the domain of $G(\nu)$ is smaller than that of $rect\ h\nu$, $[-1/2h, +1/2h]$, the multiplication by $rect\ h$ amounts to multiplying by 1, so that $H(\nu)$is finally written as:

$$H(\nu) = TF\left(\sum_{-\infty}^{+\infty} f(ph)\,.\,\delta(x - ph)\right).\,h.\,G(\nu)$$

After inverse Fourier transform, this expression yields:

$$h(x) = \sum_{-\infty}^{+\infty} f(ph)\,.\,\delta(x - ph) * hg(x)$$

Since the convolution of the Dirac distribution $\delta(x - ph)$ and $g(x)$ corresponds to a translation of the function $g(x)$ such that $\delta\ (x - ph)\ *\ g(x) = g\ (x - ph)$, the expression of $h(x)$ is given by:

$$h(x) = \sum_{-\infty}^{+\infty} hf(ph)\,.\,g(x - ph)$$

The numerical process relies on this result to build equivalent convolution products from an expansion limited to a maximum value N of the number of sampling points. $f(x)$ is rewritten as:

$$f(x) = \sum_{-\infty}^{+\infty} hf(ph)\,.\,\delta(x - ph) * {}^{1}/_{h}\, sinc\ ({}^{x}/_{h})$$

which leads to:

$$f(x) = \sum_{-\infty}^{+\infty} f(ph)\,.\,sinc\left({}^{(x - ph)}/_{h}\right), \quad h < h_0$$

Limiting this expansion to a finite number N of terms, the following function is calculated:

$$f^N(x) = \sum_{-N}^{+N} f(ph).sinc\left({(x-ph)}/{h}\right), \quad h < h_0$$

which is a quadratic mean approximation over $]-\infty, +\infty[$, using an adequate expansion basis such that:

$$\int_{-\infty}^{+\infty} |f(x) - f^N(x)|^2 \, dx = min$$

The following convolution product is then calculated:

$$h(x) = \sum_{-\infty}^{+\infty} hf^N(ph).g(x-ph)$$

of the function $f^N(x)$ and the function $g(x)$ whose spectrum does not contain the frequencies that are higher than those of $F(\nu)$. The best approximation is obtained for $h = h_0$.

In practice, if $A_{exp}(\sigma) = E_a(i)/E_0(i)$ denotes the recorded spectrum, where $E_a(i)$ is the value of the signal detected as a function of the wave number σ and $E_0(i)$ is the signal detected in the absence of absorbing gas, this spectrum is compared to a calculated spectrum $A_{th}(\sigma) = A(\sigma) * f(\sigma)$. For example, $A(\sigma) = 1 - exp(-k(\sigma,p)l)$, for an isolated line where P is the gas pressure, l is the length traveled in the gas and k is the profile of the line that can be that of a Doppler broadened line, or of a pressure broadened line (Chapter 3, Volume 1, [DAH 17]).

A minimization process is applied to:

$$\int_{-\infty}^{+\infty} \left|A_{exp}(\sigma) - A_{th}(\sigma)\right|^2 d\sigma = min$$

The nonlinear simplex method, also referred to as the Nelder–Mead method [NEL 65, PRE 86, pp. 521–528], is often used as it is a zeroth-order method that does not require a gradient calculation. Other algorithms based on the calculation of the gradient of the function to be minimized for the evaluation of second-order derivatives are nevertheless also used, such as the LM [LEV 44, MAR 63] or Broyden–Fletcher–Goldfarb–Shanno (BFGS) [BRO 69, FLE 70, GOL 70, SHA 70] algorithms.

The minimization process is conducted as follows. Consider an N-dimensional space whose points M have the values $(x_1, x_2, ...x_N)$ as coordinates. Each point M in space is associated with the value $f(x_1, x_2, ...x_N)$ of the function to minimize. Starting from an arbitrary point M_1, of value $f(M_1)$, the coordinates of M_1 are increased by $(\Delta x_1, \Delta x_2, ...\Delta x_N)$. $N + 1$ points are defined, as well as the values associated by f of N new coordinates:

Coordinates of M_1: $x_1, x_2, ...x_N$ and $f(M_1)$

Coordinates of M_2: $x_1+\Delta x_1, x_2, ...x_N$ and $f(M_2)$

Coordinates of M_3: $x_1, x_2+\Delta x_2, ...x_N$ and $f(M_3)$

...

Coordinates of M_{N+1}: $x_1, x_2, ...x_N+\Delta x_N$ and $f(M_{N+1})$

Proceeding from these points, which are classified in ascending order of the value of $f(M_i)$, the worst point, the one that has the maximum value, is replaced by a better point, and so on, using the procedure of the chosen method. In the end, the expected profile is reached such that the intensity of the line can be calculated, for example.

The nonlinear optimization algorithm of Nelder and Mead [NEL 65] relies on Dantzig's simplex model [DAN 90], a geometric method that uses N parameters to build a simplex of $N + 1$ vertices to which a certain number of operations is applied. If the function to minimize has two variables, the simplex is a triangle, if it has three variables, the simplex is a tetrahedron. An algorithm that involves the comparison of the values of the function $f(x_1, x_2, ...x_N)$ to be determined is applied to each vertex of the simplex. The vertex that yields the worst value (the highest) is replaced by a new vertex. The values of $f(x'_1, x'_2, ...x'_N)$ are recalculated, and the search for the minimum continues. This process generates a sequence of simplexes for which the values of the function at vertices are increasingly small. In principle, the hypersurface of simplexes gradually diminishes and the coordinates of the minimum are thus obtained.

This criterion is not necessarily achieved at each stage where a point is replaced by a better one. In this case, the simplex is subjected to other operations: reflection (volume preservation), expansion (volume increase) or contraction or multidimensional contraction (volume decrease near solution). At each transformation, the criterion function is evaluated to find a better point than the one calculated in the previous stage. The process is continued until the stage corresponding to the stop criterion (value below or equal to the fixed threshold).

If the number of variables is n, a general simplex with $n+1$ vertices is built from n initial values by successively applying an increase to each starting value. These operations can be implemented on the simplex, for example, by sorting the vertices of the simplex following the value of the cost function in order to select three vertices having, respectively, the best score, the worst score and the second last. The vertex with the worst score is replaced by symmetry of the simplex with respect to the hyperplane defined by the other vertices, which amounts to conducting a research in a direction opposite to the one that yielded the bad score. If the cost function is improved, then an expansion occurs; otherwise, a one-dimensional contraction takes place. If despite this, the cost function does not improve, the algorithm makes a multidimensional contraction in several directions. The vertex associated with the maximum value of $f(x_1, x_2, \ldots x_N)$ is replaced by a point in the space of n variables, which leads to a lower corresponding value of $f(x'_1, x'_2, \ldots x'_N)$.

Considering the example of a function with two variables $f(x, y)$, the simplex is a triangle. $S_1(x_1, y_1)$, $S_2(x_2, y_2)$ and $S_3(x_3, y_3)$ denote the vertices of the simplex and $z_k = f(S_k)$, the values calculated at vertices S_k, $k = 1, 2, 3$. The values calculated at vertices S_k, $k = 1, 2, 3$ in ascending order are denoted by m (minimum), μ (mean) and M (maximum). The vertex corresponding to M must be replaced. The algorithm of the simplex is achieved by calculating:

1) the mid-point:

$$S_M = (f^{-1}(m) + f^{-1}(\mu))/2;$$

2) the vertex reached by reflection:

$$S_R = S_M + (S_M - f^{-1}(M)) = 2S_M - f^{-1}(M);$$

3) the vertex reached by expansion:

$$S_D = S_R + (S_R - S_M) = 2\, S_R - S_M;$$

4) the vertex reached by contraction: S_C from the two possible vertices for contraction: $C_1 = (f^{-1}(M) + S_M)/2$ and $C_2 = (S_M + S_R)/2$ and of the condition: if $f(C_1) < f(C_2)$, then $S_C = C_1$, else $S_C = C_2$;

5) the best vertex:

$$S_F = (f^{-1}(M) + f^{-1}(m))/2.$$

The following algorithm is then applied, if $f(S_R) < \mu$, then apply case I, else case II:

– case I: if ($m < f(S_R)$), then replace $f^{-1}(M)$ by S_R, else calculate S_D and if ($f(S_D) < \mu$), replace $f^{-1}(M)$ by S_D, else replace $f^{-1}(M)$ by S_R;

– case II: if ($f(S_R) < M$), then replace $f^{-1}(M)$ by S_R. If $f(S_R) \geq M$, then calculate S_C and if $f(S_C) < M$, then replace $f^{-1}(M)$ by S_C, else calculate S_F replace $f^{-1}(M)$ by S_F and replace $f^{-1}(\mu)$ by S_M.

2

Interactions Between a Molecule and Its Solid Environment

Volumes 1 and 2 of this series [DAH 17, DAH 19a], describe the theoretical models developed for determining the absorption spectra of (homonuclear or heteronuclear) diatomic molecules and (linear or nonlinear, symmetric or non-symmetric) triatomic molecules trapped in nanocages of inert matrices of rare gases or clathrate hydrates at very low temperatures.

In rare gas matrices, the trapped molecule can occupy, depending on its size, composition and geometry, a simple substitution site (single substituted atom) or double substitution site (two substituted atoms). The distortion of the nanocage is determined by the iterative method, based on the Green functions of the perfect crystal. In clathrate hydrates, the molecule can occupy a small cage and/or a large cage in structure I or II.

In this volume, the theoretical models developed in Volumes 1 and 2 are applied to the NH_3 symmetric and/or CH_4 spherical tops trapped in a nanocage of rare gas matrix, clathrate, fullerene or adsorbed on a graphite substrate.

The perturbed motions of the molecule are studied using an atom–atom potential for describing the interaction between the trapped molecule and its environment. The frequency shift and the line broadening are determined by the Liouville formalism and using the cumulant expansion techniques of the evolution operator of the optically active system in a solid or liquid constrained environment known as a heat bath.

2.1. Introduction

The profile, shape and width of the spectral lines of a molecule trapped in a condensed environment are determined when it is possible to compare the theoretical results to the high-resolution spectroscopic measurements.

The *a priori* analyses of the behavior of molecules trapped in simple matrices, such as rare gas matrices with octahedral symmetry, make it possible to test the quality of the molecule–matrix interaction potential, the symmetry of the inclusion site and the quality of approximations used in the motion decoupling, by reconstructing the bar spectrum (frequencies and integrated intensities of the lines) and comparing it to experimental spectra.

The first models proposed for interpreting the experimental spectra were developed by Pauling [PAU 30] and Devonshire [DEV 36] for diatomic molecules (hydracids such as HCl and CO) trapped in rare gas matrices. The first model was based on the librational motion (locked rotation) of the molecule in the crystalline inclusion site, and the second model was based on the quasi-free rotational motion of the molecule.

Much later, the models were improved and generalized to other molecules and other types of site symmetry [FLY 63, SAU 66, BEY 74]. These models were used to determine frequency shifts that were in satisfactory agreement with the experimental results; however, they could not be used to determine the infrared spectral profiles of the above-mentioned molecules trapped in rare gas matrices.

A more successful model, known as RTC (Rotation–Translation Coupling) [FRI 65, BAU 66, FRI 67], took into account the coupling between the orientational motion of the diatomic molecule and its translational motion, where the matrix was assumed to be static and a non-deformed "cage model". The basic hypothesis is that the center of interaction to which the crystal field applies differs from the molecule's center of mass. Although it enabled the interpretation of certain spectral features for the HCl molecule trapped in the argon matrix, this model could not be generalized. Indeed, the cage model cannot be applied unless the translational vibration frequency ω_T of the molecule is much higher than the cut-off frequency of the vibrational modes (phonons) of the crystal environment.

The first model, which took into account the coupling between the motions of the molecule and those of the non-deformed crystal lattice of octahedral symmetry, was proposed by Kono and Lin [KON 83] for HCl and DCl trapped in rare gas matrices. First, the Born–Oppenheimer approximation is used to separate the high-frequency internal vibrational mode of the molecule from the low-frequency modes, including the orientational motion of the molecule, the vibrational motion of its center of mass and the vibrational motion of the crystal lattice (phonons). The coupling between these modes is then introduced, to calculate the frequency shifts and linewidths for reconstructing the spectral profile. The interaction potential energy between the trapped molecule and the crystal atoms limits the use of the model to the systems in which the size of the molecule is fairly comparable to that of the site it fills.

Concerning the polyatomic molecules trapped in rare gas matrices, the first theoretical model that took into account the dynamic interaction between the optical system and the heat bath was developed for the spectroscopic study of the vibration–inversion mode ν_2 of the NH_3 molecule [LAK 87a].

The basic model was described in Chapter 4 of Volume 1 [DAH 17] for the study of diatomic molecules trapped in rare gas matrices, then in Volume 2 [DAH 19a] for the study of triatomic molecules trapped in rare gas matrices and clathrate hydrates.

This volume is devoted to the construction of infrared absorption spectra, at very low temperatures, of the NH_3 molecule trapped in rare gas matrices, clathrate hydrate matrices, fullerene nanocages or adsorbed on a graphite substrate.

This chapter presents the model of the interaction potential energy between the studied molecule and the condensed medium in which it evolves. It is important to remember that this model relies on the following four hypotheses:

1) The interaction potential energy between the molecule and the solid environment system can be considered as the result of binary interactions between the molecule and an atom or molecule of this environment.

2) These binary interactions can be modeled by an atom–atom potential.

3) The distortion of atoms or molecules of the medium surrounding the active molecule can be determined using the Green functions of the perfect crystal (by minimizing the potential energy of the system).

4) Slow modes of low frequencies ω_Q and rapid modes of high frequencies ω_q of the active molecule ($\omega_Q/\omega_q \ll 1$) can be decoupled using the Born–Oppenheimer approximation.

In the case of nanocages of rare gas matrices and fullerene or even a graphite substrate, the polarizabilities of atoms of the environment give rise to the electric field. The polarization field is induced by the electrical moments of the trapped molecule.

In the case of clathrate nanocages, the electrical field of the environment surrounding the molecule is determined by the Lakhlifi–Dahoo extended inclusion model, by placing effective charges on each atom of the trapped molecule and of the molecules in the nanocage. These effective charges are determined by quantum chemistry methods.

After its application to the determination of infrared spectra of linear and nonlinear triatomic molecules [DAH 19a], the theoretical model described in Chapter 4 of Volume 1 [DAH 17] is applied to the symmetric top molecule of NH_3 perturbed by a solid environment. This is why the various stages of calculation are only briefly presented. For more details, the reader is referred to Chapter 4 of Volume 1 as well as to Chapter 6 of Volume 2 [DAH 19a] of this set.

2.2. Active molecule – solid environment system

The quantitative interpretation of spectroscopic phenomena requires that the interaction potential energy between the active molecule and its environment be known. This energy is generally considered as the result of binary interactions between the trapped molecule and the atoms or molecules of the solid environment. In fact, it often takes into account the effects of the environment through effective parameters of this binary potential energy, obtained from experimental measurements in the condensed phase, and not in the gas phase.

Moreover, this energy must have an analytical form, with an explicit dependency on the various degrees of freedom of the optically active molecule and those of the atoms or molecules of its environment (the heat bath).

Finally, this binary potential energy is the sum of long-range electrostatic, inductive and dispersive contributions (van der Waals interactions) and of short-range repulsive contributions describing, on the one hand, the overlapping of electronic clouds of two partner molecules and, on the other hand, the potential charge transfer phenomena.

2.2.1. *Binary interaction energy*

Let us consider a polyatomic molecule, for example, NH_3 (denoted by 0), in interaction with a molecule (denoted by k) of the solid environment. The position and the orientation of each of these molecules are defined with respect to a fixed reference frame $(0, \vec{X}, \vec{Y}, \vec{Z})$, generally attached to the solid environment in which the molecule is found. The molecules are assumed to be in their fundamental electronic state.

The distribution of charges in the molecules is such that they feature a set of permanent multipolar moments (dipole, quadrupole, etc.). Under the influence of an electric field, this distribution of charges can undergo a non-permanent deformation. The molecules are referred to as "polarizable" and have (dipolar, quadripolar, etc.) polarizability. This deformation is modeled by the introduction of induced multipolar moments.

Figure 2.1 presents the geometric characteristics of the system formed by the NH_3 molecule and its partner. The latter can be, for example, an atom of a rare gas matrix, a N_2 molecule of a nitrogen matrix, a H_2O molecule of a clathrate hydrate nanocage, a carbon atom of a spherical nanocage of fullerene C_{60} or even a carbon atom in a graphite substrate.

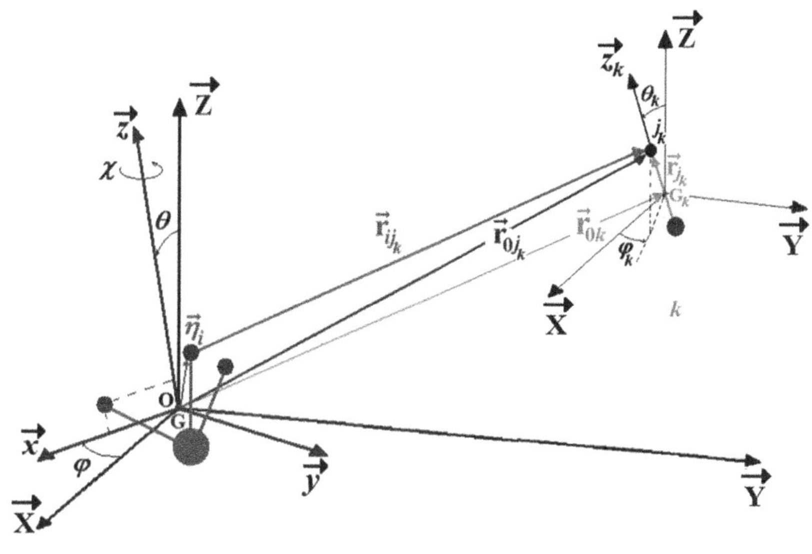

Figure 2.1. *Geometric characteristics of a symmetric top NH_3 molecule in interaction with a (diatomic, for example) molecule of its environment.* $(G,\vec{x},\vec{y},\vec{z})$ *and* $(O,\vec{X},\vec{Y},\vec{Z})$ *are, respectively, the reference frame linked to the equilibrium configuration of the molecule and the fixed reference frame (or laboratory reference frame). For a color version of this figure, see www.iste.co.uk/dahoo/infrared2.zip*

In this figure:

– $\vec{r}_{ij_k}$ and $\vec{r}_{0j_k}$ are the distance vectors, respectively, between atom i of the NH_3 molecule and, on the one hand, its center of mass and, on the other hand, atom j of molecule k;

– $\vec{r}_{0k}$ is the distance vector between the center of mass of the NH_3 molecule and that of molecule *k*;

– $\vec{\eta}_i$ and $\vec{r}_{j_k}$ are the position vectors of atom *i* of the NH_3 molecule and atom *j* of the molecule *k*, with respect to their respective attached reference frame;

– (φ, θ, χ) and (φ_k, θ_k) are the Euler angles describing the orientation of each molecule, with respect to the fixed reference frame, whose origin is in its center of mass.

Finally, in order to highlight the various degrees of freedom of two partner molecules, the vector $\vec{r}_{ij_k}$ can be written as:

$$\vec{r}_{ij_k} = \vec{r}_{0k} + \vec{r}_{j_k} - \vec{\eta}_i = \vec{r}_{0j_k} - \vec{\eta}_i \qquad [2.1]$$

2.2.2. *Dispersion–repulsion contribution*

This contribution is generally modeled by an atom–atom potential energy of Lennard-Jones (LJ) type 6–12, expressed in the following form:

$$V_{LJ} = \sum_i \sum_{j_k} 4\varepsilon_{ij} \left\{ \left(\frac{\sigma_{ij}}{\left| \vec{r}_{ij_k} \right|} \right)^{12} - \left(\frac{\sigma_{ij}}{\left| \vec{r}_{ij_k} \right|} \right)^{6} \right\} \qquad [2.2]$$

The first term of this expression (in $1/r^{12}$) is a repulsion term, dominant at the short range, and the second term (in $1/r^6$) is an attractive term, dominant at the long range.

However, the LJ potential parameters ε_{ij} and σ_{ij} are determined by the following Lorentz–Berthelot combination rules:

$$\begin{aligned} \varepsilon_{ij} &= \sqrt{\varepsilon_{ii}\varepsilon_{jj}} \\ 2\sigma_{ij} &= \sigma_{ii} + \sigma_{jj} \end{aligned} \qquad [2.3]$$

2.2.3. *Electrostatic contribution*

This contribution is due to the interaction between the permanent multipole moments of two molecules assumed to be electrically neutral. Nevertheless, depending on the size of the two molecules and the distance between them, they can interact through their charge distributions or their permanent multipole moments.

Indeed, if the distance between the two molecules is of the order of their size, each of them appears to the other as a distribution of electrical charges, and the electrostatic energy is written as the sum of charge–charge interaction energies:

$$V_{\mathrm{E}} = \frac{1}{4\pi\varepsilon_0}\sum_i\sum_{j_k}\frac{q_i q_{j_k}}{\left|\vec{r}_{ij_k}\right|} \qquad [2.4]$$

where q_i and q_{jk} are, respectively, the charge of site i of the studied molecule and the charge of site j of molecule k of the solid environment, and $\vec{r}_{ij_k}$ is the distance vector between these two sites. It is important to note that, depending on the molecule, the sites carrying the charges can be very different from the atoms making part of this molecule.

However, when the distance between the two molecules is large enough, compared to their size, each appears, instead, neutral to the other and they interact through their multipole moments. The electrostatic energy can then be written as:

$$V_{\mathrm{E}} = \frac{1}{4\pi\varepsilon_0}\left\{\begin{array}{l} -\vec{\mu}_0^A.\vec{\nabla}\vec{\nabla}\left(\frac{1}{\left|\vec{r}_{0k}\right|}\right).\vec{\mu}_k^A - \frac{1}{3}\vec{\mu}_0^A.\vec{\nabla}\vec{\nabla}\vec{\nabla}\left(\frac{1}{\left|\vec{r}_{0k}\right|}\right):\vec{\vec{\Theta}}_k^A \\ -\frac{1}{15}\vec{\mu}_0^A.\vec{\nabla}\vec{\nabla}\vec{\nabla}\vec{\nabla}\left(\frac{1}{\left|\vec{r}_{0k}\right|}\right)\vdots\vec{\vec{\vec{\Omega}}}_k^A + \frac{1}{3}\vec{\vec{\Theta}}_0^A:\vec{\nabla}\vec{\nabla}\vec{\nabla}\left(\frac{1}{\left|\vec{r}_{0k}\right|}\right).\vec{\mu}_k^A \\ +\frac{1}{9}\vec{\vec{\Theta}}_0^A:\vec{\nabla}\vec{\nabla}\vec{\nabla}\vec{\nabla}\left(\frac{1}{\left|\vec{r}_{0k}\right|}\right):\vec{\vec{\Theta}}_k^A - \frac{1}{45}\vec{\vec{\Theta}}_0^A:\vec{\nabla}\vec{\nabla}\vec{\nabla}\vec{\nabla}\vec{\nabla}\left(\frac{1}{\left|\vec{r}_{0k}\right|}\right)\vdots\vec{\vec{\vec{\Omega}}}_k^A \\ -\frac{1}{15}\vec{\vec{\vec{\Omega}}}_0^A\vdots\vec{\nabla}\vec{\nabla}\vec{\nabla}\vec{\nabla}\left(\frac{1}{\left|\vec{r}_{0k}\right|}\right).\vec{\mu}_k^A + \frac{1}{45}\vec{\vec{\vec{\Omega}}}_0^A\vdots\vec{\nabla}\vec{\nabla}\vec{\nabla}\vec{\nabla}\vec{\nabla}\left(\frac{1}{\left|\vec{r}_{0k}\right|}\right):\vec{\vec{\Theta}}_k^A \\ -\frac{1}{225}\vec{\vec{\vec{\Omega}}}_0^A\vdots\vec{\nabla}\vec{\nabla}\vec{\nabla}\vec{\nabla}\vec{\nabla}\vec{\nabla}\left(\frac{1}{\left|\vec{r}_{0k}\right|}\right)\vdots\vec{\vec{\vec{\Omega}}}_k^A + \ldots \end{array}\right\} \qquad [2.5]$$

where $\vec{r}_{0k}$ is the distance vector between the two molecules; $\vec{\mu}^A$, $\vec{\vec{\Theta}}^A$, $\vec{\vec{\vec{\Omega}}}^A$, etc. are their dipole moments (vector), quadrupole moments (second-rank tensor), octupole moments (third-rank tensor), etc. defined with respect to the fixed reference frames $(G,\vec{X},\vec{Y},\vec{Z})$ attached to the molecule and $(G_k,\vec{X},\vec{Y},\vec{Z})$ attached to the partner molecule k (see Appendix 1 (section 2.5.1)); ε_0 is the electric permittivity of the vacuum; and the symbol $\vec{\nabla}$ represents the gradient operator that, applied to a function $f(X,Y,Z)$, yields the vector:

$$\vec{\nabla} f = \left| \begin{array}{l} \nabla_X f = \dfrac{\partial f}{\partial X} \\ \nabla_Y f = \dfrac{\partial f}{\partial Y} \\ \nabla_Z f = \dfrac{\partial f}{\partial Z} \end{array} \right. \qquad [2.6]$$

2.2.4. *Induction contribution*

Since the molecules are polarizable, the permanent multipole moments of each molecule generate an electric field at the location of the other molecule, the latter being thus subjected to induced multipole moments. The induction energy is due to the interaction between the induced and permanent moments. According to the dipole polarization approximation (first order of the expansion), this energy is defined by:

$$V_{\mathrm{I}} = -\frac{1}{2}\vec{E}_{0\to k}.\vec{\vec{\alpha}}_k^A.\vec{E}_{0\to k} - \frac{1}{2}\vec{E}_{k\to 0}.\vec{\vec{\alpha}}_0^A.\vec{E}_{k\to 0} \qquad [2.7]$$

where $\vec{\vec{\alpha}}^A$ are dipole polarizabilities (second-rank tensors) of partner molecules and $\vec{E}$ are electric field vectors generated by each molecule on the other, and defined by:

$$\vec{E}_{0\to k} = \frac{1}{4\pi\varepsilon_0}\left\{ \begin{array}{l} \vec{\nabla}\vec{\nabla}\left(\dfrac{1}{|\vec{\mathbf{r}}_{0k}|}\right).\vec{\mu}_0^A - \dfrac{1}{3}\vec{\nabla}\vec{\nabla}\vec{\nabla}\left(\dfrac{1}{|\vec{\mathbf{r}}_{0k}|}\right):\vec{\vec{\Theta}}_0^A \\ +\dfrac{1}{15}\vec{\nabla}\vec{\nabla}\vec{\nabla}\vec{\nabla}\left(\dfrac{1}{|\vec{\mathbf{r}}_{0k}|}\right)\vdots\vec{\vec{\vec{\Omega}}}_0^A + \ldots \end{array} \right\} \qquad [2.8a]$$

$$\vec{E}_{k\to 0} = \frac{1}{4\pi\varepsilon_0}\left\{ \begin{array}{l} \vec{\nabla}\vec{\nabla}\left(\dfrac{1}{|\vec{\mathbf{r}}_{k0}|}\right).\vec{\mu}_k^A - \dfrac{1}{3}\vec{\nabla}\vec{\nabla}\vec{\nabla}\left(\dfrac{1}{|\vec{\mathbf{r}}_{k0}|}\right):\vec{\vec{\Theta}}_k^A \\ +\dfrac{1}{15}\vec{\nabla}\vec{\nabla}\vec{\nabla}\vec{\nabla}\left(\dfrac{1}{|\vec{\mathbf{r}}_{k0}|}\right)\vdots\vec{\vec{\vec{\Omega}}}_k^A + \ldots \end{array} \right\} \qquad [2.8b]$$

2.3. Two-center expansion of the term $\left(\dfrac{1}{\left|\vec{\mathrm{r}}_{ij_k}\right|}\right)$

As previously mentioned, the interaction energy between the trapped molecule and the atoms or molecules of the solid environment must have an analytical form containing an explicit dependency on the various degrees of freedom, namely, their translational, orientational and internal vibrational motions. This is essentially important when dealing with the dynamic coupling between the degrees of freedom of the active molecule and those of the solid environment.

In fact, very often this amounts to determining the successive derivatives of the characteristic molecular parameters. An *exact* analytical form can only be established by using approximations and truncations in the mathematical expansions at a level providing a certain convergence. The analytical expansions should be consistent with the numerical calculations.

A two-center expansion [HIR 64] is used to write the term $1/\left|\vec{\mathrm{r}}_{ij_k}\right|$ as a function of various variables of the system:

$$\frac{1}{\left|\vec{\mathrm{r}}_{ij_k}\right|}=\frac{1}{\left|\vec{\mathrm{r}}_{\mathbf{0}k}\right|}\left\{1+\sum_{\substack{l_0,l_k\\p_0,q_0,p_k}}^{*}\left[F^{l_0,l_k}_{p_0,q_0,p_k}(\eta_i,\beta_i,\chi_i,r_{j_k},\beta_{j_k})D^{l_0}_{p_0,q_0}(\varphi,\theta,\chi)Y^{p_k}_{l_k}(\varphi_k,\theta_k)\right]\right\} \quad [2.9]$$

with

$$F^{l_0,l_k}_{p_0,q_0,p_k}(\eta_i,\beta_i,\chi_i,r_{j_k},\beta_{j_k})=\frac{(4\pi)^2(-1)^{l_k+p_k+p_0}}{(2l_0+1)(2l_k+1)}\left(\frac{(\eta_i)^{l_0}(r_{j_k})^{l_k}}{\left|\vec{\mathrm{r}}_{\mathbf{0}k}\right|^{l_0+l_k}}\right)Y^{q_0}_{l_0}(\beta_i,\chi_i)\times$$

$$Y^{0}_{l_k}(\beta_{j_k},0)\sum_{m=-l_{<}}^{+l_{<}}\left\{\begin{array}{c}\dfrac{(l_0+l_k)!}{\left[(l_0+|m|)!(l_0-|m|)!(l_k+|m|)!(l_k-|m|)!\right]^{1/2}}\times\\ \displaystyle\sum_{l=|l_0-l_k|}^{l_0+l_k}\sqrt{\frac{4\pi}{(2l+1)}}C(l_0l_kl;m,-m)C(l_0l_kl;p_0,-p_k)\\ C(l_0l_kl;q_0,0)Y^{p_0-p_k}_{l}(\Phi_{0k},\Theta_{0k})\end{array}\right\} \quad [2.10]$$

In expression [2.9], l_0 and l_k are natural integers; p_0, q_0 and p_k are relative integers such that: $-l_0\le p_0\le l_0$, $-l_0\le q_0\le l_0$ and $-l_k\le p_k\le l_k$; the symbol (*) on the sum symbol indicates that $l_0+l_k\ne 0$; (β_i,χ_i) and β_{jk} are the angles identifying, respectively, the atom i of the NH_3 molecule and the atom j of the molecule k in their respective reference frame; Y represents the spherical harmonics; and D are the elements of the rotation matrix [ROS 67], whose expression and calculation subroutine in FORTRAN are presented in Appendix 2 (section 2.5.2). On the other hand, in expression [2.10], $l_{\langle} = \inf\{l_0, l_k\}$ represents the smallest of two values l_0 and l_k; C are Clebsch–Gordan coefficients whose expression and calculation subroutine in FORTRAN are also presented in Appendix 3 (section 2.5.3); and (Φ_{0k}, Θ_{0k}) are the Euler angles identifying the position of the center of mass of molecule k in the solid environment with respect to the fixed reference frame $(\mathrm{O}, \vec{\mathrm{X}}, \vec{\mathrm{Y}}, \vec{\mathrm{Z}})$.

It is important to note that the dispersion–repulsion contribution V_{LJ} [2.2] to the potential energy depends on $\left(1/\left|\vec{\mathrm{r}}_{ij_k}\right|\right)^6$ and $\left(1/\left|\vec{\mathrm{r}}_{ij_k}\right|\right)^{12}$, which leads us to making the binomial expansions of expression [2.9]. Moreover, in expression [2.10] the ratio $\frac{(\eta_i)^{l_0}(r_{j_k})^{l_k}}{\left|\vec{\mathrm{r}}_{0k}\right|^{l_0+l_k}}$ appears, whose importance is essential in the convergence and truncation of the series and binomial expansions.

Finally, the following section considers, as an example, the inclusion of an ammonia molecule in an atomic matrix of rare gas. The molecule k is therefore an atom denoted by j; the two-center expansion of expression [2.9] is replaced by a single-center expansion. At the fourth order of truncation, the resulting analytical expression of the Lennard-Jones dispersion–repulsion contribution to the interaction potential energy is:

$$\begin{aligned} V_{\mathrm{LJ}} &= 4\sum_j \sum_{\mathrm{n}=6,12} (-1)^{\mathrm{n}/2} \sum_i \varepsilon_{ij} \left(\frac{\sigma_{ij}}{\left|\vec{\mathrm{r}}_{0j}\right|}\right)^{\mathrm{n}} \\ &\times \sum_{l_0, p_0, q_0} \left\{ \frac{4\pi}{(2l_0+1)} \, {}^{\mathrm{n}}F_{q_0}^{l_0}(\eta_i, \beta_i, \chi_i) Y_{l_0}^{p_0}(\Theta_{0j}, \Phi_{0j}) D_{p_0, q_0}^{l_0}(\varphi, \theta, \chi) \right\} \end{aligned} \qquad [2.11]$$

with

$${}^{\mathrm{n}}F_0^0(\eta_i, \beta_i, \chi_i) = \left\{ \begin{aligned} & 1 + \frac{\mathrm{n}(\mathrm{n}-1)}{6}\left(\frac{\eta_i}{\left|\vec{\mathrm{r}}_{0j}\right|}\right)^2 + \\ & \frac{(\mathrm{n}-1)\mathrm{n}(\mathrm{n}+1)(\mathrm{n}+2)}{120}\left(\frac{\eta_i}{\left|\vec{\mathrm{r}}_{0j}\right|}\right)^4 \end{aligned} \right\} \sqrt{4\pi}\, Y_0^0(\beta_i, \chi_i)$$

$$ {}^{n}F_{0}^{1}(\eta_i,\beta_i,\chi_i) = n\left(\frac{\eta_i}{\left|\vec{r}_{0j}\right|}\right)\left\{1+\frac{(n-1)(n+2)}{10}\left(\frac{\eta_i}{\left|\vec{r}_{0j}\right|}\right)^2\right\}\sqrt{\frac{4\pi}{3}}Y_1^0(\beta_i,\chi_i) $$

$$ {}^{n}F_{0}^{2}(\eta_i,\beta_i,\chi_i) = n\left(\frac{\eta_i}{\left|\vec{r}_{0j}\right|}\right)^2\left\{\begin{matrix}\frac{n+2}{3}+\\ \frac{(n-1)(3n^2+14n+32)}{126}\left(\frac{\eta_i}{\left|\vec{r}_{0j}\right|}\right)^2\end{matrix}\right\}\sqrt{\frac{4\pi}{5}}Y_2^0(\beta_i,\chi_i) $$

$$ {}^{n}F_{0}^{3}(\eta_i,\beta_i,\chi_i) = n\left(\frac{\eta_i}{\left|\vec{r}_{0j}\right|}\right)^3\frac{(n+2)(n+4)}{15}\sqrt{\frac{4\pi}{7}}Y_3^0(\beta_i,\chi_i) $$

$$ {}^{n}F_{\pm3}^{3}(\eta_i,\beta_i,\chi_i) = n\left(\frac{\eta_i}{\left|\vec{r}_{0j}\right|}\right)^3\frac{(n+2)(n+4)}{15}\sqrt{\frac{4\pi}{7}}Y_3^0(\beta_i,\chi_i) $$

$$ {}^{n}F_{0}^{4}(\eta_i,\beta_i,\chi_i) = n\left(\frac{\eta_i}{\left|\vec{r}_{0j}\right|}\right)^4\frac{(n^3+12n^2+44n+48)}{105}\sqrt{\frac{4\pi}{9}}Y_4^0(\beta_i,\chi_i) $$

$$ {}^{n}F_{\pm3}^{4}(\eta_i,\beta_i,\chi_i) = n\left(\frac{\eta_i}{\left|\vec{r}_{0j}\right|}\right)^4\frac{(n^3+12n^2+44n+48)}{105}\sqrt{\frac{4\pi}{9}}Y_4^0(\beta_i,\chi_i) $$

It is important to note that the coefficients of the spherical harmonics in the functions ${}^{n}F_{q_0}^{l_0}(\eta_i,\beta_i,\chi_i)$ (with $q_0 \neq 0$) are identical to those of the spherical harmonics in ${}^{n}F_{0}^{l_0}(\eta_i,\beta_i,\chi_i)$. Moreover, with the ammonia molecule having a symmetry axis C_3, the non-zero values of the relative number q_0 are multiples of 3 respecting the condition $\left|q_0\right| \leq l_0$.

2.4. Conclusion

This chapter presents the various contributions to the interaction potential energy between the studied molecule and an atom or a molecule in the solid environment. This potential energy will be used in the following chapters to determine the modifications of the infrared absorption spectrum, at very low temperatures, of the ammonia molecule trapped in a nanocage of a rare gas matrix, clathrate hydrates, fullerene or adsorbed on a graphite substrate.

2.5. Appendices

2.5.1. *Appendix 1: Multipole moments and dipole polarizability of a molecule with respect to its fixed reference frame* $(G,\vec{X},\vec{Y},\vec{Z})$

$$\vec{\mu}^A = \mathbf{M}^{-1}(\varphi,\theta,\chi)\vec{\mu}$$

$$\vec{\vec{\Theta}}^A = \mathbf{M}^{-1}(\varphi,\theta,\chi)\mathbf{M}^{-1}(\varphi,\theta,\chi)\vec{\vec{\Theta}}$$

$$\vec{\vec{\vec{\Omega}}}^A = \mathbf{M}^{-1}(\varphi,\theta,\chi)\mathbf{M}^{-1}(\varphi,\theta,\chi)\mathbf{M}^{-1}(\varphi,\theta,\chi)\vec{\vec{\vec{\Omega}}}$$

$$\vec{\vec{\alpha}}^A = \mathbf{M}^{-1}(\varphi,\theta,\chi)\mathbf{M}^{-1}(\varphi,\theta,\chi)\vec{\vec{\alpha}}$$

In these expressions, the multipole moments $\vec{\mu}$, $\vec{\vec{\Theta}}$, $\vec{\vec{\vec{\Omega}}}$ and the dipole polarizability $\vec{\vec{\alpha}}$ are defined with respect to the mobile reference frame $(G,\vec{x},\vec{y},\vec{z})$ attached to the equilibrium configuration of the molecule, and $\mathbf{M}(\varphi,\theta,\chi)$ is the unitary rotational matrix transformation, that is, matrix of transition from the fixed reference frame $(G,\vec{X},\vec{Y},\vec{Z})$ to the mobile reference frame (see Appendix 1 (section 2.7.1) in Chapter 2 of Volume 2):

$$\mathbf{M}(\varphi,\theta,\chi) = \begin{pmatrix} \cos\varphi\cos\theta\cos\chi - \sin\varphi\sin\chi & \sin\varphi\cos\theta\cos\chi + \cos\varphi\sin\chi & -\sin\theta\cos\chi \\ -\cos\varphi\cos\theta\sin\chi - \sin\varphi\cos\chi & -\sin\varphi\cos\theta\sin\chi + \cos\varphi\cos\chi & \sin\theta\sin\chi \\ \cos\varphi\sin\theta & \sin\varphi\sin\theta & \cos\theta \end{pmatrix}$$

NOTE.– If the angle of proper rotation χ does not exist (linear molecule), the unitary matrix transformation $\mathbf{M}(\varphi,\theta)$ readily results by putting $\chi = 0$ in the above matrix.

2.5.2. *Appendix 2: Elements of the rotational matrix*

The expression of the rotational matrix elements $D^l_{p,q}(\varphi,\theta,\chi)$ as a function of Euler angles is [ROS 67]:

$$D^l_{p,q}(\varphi,\theta,\chi) = e^{-ip\phi} d^l_{p,q}(\theta) e^{-iq\chi}$$

where l is a natural integer, $-l \le p \le l$, $-l \le q \le l$ and

$$d^l_{p,q}(\theta) = \left[(l+p)!(l-p)!(l+q)!(l-q)!\right]^{\frac{1}{2}} \times$$
$$\sum_\lambda \frac{(-1)^\lambda}{(l-p-\lambda)!(l+q-\lambda)!(p-q+\lambda)!\lambda!} \times \left[\cos(\frac{\theta}{2})\right]^{2l-p+q-2\lambda} \times \left[-\sin(\frac{\theta}{2})\right]^{p-q+2\lambda}$$

ROTATIONAL MATRIX ELEMENTS CALCULATION SUBROUTINE

$d^l_{p,q}(\theta)$ with $l = J, p = MP, q = M$ and $\theta = THETA$

```
C*****************************************************************************
      FUNCTION DJMPM(J,MP,M,THETA)
      IMPLICIT DOUBLE PRECISION (A-H,O-Z)
      IF(IABS(MP).GT.J.OR.IABS(M).GT.J)GO TO 100
      COST2 = DCOS(THETA/2.)
      SINT2 = -DSIN(THETA/2.)
      JPM = J+M
      JMM = J-M
      JPMP = J+MP
      JMMP = J-MP
      MMMP = M-MP
      NC = JPM+JMMP
      NS = -MMMP
      RFJPM = DSQRT(FACT(JPM))
      RFJMM = DSQRT(FACT(JMM))
      RFJPMP = DSQRT(FACT(JPMP))
      RFJMMP = DSQRT(FACT(JMMP))
C**************************************************************************
      NUMAX = MIN0(JPM,JMMP)
      NUMIN = MAX0(0,MMMP)
      DJMPM = 0.
      DO 1 NU=NUMIN,NUMAX
      FJMMPMNU = FACT(JMMP-NU)
      FJPMMNU = FACT(JPM-NU)
      FNUMMMMP = FACT(NU-MMMP)
      FNU = FACT(NU)
      COEFJM = (RFJPM/FJPMMNU)*RFJMM
      COEFJMP = (RFJPMP/FJMMPMNU)*RFJMMP
      COEF = (-1.)**NU *(COEFJM/FNU)*(COEFJMP/FNUMMMMP)
```

```
      DJMPM = DJMPM + COEF*(COST2**(NC-2*NU))*(SINT2**(NS+2*NU)
 1    CONTINUE
      GO TO 101
100 WRITE(*,*)'ATTENTION: VALEUR ABSOLUE DE M OU MP > J'
101 RETURN
      END
C
C*****************************************************************
C-CALCULTION OF THE FACTORIAL FUNCTION
C*****************************************************************
      FUNCTION FACT(NF)
      IMPLICIT DOUBLE PRECISION (A-H,O-Z)
      IF(NF)121,122,123
123 FACT=1.
      DO 125 IF=1,NF
125 FACT=FACT*IF
      GOTO 124
122 FACT=1.
      GOTO 124
121 FACT=0.
124 RETURN
      END
C**********************************************************************
```

2.5.3. *Appendix 3: Clebsch–Gordan coefficients*

The values of the Clebsch–Gordan coefficients $C(J_1J_2J_3;M_1,M_2)$ can be calculated with the following expression [ROS 67]:

$$C(J_1J_2J_3;M_1,M_2) = \delta_{M_3,M_1+M_2}\sqrt{2J_3+1}\times$$

$$\left[\frac{(J_1+J_2-J_3)!(J_3+J_1-J_2)!(J_2+J_3-J_1)!}{(J_1+J_2+J_3+1)!}\times \right.$$
$$\left.(J_1+M_1)!(J_1-M_1)!(J_2+M_2)!(J_2-M_2)!(J_3+M_3)!(J_3-M_3)!\right]^{1/2}\times$$

$$\sum_\lambda \frac{(-1)^\lambda}{\lambda!}\left[\frac{(J_1+J_2-J_3-\lambda)!(J_1-M_1-\lambda)!(J_2+M_2-\lambda)!}{(J_3-J_2+M_1+\lambda)!(J_3-J_1-M_2+\lambda)!}\right]^{-1}$$

CLEBSCH–GORDAN COEFFICIENTS CALCULATION SUBROUTINE:

$$C = C(J_1J_2J_3;M_1,M_2)$$

```
C*******************************************************************
      WRITE(0,500)
500  FORMAT(4X, 'J1',4X, 'J2',4X, 'J3',4X, 'M1',4X, 'M2',
     &3X, 'M3',10X, 'COEF. CLEB.')
      DO 1 JF1=1,7
      J1=JF1-1
      MJ1=2*J1+1
      DO 2 MF1=1,MJ1
      M1=MF1-J1-1
      DO 5 JF2=1,7
      J2=JF2-1
      MJ2=2*J2+1
      DO 6 MF2=1,MJ2
      M2=MF2-J2-1
      IS3=J1+J2
      IR3=J1-J2
      IAR3=IABS(IR3)
      J3MIN=IAR3+1
      J3MAX=IS3+1
      DO 3 JF3=J3MIN,J3MAX
      J3=JF3-1
      MJ3=2*J3+1
      DO 4 MF3=1,MJ3
      M3=MF3-J3-1
      IF(M3.NE.(M1+M2))GO TO 4
      MA3=IABS(M3)
      IF(J3-IS3)100,100,101
100  IF(J3-IAR3)101,102,102
102  IF(J3-MA3)101,103,103
103  CC=((2*J3+1)*FACT(J1+J2-J3)*FACT(J3+J1-J2)/
     &FACT(J1+J2+J3+1))*FACT(J1+M1)*FACT(J1-M1)*
     &FACT(J2+M2)*FACT(J2-M2)*FACT(J3+M3)*
     &FACT(J3-M3)*FACT(J3+J2-J1)
      CC=SQRT(CC)
      JP1=2*J1
      JP2=2*J2
      IF(JP1-JP2)104,104,105
104  NUBO=JP1+1
      GO TO 106
105  NUBO=JP2+1
106  CA=0.
      DO 107 NU1=1,NUBO
      NU=NU1-1
```

```
      DNU=FACT(J1+J2-J3-NU)*FACT(NU)*FACT(J1-M1-NU)*
     &FACT(J2+M2-NU)*FACT(J3-J2+M1+NU)*FACT(J3-J1-M2+NU)
      IF(DNU)108,108,109
108   CNU=0.
      GO TO 107
109   CNU=(-1.)**NU/DNU
      CA=CA+CNU
107   CONTINUE
      C=CA*CC
      GO TO 110
101   C=0.
110   WRITE(0,400)J1,J2,J3,M1,M2,M3,C
400   FORMAT(6(3X,I3),10X, 'C=',F9.6)
 4    CONTINUE
 3    CONTINUE
 6    CONTINUE
 5    CONTINUE
 2    CONTINUE
 1    CONTINUE
      STOP
      END
C
C********************************************************************
C-CALCULATION OF THE FACTORIAL FUNCTION
C********************************************************************
      FUNCTION FACT(NF)
      IF(NF)121,122,123
123   X=1.
      DO 125 IF=1,NF
125   X=X*IF
      GO TO 124
122   X=1.
      GO TO 124
121   X=0.
124   FACT=X
      END
C********************************************************************
```

3

Nanocage of Rare Gas Matrix

The spectral profile, shape and width of the spectral lines of a molecule trapped in a condensed medium could be constructed when it was possible to compare the theoretical results to the high-resolution spectroscopy measurements.

Volumes 1 and 2 of this series [DAH 17, DAH 19a] describe the theoretical models developed in order to determine the absorption spectra of (homonuclear or heteronuclear) diatomic molecules and (linear or nonlinear, symmetric or non-symmetric) triatomic molecules trapped in nanocages of inert matrices of rare gases or hydrate clathrates at very low temperatures.

Depending on its size, composition and geometry, the molecule trapped in rare gas matrices can occupy a simple substitution site (only one atom is substituted) or a double substitution site (two atoms are substituted). The distortion of the nanocage is determined by the iterative method based on Green's functions of the perfect crystal.

3.1. Introduction

A priori analyses of the behavior of molecules trapped in simple matrices such as rare gas matrices of octahedral symmetry can be used to test the quality of the molecule–matrix interaction potential, the symmetry of the inclusion site and the quality of the approximations used in motion decoupling, by reconstructing the bar spectra (frequencies and integrated intensities of the lines) and comparing them to experimental spectra.

In this chapter, the theoretical models developed in Volumes 1 and 2 are mainly applied to the case of the symmetric top NH_3 molecule trapped in a nanocage of rare gas matrix.

The perturbed motions of the molecule are studied using an atom–atom potential to describe the interaction between the trapped molecule and its environment. Frequency shifts and line broadening are determined within the Liouville formalism and by using cumulative expansion techniques of the evolution operator of the optically active system in a solid or liquid constrained medium heat bath.

3.2. Rare gases in solid state

At very low temperatures, solid rare gases are inert crystals consisting of atoms weakly bound by van der Waals type electrostatic forces, and their cohesion energy is weak. They crystallize in a face-centered cubic (fcc) system, belonging to the octahedral symmetry group O_h. The unit cell is a cube of edge a, the lattice parameter (Table 3.1).

Matrix	Neon	Argon	Krypton	Xenon
Melting temperature (K)	25	84	115	161
Cohesion energy (kcal/mole)	0.45	1.85	2.67	3.83
Lattice parameter a (Å)	4.34	5.30	5.67	6.10
Distance between nearest neighbors R^e (Å)	3.07	3.75	4.01	4.31

Table 3.1. *Melting temperature, cohesion energy, lattice parameter and distance between nearest neighbors (van der Waals radius) of rare gas matrices in the face-centered cubic (fcc) system*

Nevertheless, in the presence of impurities, they can be stabilized in the hexagonal close-packed structure (hcp). Moreover, the rigidity of the matrix increases with the size of the atoms composing it (Ne, not rigid, → Ar → Kr → Xe).

Furthermore, the interatomic forces are generally assumed to be central and binary. Various models of potential energy were elaborated, among which is the Lennard-Jones (LJ) model, which is used in this book. In a perfect crystal of rare gas, the potential energy is written as:

$$V_C = \sum_j \sum_{j'>j} 4\varepsilon_{gr-gr} \left\{ \left(\frac{\sigma_{gr-gr}}{|\vec{r}_{jj'}|} \right)^{12} - \left(\frac{\sigma_{gr-gr}}{|\vec{r}_{jj'}|} \right)^{6} \right\} \quad [3.1]$$

The atom–atom parameters ε and σ of the LJ potential are given in Table 3.2.

Atom X	Ne	Ar	Kr	Xe	N	H
$\varepsilon_{X\text{-}X}$ (cm^{-1})	24.33	84.01	125.00	157.10	26.50	17.20
$\sigma_{X\text{-}X}$ (Å)	2.764	3.448	3.650	3.970	3.380	2.53

Table 3.2. *Atom–atom parameters of Lennard-Jones potential of rare gases, nitrogen and hydrogen*

Chapter 4 of Volume 1 [DAH 17] presents a dynamic study of the perfect fcc lattice of N atoms (N being very large). It is important to recall that the density of vibrational modes (density of phonons) of the crystal lattice is determined by solving, within harmonic approximation, the eigenvalue equation associated with the Hamiltonian of the lattice [HER 59, SCH 52]

$$H_C = T_C + V_C = \sum_{\alpha,j} \frac{P_\alpha^2(j)}{2M_j} + \frac{1}{2}\sum_{\alpha,j}\sum_{\beta,j'} A_{\alpha\beta}(j,j')u_\alpha(j)u_\beta(j') + V_C^e \qquad [3.2]$$

where T_C is the kinetic Hamiltonian of translation of the crystal atoms, V_C is its potential energy, V_C^e is its potential energy at equilibrium (rigid crystal), and:

– $u_\alpha(j)$ and $u_\beta(j')$ are the dynamic translational variables of atoms j and j' along the directions α and β (X, Y, Z), around their equilibrium position, with respect to the reference frame associated with the crystal (fixed or laboratory reference frame);

– $P_\alpha(j)$ and M_j are, respectively, the components along the direction α of the conjugate momentum of the dynamic translation variable $u_\alpha(j)$ of atom j and its mass. Indeed, the lattice being perfect, the N atoms have the same mass;

– and $A_{\alpha\beta}(j,j') = \left(\dfrac{\partial^2 V_C}{\partial u_\alpha(j)\partial u_\beta(j')}\right)_{R_j^e, R_{j'}^e}$ is the second-order force constant, namely, the force exerted along direction α on the atom j when the atom j' moves along direction β.

Appendix 1 (section 3.6.1) presents the vibrational normal modes (phonons) of a monoatomic Bravais lattice of the fcc structure of the octahedral symmetry.

The density of phonon modes $\nu(\omega)$ is defined as the number of modes whose frequency is in the interval $[\omega, \omega + d\omega]$. As an example, Figure 3.1 presents the curve of the normalized density of phonon states $\omega^2 \nu(\omega^2)$ as well as the function $S(\omega^2)$

$$S(\omega^2) = \mathrm{PP}\int_0^{\infty} \frac{\omega'^2\,\nu(\omega'^2)}{\omega^2 - \omega'^2} d\omega'^2 \qquad [3.3]$$

of the perfect crystal of argon [MAN 71].

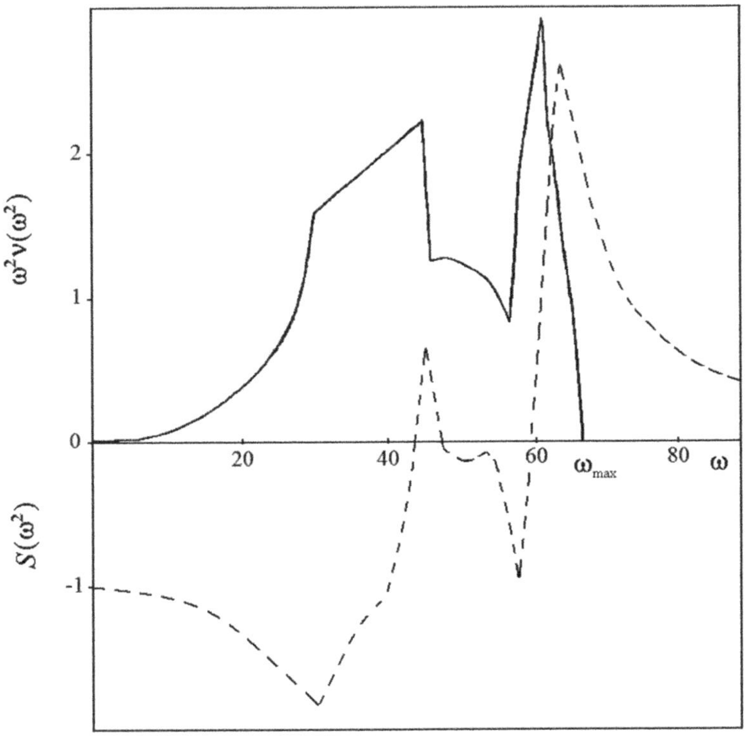

Figure 3.1. *Curves of the normalized density of phonon states $\omega^2\nu(\omega^2)$ (solid line curve) and of the function $S(\omega^2)$ (dashed line curve) of the perfect crystal of argon*

3.3. Molecule inclusion and deformation of the doped crystal

3.3.1. *Molecule inclusion*

In a rare gas (fcc) matrix, various trapping sites are possible. Depending on the nature, size and shape of the included molecule, it can replace an atom (simple site), two nearest-neighbor atoms (double site) or more of the lattice or, if its size is sufficiently small, it can be inserted into an interstitial site. This leads to a more or less significant distortion of the matrix around this molecule.

As discussed further on, since the symmetric top NH_3 molecule has a reasonable size (van der Waals diameter 3.10 Å) compared to those of rare gas atoms (Ar, Kr, Xe) and a flat pyramidal shape of small extent, it will put up with a simple substitution site. Figure 3.2 presents the geometric characteristics of the inclusion of a molecule in a rare gas matrix; $\vec{\xi}_0$ and $\vec{\xi}_j$ are, respectively, displacement vectors of the center of mass of the included molecule and of the atom *j* of the matrix due to this inclusion, and $\vec{R}_j^e$ is the position vector of the atom *j* with respect to the center of the site in the perfect crystal; it characterizes the distance between nearest-neighbor atoms (see Table 3.1).

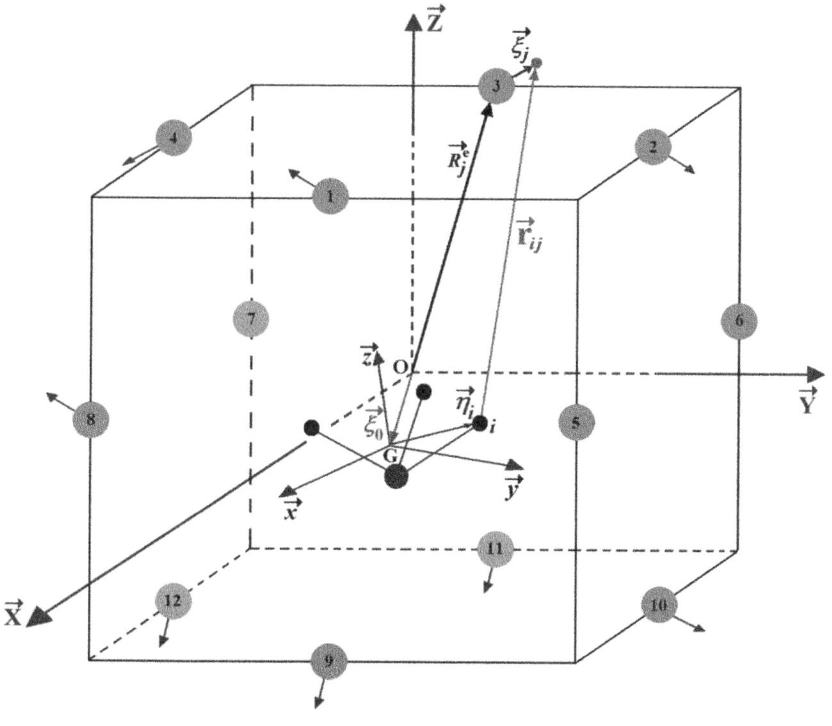

Figure 3.2. *Inclusion diagram, in a simple substitution site, of the NH_3 molecule in a rare gas matrix. $\vec{R}_j^e$ is the equilibrium position vector of the atom j in the fixed reference frame $(O,\vec{X},\vec{Y},\vec{Z})$ attached to the crystal. For a color version of this figure, see www.iste.co.uk/dahoo/infrared2.zip*

It is important to note that Figure 2.1 presents the geometric characteristics of the active molecule NH_3 and its partner, molecule k, belonging to the nanocage. In the case of the rare gas matrix, the partner k is replaced by the atom j of the rare gas, and the index k is no longer useful. The atoms of index $j = 1, \ldots 12$ are the nearest neighbors forming the first shell around the molecule. The Cartesian coordinates of the atoms of the rigid molecule, in the reference frame $(\mathrm{G}, \vec{x}, \vec{y}, \vec{z})$ attached to its equilibrium configuration, are presented in Table 3.3.

Atom coordinates	$\eta^e_{i_x}$ **(Å)**	$\eta^e_{i_y}$ **(Å)**	$\eta^e_{i_z}$ **(Å)**
N ($i = 1$)	0	0	- 0.0676
H ($i = 2$)	0.9414	0	0.3133
H ($i = 3$)	- 0.4707	0.8153	0.3133
H ($i = 4$)	- 0.4707	- 0.8153	0.3133

Table 3.3. *Cartesian coordinates of the atoms of the ammonia molecule NH_3 in the reference frame* $(\mathrm{G}, \vec{x}, \vec{y}, \vec{z})$ *attached to its equilibrium configuration*

3.3.2. *Deformation of the doped crystal*

Chapter 4 of Volume 1 describes how to determine the distortion of the matrix around the included molecule using an iterative method based on Green's function of the perfect crystal of rare gas.

It is important to recall that the internal energy of the doped crystal is written as:

$$U\left(\left\{\vec{R}^e_j, \vec{\xi}_j\right\}, \vec{\xi}_0, \vec{\Omega}, \{q_v\}\right) = V_\mathrm{C} + \hat{V}_\mathrm{MC}\left(\left\{\vec{R}^e_j, \vec{\xi}_j\right\}, \vec{\xi}_0, \vec{\Omega}, \{q_v\}\right) + V_\mathrm{vib}\left(\{q_v\}\right) \qquad [3.4]$$

where $\vec{\Omega} \equiv (\varphi, \theta, \chi)$ and $\{q_v\}$ represent, respectively, the orientation and the internal vibrations (q being the dimensionless normal coordinates) of the molecule at its inclusion site. On the other hand, V_C (equation [3.1]) and $V_\mathrm{vib}\left(\{q_v\}\right)$ (equation [4.42] of Volume 3) are the potential energies of the perfect crystal and of the internal vibrations of the molecule in gas phase, and $\hat{V}_\mathrm{MC}$ is the interaction energy between the molecule and the matrix V_MC from which the interaction energy between the substituted (ghost) atom and the rest of the matrix, except the included molecule, was deducted:

$$\hat{V}_{\text{MC}}\left(\left\{\vec{R}_j^{\text{e}},\vec{\xi}_j\right\},\vec{\xi}_0,\vec{\Omega},\{q_v\}\right)=\sum_j\left[\sum_i 4\varepsilon_{ij}\left\{\left(\frac{\sigma_{ij}}{\left|\vec{\mathrm{r}}_{ij}\right|}\right)^{12}-\left(\frac{\sigma_{ij}}{\left|\vec{\mathrm{r}}_{ij}\right|}\right)^{6}\right\}-\frac{1}{2}\vec{E}_{\mathbf{0}\to j}.\vec{\vec{\alpha}}_j^{A}.\vec{E}_{\mathbf{0}\to j}\right]$$
$$-\sum_j 4\varepsilon_{\text{gr-gr}}\left\{\left(\frac{\sigma_{\text{gr-gr}}}{\left|\vec{R}_j^{\text{e}}\right|}\right)^{12}-\left(\frac{\sigma_{\text{gr-gr}}}{\left|\vec{R}_j^{\text{e}}\right|}\right)^{6}\right\} \qquad [3.5]$$

In this expression, $\vec{\vec{\alpha}}_j^{A}$ and $\vec{E}_{\mathbf{0}\to j}$ are, respectively, the dipole polarizability tensor of the rare gas atom j and the field created by the multipole moments of the molecule at the location of this atom (see Chapter 1), $\vec{\mathrm{r}}_{ij}$ is the distance vector between the atom i of the molecule and the atom j of the matrix, and ε and σ are the LJ potential parameters presented in Table 3.2. The rules of combination of these parameters are presented in expression [2.3].

The dependency in angular and vibrational coordinates of the molecule is contained in the position vector $\vec{\eta}_i$ of the atom i of the molecule through the distance vector $\vec{\mathrm{r}}_{ij}$, written as:

$$\vec{\mathrm{r}}_{ij}=\vec{R}_j^{\text{e}}+\vec{\xi}_j-\vec{\xi}_0-\vec{\eta}_i \qquad [3.6]$$

as well as in its multipole moments. A Taylor series expansion of these quantities depending on the normal vibrational coordinates can be used to write:

$$\vec{\eta}_i=\vec{\eta}_i^{e}+\sum_v\left(\frac{\partial\vec{\eta}_i}{\partial q_v}\right)_0 q_v+\frac{1}{2}\sum_{v,v'}\left(\frac{\partial^2\vec{\eta}_i}{\partial q_v\partial q_{v'}}\right)_0 q_v q_{v'}+\ldots \qquad [3.7a]$$

$$\vec{\mu}_0=\vec{\mu}_0^{e}+\sum_v\left(\frac{\partial\vec{\mu}_0}{\partial q_v}\right)_0 q_v+\frac{1}{2}\sum_{v,v'}\left(\frac{\partial^2\vec{\mu}_0}{\partial q_v\partial q_{v'}}\right)_0 q_v q_{v'}+\ldots \qquad [3.7b]$$

$$\vec{\vec{\Theta}}_0=\vec{\vec{\Theta}}_0^{e}+\sum_v\left(\frac{\partial\vec{\vec{\Theta}}_0}{\partial q_v}\right)_0 q_v+\frac{1}{2}\sum_{v,v'}\left(\frac{\partial^2\vec{\vec{\Theta}}_0}{\partial q_v\partial q_{v'}}\right)_0 q_v q_{v'}+\ldots \qquad [3.7c]$$

These molecular quantities are defined in the reference frame $(G,\vec{x},\vec{y},\vec{z})$ attached to the equilibrium configuration of the molecule; their explicit dependency in angular coordinates $\vec{\Omega} \equiv (\varphi, \theta, \chi)$ appears when they are expressed in the fixed reference frame $(G,\vec{X},\vec{Y},\vec{Z})$, using the matrix $\mathbf{M}(\varphi, \theta, \chi)$ of transition from the fixed reference frame to the reference frame attached to the molecule (Appendix 1 of Chapter 2 (section 2.5.1)) [ROS 67].

The equilibrium of the doped crystal is obtained when the sum of its internal forces is zero, i.e. when the total energy U is minimal:

$$\vec{\nabla} U\left(\left\{\vec{R}_j^e, \vec{\xi}_j\right\}, \vec{\xi}_0, \vec{\Omega}, \{q_v\}\right) = \vec{0} \qquad [3.8]$$

The gradient operator $\vec{\nabla}$ relates to all the degrees of freedom of the system. Indeed, the internal vibrational motions of the included molecule are significantly more rapid than its external motions and those of the matrix atoms. The deformation of the matrix around the assumed rigid molecule can be determined.

In the approximation of a reasonably small distortion of the atoms of the matrix surrounding the trapped molecule, the total interaction energy U can be written as a second-order Taylor series expansion depending on the displacement vectors $\vec{\xi}$ of atoms and on the molecule's center of mass. They result from the application of the equilibrium condition of the doped crystal and the use of the elements of Green's function matrix of the perfect crystal (Table 3.4).

Distance	g_{XX}	g_{YY}	g_{ZZ}	g_{XY}	g_{XZ}	g_{YZ}
0	0.838	0.838	0.838	0	0	0
$\mathbf{a/\sqrt{2}}$	0.257	0.257	0.206	0	0	0.081
a	0.145	0.143	0.143	0	0	0
$\mathbf{a\sqrt{2}}$	0.144	0.128	0.128	0.016	0.027	0.027

Table 3.4. *Numerical values of several elements of the (dimensionless) Green's function matrix of a perfect monoatomic crystal, depending on the relative distance between an atom j and an atom j' nearest neighbor, second nearest neighbor and third nearest neighbor*

These displacement vectors depend on the orientational and vibrational motions of the molecule. Nevertheless, according to the adiabatic approximation, it is possible to calculate them for each set of coordinates $\left(\vec{\Omega},\{q_v\}\right)$ and write them as follows:

$$\vec{\xi}_k \approx \vec{\xi}_k^{\,s} + \vec{\xi}_k\left(\vec{\Omega},\{q_v\}\right) \qquad [3.9]$$

where $\vec{\xi}_k^{\,s}$ are the displacement vectors of the center of mass of the molecule ($k = 0$) and of the atoms of the matrix ($k = j$) obtained for the rigid molecule and at its equilibrium orientation (static model or cage model), and $\vec{\xi}_k\left(\vec{\Omega},\{q_v\}\right)$ characterizes the variation in the vicinity of $\vec{\xi}_k^{\,s}$, depending on the orientational and vibrational coordinates.

It is obvious that the equilibrium condition (equation [3.8]) is not verified unless $\vec{\xi}_k$ is considered in its entirety. Nevertheless, if the matrix gets deformed only slightly when the vibrational and orientational dynamics of the molecule evolves, then the displacements $\vec{\xi}_k$ can be approximated by $\vec{\xi}_k^{\,s}$. Otherwise, in the case of a more or less significant modulation of $\vec{\xi}_k$ depending on variables $\left(\vec{\Omega},\{q_v\}\right)$, the second part is no longer negligible and an inertial treatment of the Hamiltonian is required.

3.3.3. *NH_3 in an argon matrix*

The application of the above inclusion model to the case of the rigid NH_3 molecule trapped in an argon matrix led to the determination of the deformation generated by the inclusion of the molecule up to the 4th matrix shell around it, as well as its interaction energy V_{MC}.

The equilibrium of the system is reached when the molecule is at the center of the site, oriented so that its symmetry axis C_3 is collinear with one of the symmetry axes C_3 of the site, for example, $\vec{\Omega}^e \equiv (\varphi^e = \pi/4, \theta^e \approx 125.3°, \chi^e = \pi/6)$, with a very slight isotropic distortion of about +0.01 Å of the argon atoms around the molecule. Therefore, the doped crystal preserves a quasi-perfect octahedral symmetry. The minimal interaction potential energy is therefore $V^e_{MC} = -1,263$ cm^{-1} [LAK 87a]. Moreover, it was noted that during a translational motion of the molecule, the argon atoms are motionless. This makes it possible to separate the displacement vectors according to expression [3.9], in which $\vec{\xi}_k\left(\vec{\Omega},\{q_v\}\right)$ is very small.

3.3.4. *Renormalization of the system's Hamiltonian*

Chapter 4 of Volume 1 of this set presents a quite detailed approach of the total Hamiltonian of the system consisting of the trapped diatomic molecule and the matrix. It is important to recall a few significant points when the trapped molecule is polyatomic.

The Hamiltonian of the system:

$$H = H_{\text{vib}} + T_{\text{rot}} + T_{\text{trans}} + T_{\text{C}} + U \tag{3.10}$$

H_{vib}, T_{rot} and T_{trans} represent, respectively, the Hamiltonians of vibration, rotation and translation of the isolated molecule (expressions [4.7] and [4.8]), and $T_{\text{C}} = \sum_{\alpha,j} \frac{P_\alpha^2(j)}{2M_j}$ is the kinetic Hamiltonian of vibration of the lattice.

Consider $\vec{\zeta}$ the dynamic variable of lattice vibration defined from the equation of the perfect crystal. $\vec{\zeta}$ can then be written in the form of a sum of displacement $\vec{\xi}$ defined by expression [3.9] and the dynamic variable $\vec{u}$ of small amplitude, as follows:

$$\vec{\zeta} = \vec{\xi} + \vec{u} \tag{3.11}$$

The interaction potential energy U, defined by expression [3.4], can then be written in the following symbolic form:

$$U = \varphi^{(0)}\left(\vec{\Omega},\{q_v\}\right) + \underline{\varphi}^{(1)}\left(\vec{\Omega},\{q_v\}\right).\vec{u} + \frac{1}{2}\underline{\varphi}^{(2)}\left(\vec{\Omega},\{q_v\}\right)\vec{u}.\vec{u} + \dots \tag{3.12}$$

This type of expression amounts to recentering the motions around an equilibrium configuration of a crystal "prepared" by the inclusion of the molecule and not around that of the perfect crystal.

The first term $\varphi^{(0)}\left(\vec{\Omega},\{q_v\}\right)$ of expression [3.12] is the potential energy independent of lattice dynamics, but taking into account a significant part of the vibrational–orientational dynamics of the molecule. The second term $\underline{\varphi}^{(1)}\left(\vec{\Omega},\{q_v\}\right)$ linear in $\vec{u}$ characterizes the forces acting on the molecule and on the crystal atoms. The third term $\underline{\varphi}^{(2)}\left(\vec{\Omega},\{q_v\}\right)$ represents the harmonic force constants associated with the dynamics of the doped crystal. As for the higher-order terms, they represent anharmonic force constants.

The Hamiltonian of the system can then be decomposed as:

$$H=\left[H_{\text{vib}}+T_{\text{rot}}+\varphi^{(0)}\left(\vec{\Omega},\{q_{v}\}\right)\right]+\left[T_{\text{trans}}+T_{\text{C}}+\sum_{l\neq 0}\varphi^{(l)}\left(\vec{\Omega},\{q_{v}\}\right).\vec{u}^{l}\right] \quad [3.13]$$

The first part of this expression is the Hamiltonian related to the "optical system or optically active molecule" characterized by the variables of vibrations and orientation of the trapped molecule. The second part designates the Hamiltonian of the "bath", characterized by motions of the center of mass of the molecule, vibrations of lattice "phonons" and the dynamic part of the potential energy coupling the variables $\left(\vec{\Omega},\{q_{v}\}\right)$ and $\vec{u}$. It is important to note that the motion of the molecule's center of mass can have a frequency in the band of phonons of the perfect crystal, a non-localized phonon or above the cut-off frequency of this band, referred to as the localized mode.

A more appropriate separation of the Hamiltonian [3.13] requires a clarification of the dynamic conditions under which the system evolves to solve the equations of motion.

3.4. Motions of NH_3 trapped in an argon matrix

In the hypothesis that the matrix is not driven by the vibrational and orientational motions of the molecule, either because the rigidity of the matrix is sufficiently large compared to the forces binding the molecule to the matrix, or because the motions of this molecule are sufficiently rapid to annihilate the drive on average, as in the case of NH_3 trapped in an argon matrix, then $\vec{\xi}\left(\vec{\Omega},\{q_{v}\}\right)\cong\vec{\xi}^{s}$; and, using the adiabatic approximation, it is possible to obtain a separation of the total Hamiltonian (renormalization) in the form:

$$H=\tilde{H}_{\text{vib}}+\tilde{H}_{\text{orient}}+\tilde{H}_{\text{bath}}+\Delta\tilde{H}\left(\vec{\Omega},\vec{u}\right) \quad [3.14]$$

In this expression, $\tilde{H}_{\text{a}}=\tilde{H}_{\text{vib}}+\tilde{H}_{\text{orient}}$ is the Hamiltonian of the vibration–orientation optical system of the trapped molecule, $\tilde{H}_{\text{bath}}$ is the Hamiltonian of the heat bath constituted of the translational vibrational modes of the atoms of the matrix and of the molecule's center of mass. Finally, $\Delta\tilde{H}\left(\vec{\Omega},\vec{u}\right)$

characterizes the dynamic coupling between the orientational motion of the rigid molecule (low-frequency mode) and the heat bath, which is considered a perturbation Hamiltonian. This coupling is responsible for the transfer mechanisms of the orientation energy to the modes of phonons of the doped crystal (phase relaxation), which involves frequency shifts and broadenings of the spectral lines.

The following section, concerning the ammonia molecule trapped in an argon matrix at very low temperatures, focuses on the construction of the far-infrared (FIR) spectrum (rotational spectrum) and near-infrared (NIR) spectrum in the region of the vibration–inversion mode, in order to be able to compare the calculated spectra to the observed spectra [ABO 77].

3.4.1. *Vibration–inversion mode* ν_2

In a condensed medium, the intermolecular forces shift the frequencies of molecular vibrations with respect to the frequencies measured in the gas state. The shift of the vibrational frequency in matrix, with respect to that of the gas phase, is defined by:

$$\Delta\omega = \omega_{matrix} - \omega_{gas} \tag{3.15}$$

This shift $\Delta\omega$ comprises a static contribution generated by the interaction of the trapped molecule with the nearest-neighbor atoms due to the surrounding crystal field and a dynamic contribution at the origin of the energy transfer processes, corresponding to the coupling between the lattice vibrations (phonons) and the vibrations of the molecule. In terms of the relative value, $\Delta\omega/\omega_{gas}$ does not generally exceed 1–2%, which corresponds to a weak coupling between the atoms of the molecule and those of the matrix. The coefficients of the expansion of the vibration potential energy, depending on the dimensionless normal coordinates characterizing the normal modes of vibration, are in this case weakly modified by the crystal field.

However, in the case of the vibration–inversion mode ν_2 of the ammonia molecule, the splitting of the vibrational levels due to the inversion phenomenon, which, moreover, is modified by the matrix effect, must be taken into account.

In expression [3.14], the first term $\tilde{H}_{\text{vib}}$ is the renormalized Hamiltonian of the vibration of the molecule in its inclusion site in the cage model:

$$\tilde{H}_{\text{vib}} = H_{\text{vib}} + V_{\text{MC}}^{\text{e}}(q_2) \tag{3.16}$$

where $V_{\text{MC}}^{\text{e}}(q_2)$ contains the dependency of the potential energy of the molecule–matrix interaction in terms of the normal coordinate of vibration of mode ν_2, the molecule being fixed at the center of the site, to its equilibrium orientation. This contribution is small and considered as a perturbation, and written as:

$$V_{\text{MC}}^{\text{e}}(q_2) = \left(\frac{\partial V_{\text{MC}}^{\text{e}}}{\partial q_2}\right)_0 q_2 + \frac{1}{2}\left(\frac{\partial^2 V_{\text{MC}}^{\text{e}}}{\partial q_2^2}\right)_0 q_2^2 = \beta_1 q_2 + \frac{1}{2}\beta_2 q_2^2 \qquad [3.17]$$

The calculation by the stationary perturbation method leads to a frequency shift by the matrix effect $\Delta\omega_2 \approx +15\ \text{cm}^{-1}$ (+18 cm^{-1} observed), which takes the frequency from 950 cm^{-1} (see Table 4.7 of Volume 3) in the gas phase to 965 cm^{-1} in an argon matrix.

Moreover, given the modification of the double potential well by the matrix effect, the associated Hamiltonian is written as (see expressions [4.44] and [4.61] of Volume 3):

$$\begin{aligned}\tilde{H}_{\nu_2} &= H_{\nu_2} + \Delta V_{\text{MC}}^{\text{e}}(\theta_2) \\ &= -\frac{\hbar^2}{2\text{r}^{\text{e}\,2}\mu_2}\frac{\partial^2}{\partial\theta_2^2} + V(\theta_2) + \Delta V_{\text{MC}}^{\text{e}}(\theta_2)\end{aligned} \qquad [3.18]$$

where μ_2 is the reduced mass of the molecule, $V(\theta_2) = \frac{1}{2}k_0\theta_2^2 + A_0\left(e^{-a_0\theta_2^2} - 1\right)$ is the potential energy function (double potential well) associated with the vibration–inversion mode in the gas phase (Figure 4.9 of Volume 3) and $\Delta V_{\text{MC}}^{\text{e}}(\theta_2)$ is the variation of the interaction potential energy of molecule–matrix during the inversion motion, the other degrees of freedom being at their equilibrium value. The addition of $\Delta V_{\text{MC}}^{\text{e}}(\theta_2)$ and $V(\theta_2)$ makes it possible to obtain a double potential well of the same type as in Figure 4.9 of Volume 3, and constants k_0, A_0 and a_0 modified as shown in Table 3.5.

The height of the barrier between two potential wells passes from 2,047 cm^{-1} in the gas phase to 2,254 cm^{-1} in an argon matrix, which is an increase of 205 cm^{-1}, which induces a decrease in the splitting of v_2 levels.

The calculated values of splitting $\delta(\text{v}_2 = 0)$ of the fundamental level and $\delta(\text{v}_2 = 1)$ of the first excited level are, respectively, 0.54 cm^{-1} and 23.05 cm^{-1}. The observed value of the total splitting of the two levels is ~ 24 cm^{-1}, obtained for NH_3 in an argon matrix at a temperature of 8 K [ABO 77]. It is important to note that the wave functions and the transitions obtained are of the same type as those in Figure 4.9 of Volume 3.

	k_0 (cm^{-1}.rad^{-2})	A_0 (cm^{-1})	a_0 (rad^{-2})
NH_3 gas	60,137	12,469	4.974
NH_3 in Ar matrix	61,380	12,704	5.185

Table 3.5. *Values of the constants associated with the double well of vibration–inversion potential of NH_3 in gas phase and trapped in an argon matrix*

3.4.2. *Orientational motion*

In expression [3.14], $\tilde{H}_{\text{orient}}$ is the renormalized Hamiltonian of the orientational motion of the rigid molecule, fixed at its equilibrium position at the center of the site. It is written as:

$$\tilde{H}_{\text{orient}} = H_{\text{rot}} + V^{\text{e}}_{\text{MC}}(\varphi,\theta,\chi) \qquad [3.19]$$

where H_{rot} is the rotational Hamiltonian of the rigid oblate symmetric top molecule (expression [4.49] of Volume 3) and $V^{\text{e}}_{\text{MC}}(\varphi,\theta,\chi)$ is the perturbed orientational potential energy surface of the molecule in its inclusion site at its equilibrium position.

As previously mentioned, eight equilibrium orientations of the included molecule are possible (axis C_3 of the molecule along the diagonals of the cubic site of the matrix), for example, $\vec{\Omega}^{\text{e}} \equiv (\varphi^{\text{e}} = \pi/4, \theta^{\text{e}} \approx 125.3°, \chi^{\text{e}} = \pi/6)$. These are configurations corresponding to the highest symmetry. The minimal value of the energy is $V^{\text{e}}_{\text{MC}} = -1{,}263$ cm^{-1}. A detailed study of the potential energy surface $V_{\text{MC}}(\varphi,\theta,\chi)$ shows that this motion is weakly perturbed. Indeed, the height of the energy barrier between two equivalent orientations is ~ 60 cm^{-1}, as shown by the level curves $V^{\text{e}}_{\text{MC}}(\varphi,\theta)$ (on the left, for χ^{e}) and $V^{\text{e}}_{\text{MC}}(\theta,\chi)$ (on the right, for φ^{e}) of Figure 3.3.

Hence, the potential energy hypersurface $V_{\text{MC}}(\varphi,\theta,\chi)$ can be numerically adjusted on the basis of the rotation matrix elements $D^{l}_{p,q}(\varphi,\theta,\chi)$ in the form:

$$V_{\text{MC}}(\varphi,\theta,\chi) = \sum_{l=0}^{l_{\max}} \sum_{\substack{p=-l \\ q=-l}}^{+l} A^{l}_{p,q} \times D^{l}_{p,q}(\varphi,\theta,\chi) \qquad [3.20]$$

where $A^l_{p,q}$ are the expansion coefficients, which can be real or complex (see Appendix 2 (section 3.6.2)), l is an integer number, and p and q are the relative integer numbers such that: $-l \leq p \leq +l$ and $-l \leq q \leq +l$.

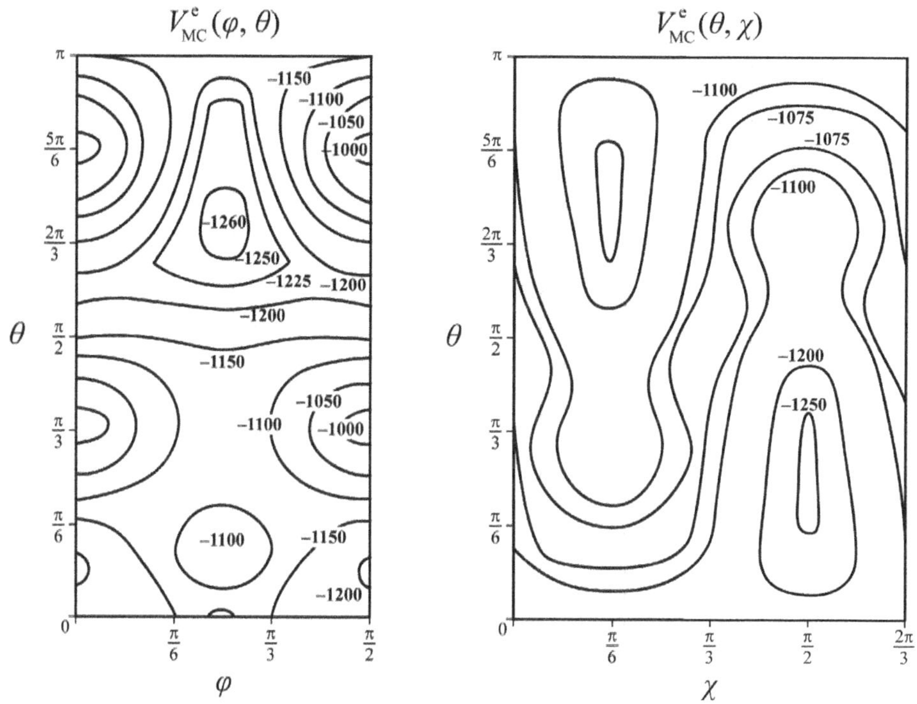

Figure 3.3. *Level curves of the potential energy surfaces (cm^{-1}): $V^e_{MC}(\varphi,\theta)$ (on the left, for χ^e) and $V^e_{MC}(\theta,\chi)$ (on the right, for φ^e)*

However, given the C_3 symmetry of the molecule, the relative integer number q associated with the spinning angle χ can only take the values $q = 0, \pm 3, \pm 6, \ldots$ Moreover, assuming that the inclusion site is octahedral, the relative integer number p associated with the angle φ can only take the values $p = 0, \pm 4, \pm 8, \ldots$ Finally, the natural integer number l only takes even values ≥ 4.

Therefore, the expansion up to the order l_{max} = 8 of the function $V_{MC}(\varphi, \theta, \chi)$ leads to the values of the coefficients A (in cm^{-1}) in expression [3.20], presented in Table 3.6. The coefficients associated with values $l_{max} > 8$ are negligible.

q	l / p	0	4	6	8
0	0	– 1097.7	– 42.9	+ 10.3	– 0.5
	4	–	+ 18.4	+ 58.0	~ 0
3	0	–	+ 74.3	+ 25.1	~ 0
	4	–	+ 42.6	+ 49.6	~ 0
6	0	–	–	– 60.5	~ 0
	4	–	–	+ 32.3	~ 0

Table 3.6. *Values of expansion coefficients A (cm^{-1}) [3.20] of the hypersurface of potential energy of the orientation of NH_3 trapped in an argon matrix*

Solving the Schrödinger eigenvalue equation associated with the Hamiltonian $\tilde{H}_{\text{orient}}$:

$$\tilde{H}_{\text{orient}} \left| jmk \right\rangle = \left[H_{\text{rot}} + V_{\text{MC}}^{\text{e}} (\varphi, \theta, \chi) \right] \left| jmk \right\rangle = E_{jmk} \left| jmk \right\rangle \qquad [3.21]$$

yields the eigenvalues of energies E_{jmk} and the eigenvectors $\left| jmk \right\rangle$ of the perturbed orientational motion, where j, m and k are the quantum numbers associated, respectively, with the angular variables of nutation θ, precession φ and proper rotation (spinning motion) χ.

In order to solve equation [3.21], a perturbation method is used, writing the eigenvector $\left| jmk \right\rangle$ as a linear combination of eigenvectors $\left| JMK \right\rangle$ of free rotation of the molecule in the gas phase:

$$\left| jmk \right\rangle = \sum_{J,M,K} \Lambda_{JMK}^{jmk} \left| JMK \right\rangle \qquad [3.22]$$

where Λ_{JMK}^{jmk} are complex coefficients determined by solving equation [3.21], with

$$\text{H}_{rot} \left| \text{JMK} \right\rangle = E_{JK} \left| \text{JMK} \right\rangle = \left[\text{B}^{\text{e}} \text{J}(\text{J}+1) + (\text{C}^{\text{e}} \text{-} \text{B}^{\text{e}}) \text{K} + ... \right] \left| \text{JMK} \right\rangle \quad \text{, becomes :}$$

$$\text{H}_{rot} \left| \text{JMK} \right\rangle = E_{JK} \left| \text{JMK} \right\rangle = \left[\text{B}^{\text{e}} \text{J}(\text{J}+1) + (\text{C}^{\text{e}} \text{-} \text{B}^{\text{e}}) \text{K}^2 + ... \right] \left| \text{JMK} \right\rangle \qquad [3.23]$$

and E_{JK} is the eigenenergy of free rotation containing the corrections due to centrifugal distortion. The rotational constants of the molecule are $B^{\text{e}} = 9.941\ \text{cm}^{-1}$ and $C^{\text{e}} = 6.309\ \text{cm}^{-1}$.

Introducing expressions [3.20], [3.22] and [3.23] into Schrödinger eigenvalue equation [3.21], the latter becomes:

$$\sum_{J,M,K} \Lambda_{JMK}^{jmk} \left[\left(E_{JK} - E_{jmk} \right) + \sum_{l=4}^{l_{\max}} \sum_{\substack{p=-l \\ q=-l}}^{+l} A_{p,q}^{l} \times D_{p,q}^{l}(\varphi,\theta,\chi) \right] \left| JMK \right\rangle = 0 \qquad [3.24]$$

By multiplying this equation by $\left\langle J'M'K' \right|$ and using the equation

$$\begin{aligned} \left\langle J'M'K' \middle| JMK \right\rangle &= (-1)^{M'-K'}(-1)^{M-K}\sqrt{\frac{2J'+1}{8\pi^2}}\sqrt{\frac{2J+1}{8\pi^2}} \\ &\times \int_0^{2\pi}\int_0^{\pi}\int_0^{2\pi} D_{-M',-K'}^{J'*}(\varphi,\theta,\chi) D_{-M,-K}^{J}(\varphi,\theta,\chi) d\varphi \sin\theta d\theta d\chi \\ &= \delta_{JJ'}\delta_{MM'}\delta_{KK'} \end{aligned} \qquad [3.25]$$

the secular equation is written as:

$$\sum_{J,M,K} \left[\begin{array}{l} \left(E_{JK} - E_{jmk} \right) \delta_{JJ'}\delta_{MM'}\delta_{KK'} + \\ \sum_{l,p,q} A_{p,q}^{l} \times \left\langle J'M'K' \right| D_{p,q}^{l}(\varphi,\theta,\chi) \left| JMK \right\rangle \end{array} \right] \Lambda_{JMK}^{jmk} = 0 \qquad [3.26]$$

where δ is the Kronecker symbol, and the matrix element $\left\langle J'M'K' \right| D_{p,q}^{l}(\varphi,\theta,\chi) \left| JMK \right\rangle$ can be expressed by means of Clebsch–Gordan coefficients (see Appendix 3 of Chapter 2 (section 2.5.3)) as:

$$\begin{aligned} \left\langle J'M'K' \right| D_{p,q}^{l}(\varphi,\theta,\chi) \left| JMK \right\rangle &= \sqrt{\frac{(2J'+1)}{(2J+1)}} \times \\ & C(JlJ';-M,p)C(JlJ';-K,q) \end{aligned} \qquad [3.27]$$

Hence, the secular equation becomes:

$$\sum_{J,M,K} \left[\begin{array}{l} \left(E_{JK} - E_{jmk} \right) \delta_{JJ'}\delta_{MM'}\delta_{KK'} + \sqrt{\frac{(2J'+1)}{(2J+1)}} \sum_{l,p,q} A_{p,q}^{l} \times \\ \times C(JlJ';-M,p)C(JlJ';-K,q) \end{array} \right] \Lambda_{JMK}^{jmk} = 0 \qquad [3.28]$$

and the determination of rotation eigenvalues and eigenvectors of the molecule involves the diagonalization of a symmetric and scattered square matrix of elements $M_{J'M'K'}^{JMK}$:

$$M_{J'M'K'}^{JMK} = \left(E_{JK} - E_{jmk}\right)\delta_{JJ'}\delta_{MM'}\delta_{KK'} + \sqrt{\frac{(2J'+1)}{(2J+1)}}\sum_{l,p,q} A_{p,q}^{l} \times \times C(JlJ';-M,p)C(JlJ';-K,q) \quad [3.29]$$

NOTE.– Clebsch–Gordan coefficients are not zero only if $|J-l| \leq J' \leq J+l$, $M'=M+p$ and $K'=K+q$.

The dimension of this matrix depends on the maximal value taken by the number l in expression [3.20] and on the rotational energy levels involved in the spectral lines. Indeed, levels j and j' perturb (contaminate) each other through the number l. For example, if $l_{\max} = 8$, then a state of number j can be perturbed by states of numbers j' such that $j' \leq j + 8$. Moreover, it can be noted that, depending on the non-zero values of the coefficients of expression [3.20], this matrix contains many zero elements and could be arranged in independent submatrices attached to species A ($k = 0$), E ($k = \pm 1$), etc., of dimensions 230x230, 225x225, etc., respectively.

Considering the splitting of the fundamental vibrational level $v_2 = 0$, due to the tunneling effect-based inversion motion, the orientation level scheme obtained after the diagonalization of matrix [3.28] consists of levels $j = 0$ and $j = 1$, very similar to those of the isolated molecule (free rotation), with a weak degeneracy lifting on the magnetic quantum number m of level $j = 1$. On the other hand, level $j = 2$ undergoes a quite strong degeneracy lifting in m, as shown in Figure 3.4. This figure also shows orientational transitions, denoted by $R(j_k)$ (R for the transitions $\Delta j = +1$), whose elements are non-zero, and taking into account the selection rules due to inversion: $s \leftrightarrow a$ (see Chapter 4 of Volume 3).

At very low temperatures (~10–40 K or ~7–28 cm^{-1}), the predominantly populated levels are $j = 0$, 1, and the orientational motion of the molecule is quasi-free.

It must be noted that during the orientational motion, the molecule is not rigorously rigid. Then, by using the adiabatic approximation, its rotational constants, on the one hand, and the coefficients of the expansion of expression [3.20], on the other, depend on the state of vibration of the molecule. The Hamiltonian $\tilde{H}_{\text{orient}}$ can be written as $\tilde{H}_{\text{orient}}^{\text{v}}$, and the previous matrices must be built and diagonalized in each vibrational state (here, the vibration–inversion state). However, the modifications brought to the level scheme remain negligible.

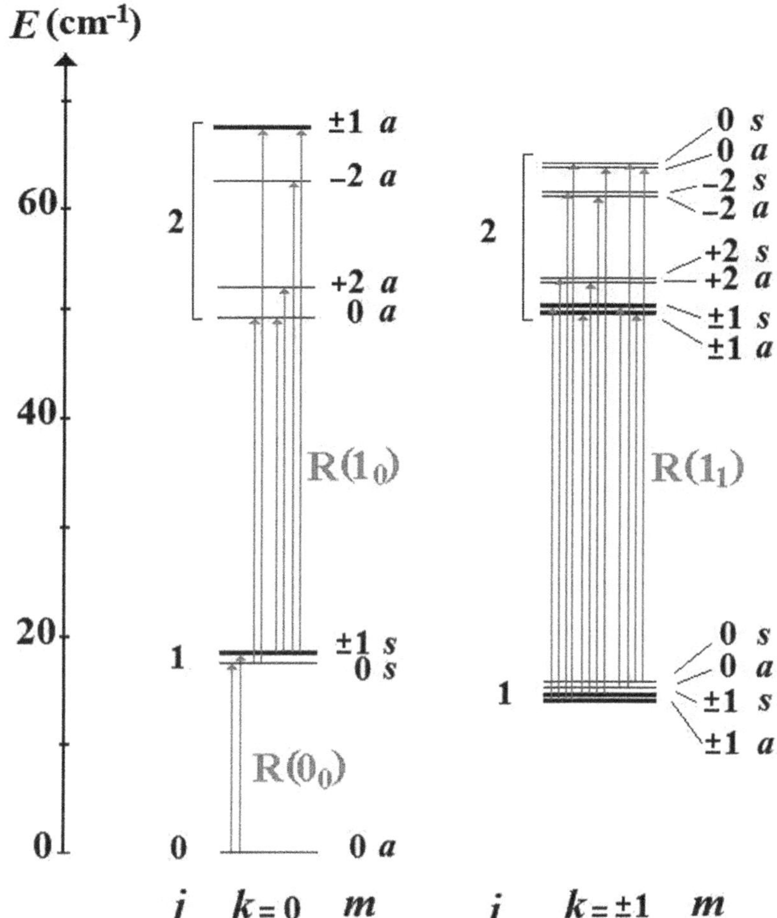

Figure 3.4. *Orientational level scheme attached to species A (k = 0) and E (k = 1) of NH_3 trapped in an argon matrix, considering the splitting of levels (s for symmetric and a for antisymmetric) due to inversion. The allowed orientational transitions are also presented. For a color version of this figure, see www.iste.co.uk/dahoo/infrared2.zip*

Hence, in the adiabatic approximation, the eigenenergies and eigenvectors of the Hamiltonian of the optical system are written as:

$$\begin{aligned} |\mathrm{v};jmk\rangle &= |\mathrm{v}\rangle \otimes |jmk\rangle \\ E_{\mathrm{v};jmk} &= E_{\mathrm{v}} + E_{jmk} \end{aligned} \quad [3.30]$$

3.4.3. *Translational motion*

The approximation of the orientational motion of the molecule by a free rotation makes it possible to consider it as a point defect in a crystal with octahedral symmetry and to determine the frequency of the vibrational (translational) motion of its center of mass as a function of the defects of mass and force constant in relation to the perfect crystal (see section 4.5.6 of Volume 1).

In expression [3.14], the Hamiltonian $\tilde{H}_{\text{bath}}$ characterizes the heat bath as susceptible to absorb the energy generated by the optical (vibration–orientation) modes of the molecule. It involves the translational vibration modes of the atoms of the matrix and of the rigid molecule's center of mass (phonons). This Hamiltonian is written as:

$$\tilde{H}_{\text{bath}} = H_{\text{C}} + T_{\text{trans}} + \hat{V}^{\text{e}}_{\text{MC}}(\vec{u}) \qquad [3.31]$$

where H_{C} is the Hamiltonian of the perfect crystal defined by expression [3.2], T_{trans} is the kinetic Hamiltonian of translation of the isolated rigid molecule's center of mass, and $\hat{V}^{\text{e}}_{\text{MC}}(\vec{u})$ is the potential energy of translation of the lattice containing the rigid molecule, assumed to be a point defect from which the interaction energy between the substituted (ghost) atom and the rest of the matrix, except the included molecule, is deducted.

The frequencies associated with the vibrational motions of the perfect crystal are contained in the function of the normalized density of phonon states $\omega^2 \nu(\omega^2)$, while those associated with vibrational motions of the molecule's center of mass are obtained by solving the equation [MAN 68, MAN 71, MAR 65]:

$$1 - \rho(\omega^2) S(\omega^2) = 0 \qquad [3.32]$$

where $S(\omega^2)$ is the function defined by expression [3.3] and presented in Figure 3.1, and

$$\rho(\omega^2) = \left(-1 + \frac{M_j}{M_0}\right) + M_j \left(\frac{\partial^2 V^{\text{e}}_{\text{C}}}{\partial u^2(j)}\right)^{-1} \left(1 - \left(\frac{\partial^2 V^{\text{e}}_{\text{C}}}{\partial u^2(j)}\right) \Big/ \left(\frac{\partial^2 V^{\text{e}}_{\text{MC}}}{\partial u^2(0)}\right)\right) \times \omega^2 \qquad [3.33]$$

The mass and force constant defects are, respectively, $M_j - M_0 = 39.95 - 17.03 = 22.92$ g/mole and $\left(\frac{\partial^2 V^{\text{e}}_{\text{C}}}{\partial u^2(j)}\right) - \left(\frac{\partial^2 V^{\text{e}}_{\text{MC}}}{\partial u^2(0)}\right) = 2308 - 1720 = 588 \ \text{cm}^{-1}.\text{Å}^{-2}$.

Solving equation [3.32] makes it possible to obtain the frequency $\omega_L \approx 68.5 \ \mathrm{cm}^{-1}$ apart from the density of phonon states $\omega^2 \nu(\omega^2)$ of the perfect crystal of argon, whose cut-off frequency is $\omega_{\max} = 68 \ \mathrm{cm}^{-1}$. It is therefore a localized mode that will be susceptible to participating to the energy transfer between the rotational motion of the molecule and the vibrational motions of the doped crystal lattice.

3.4.4. *Orientational motion–heat bath coupling*

In the harmonic description of the dynamics of the doped crystal lattice, the perturbation Hamiltonian $\Delta\tilde{H}\left(\vec{\Omega},\vec{u}\right)$ can be expanded in the form of a two-term sum:

$$\begin{aligned}\Delta\tilde{H}\left(\vec{\Omega},\vec{u}\right) &= \sum_j \left[\vec{\nabla} V_{\mathrm{MC}}(\vec{\Omega})\right]\vec{u}(j) + \frac{1}{2}\sum_{j,j'}\left[\vec{\nabla}\vec{\nabla} V_{\mathrm{MC}}(\vec{\Omega})\right]\vec{u}(j)\vec{u}(j') \\ &= \Delta\tilde{H}_{1\mathrm{p}}\left(\vec{\Omega},\vec{u}\right) + \Delta\tilde{H}_{2\mathrm{p}}\left(\vec{\Omega},\vec{u}\right)\end{aligned} \qquad [3.34]$$

implying a phase relaxation, respectively, with one phonon and two phonons.

A long and tedious analytical processing (which can be found in [LAK 87a] and [GIR 85]) can be used to write these couplings as products of functions depending on dynamic coordinates $u_\alpha(j)$ (including $j = 0$ for the molecule) and the rotation matrix elements (truncated at the order $l_1 = 2$ and $l_2 = 2$) as:

$$\begin{aligned}\Delta\tilde{H}_{1\mathrm{p}}\left(\vec{\Omega},\vec{u}\right) &= \sum_{l_1=1}^{2}\sum_{p_1=-l_1}^{+l_1} X_{l_1,p_1}\left(\{R_j^{\mathrm{e}}\};\{u_\alpha(j)\}\right)\times D^{l_1}_{-p_1,0}(\varphi,\theta,\chi) \\ \Delta\tilde{H}_{2\mathrm{p}}\left(\vec{\Omega},\vec{u}\right) &= \sum_{l_2=1}^{2}\sum_{p_2=-l_2}^{+l_2} X_{l_2,p_2}\left(\{R_j^{\mathrm{e}}\};\{u_\alpha(j)u_\beta(j')\}\right)\times D^{l_2}_{-p_2,0}(\varphi,\theta,\chi)\end{aligned} \qquad [3.35]$$

j and j' taking values of 0 for the molecule at 12 for the nearest-neighbor atoms of the latter.

The functions X are then expressed in terms of normal modes of vibration of the doped crystal. They are presented in Appendix 3 (section 3.6.3) for the one phonon coupling $\Delta\tilde{H}_{1\mathrm{p}}\left(\vec{\Omega},\vec{u}\right)$. The (fairly long) expressions associated with the two phonon coupling $\Delta\tilde{H}_{2\mathrm{p}}\left(\vec{\Omega},\vec{u}\right)$ can be found in [LAK 87a], which is freely available on open archive sites (HAL, for example).

3.5. Infrared spectra

3.5.1. *Infrared absorption coefficient*

The infrared absorption coefficient attached to a number $\mathcal{N}$ of optically active molecules, per unit volume and without interactions between them, in a matrix, is defined as the real part of the spectral density, i.e. the temporal Fourier transform of the autocorrelation function $\Phi(t)$

$$\begin{aligned} I(\omega) &= \frac{4\pi\mathcal{N}\omega}{3hc}\mathrm{Re}\int_0^\infty dt e^{i\omega t}\Phi(t) \\ \Phi(t) &= \mathrm{Tr}\left[\rho(0)\vec{\mu}_0^A(0)\vec{\mu}_0^A(t)\right] \end{aligned} \qquad [3.36]$$

In this expression, $\vec{\mu}_0^A$ is the dipole moment operator of the active molecule defined in the reference frame $(\mathrm{G},\vec{\mathrm{X}},\vec{\mathrm{Y}},\vec{\mathrm{Z}})$ attached to it, $\rho(0)$ is the initial canonical density matrix of the total system, ω is the frequency (expressed in wave number, cm^{-1}), and c is the speed of light. The symbol Tr (Trace) represents the average over the initial conditions at instant $t = 0$ and over all the possible evolutions of the system between instant 0 and instant t.

Chapter 4 of Volume 1 (section 4.6.2) analyzes the autocorrelation function $\Phi(t)$ according to the Heisenberg representation.

It is important to recall, however, that only a semi-analytical processing makes it possible to determine this autocorrelation function, through certain approximations:

– the permanent dipole moment of the molecule is much higher than the induced dipole moment due to its inclusion in the matrix and does not depend on the bath modes;

– thanks to the process of renormalization of the total Hamiltonian, the orientation–heat bath coupling Hamiltonian $\Delta\tilde{H}$ is small and remains so during the time, justifying the use of the *initial chaos* hypothesis (initial decoupling of the optical system and bath);

– the spectral lines are assumed to be ones isolated from the others.

Hence, the infrared absorption coefficient can be written as:

$$\begin{aligned} I(\omega) = \frac{4\pi\mathcal{N}\omega}{3hc}\sum_{i,f}(\rho_{\mathrm{a}}^i - \rho_{\mathrm{a}}^f)\left|\left\langle i\left|\vec{\mu}_0^A\right|f\right\rangle\right|^2 \mathrm{Re}\int_0^\infty dt \exp\left\{i(\omega - \omega_{if} - \Delta\omega_{\mathrm{v}_i\mathrm{v}_f})t\right\} \\ \exp\left\{-i\Delta^{(1)}t\right\}\exp\left\{-i\Delta^{(2)}t\right\}\exp\left\{-\Gamma^{(2)}t\right\} \end{aligned} \qquad [3.37]$$

In this expression, $|i\rangle \equiv |v_i;j_i m_i k_i\rangle$ and $|f\rangle \equiv |v_f;j_f m_f k_f\rangle$ represent the initial and final optical states between which the transition of frequency ω_{if} takes place, ρ_a^i and ρ_a^f are their respective population at instant $t=0$ and $\Delta\omega_{v_i v_f}$ is the vibrational frequency shift obtained in section 2.4.1. Finally, ($\Delta^{(1)}$, $\Delta^{(2)}$) and $\Gamma^{(2)}$ are, respectively, the frequency shifts and the spectral line broadening associated with $i \rightarrow f$ transition (see section 4.6 of Volume 1). The temperature dependency of these shifts and widths is contained in the averages and the correlation functions of the orientation–bath coupling Hamiltonian.

In expression [3.37], the population ρ_a^i (respectively, ρ_a^f) of the initial (respectively, final) state of the optically active system is expressed as a function of its energy E_i (respectively, E_f) and of the partition function Z, as:

$$\rho_a^i = \frac{e^{-E_i/k_BT}}{Z} = \frac{e^{-E_i/k_BT}}{\sum_n e^{-E_n/k_BT}} \qquad [3.38]$$

and the transition elements of the dipole moment in the fixed reference frame are given in equations [4.48] and [4.55] of Volume 3.

3.5.2. *Bar spectrum*

This is an important stage in the construction of the spectrum, since it offers information on the quality of the chosen basis and gives an idea on the initial amplitude of the coupling between the optical system and the bath. Indeed, to the extent that the bar spectrum can be used to reconstruct the majority of signals of the experimental spectrum, it can be admitted that the coupling is, and remains, a small perturbation during the evolution of the total system.

For the vibration–inversion mode v_2, of states $|v_2\alpha\rangle$ (with $\alpha = s$ for symmetric and a for antisymmetric), of the NH_3 molecule trapped in an argon matrix, and after renormalization of the total Hamiltonian, the bar spectrum is obtained by ignoring the dynamic coupling $\Delta\tilde{H}$.

3.5.2.1. *Far-infrared (FIR) spectrum*

The integrated absorption coefficient (bar lines) involving only inversion–orientation states of the fundamental vibrational level $v_2 = 0$ is written as:

$$I^0(\omega) = \frac{8\pi^2 \mathcal{N} \omega}{3hc} \sum_{i,f,\alpha_i,\alpha_f}^{\alpha_i \neq \alpha_f} (\rho_a^i - \rho_a^f) \left| \left\langle 0\alpha_i; j_i m_i k_i \left| \vec{\mu}_0^{eA} \right| 0\alpha_f; j_f m_f k_f \right\rangle \right|^2 \times \delta(\omega - \omega_{0\alpha_i;j_i m_i k_i \to 0\alpha_f;j_f m_f k_f}) \quad [3.39]$$

Figure 3.5 presents the calculated bar spectrum and the experimental spectrum obtained with a resolution of 1.1 cm^{-1} by Cugley and Pullin [CUG 72, CUG 73] at temperature $T = 13$ K.

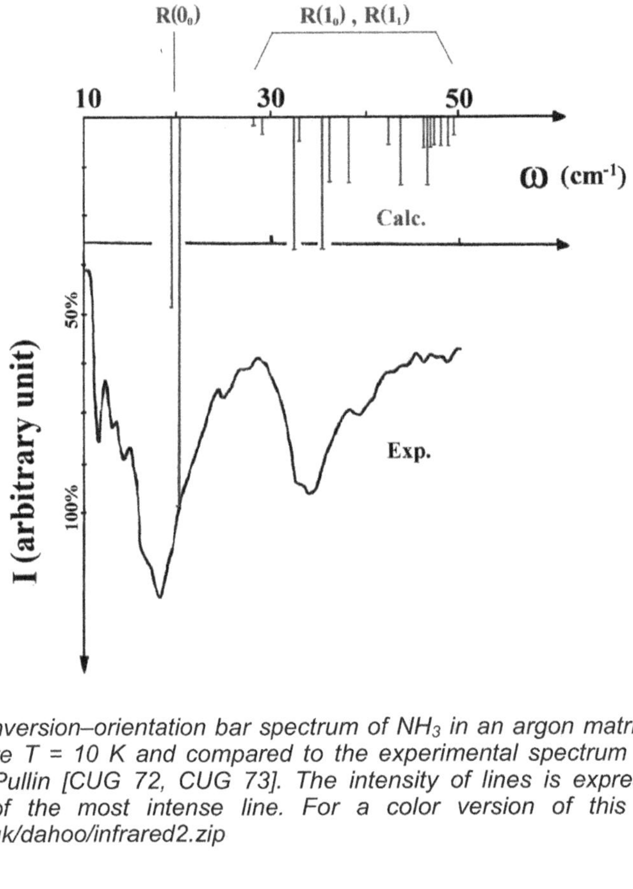

Figure 3.5. *Inversion–orientation bar spectrum of NH_3 in an argon matrix calculated at temperature T = 10 K and compared to the experimental spectrum obtained by Cugley and Pullin [CUG 72, CUG 73]. The intensity of lines is expressed as the percentage of the most intense line. For a color version of this figure, see www.iste.co.uk/dahoo/infrared2.zip*

The branch R(0_0) consists of two lines:

– one line at 19.4 cm^{-1} associated with the transition $|0a;000\rangle \to |0s;100\rangle$;

– the other line at 20.3 cm^{-1} associated with transition $|0a;000\rangle \to |0s;1{\pm}10\rangle$.

In the range from 28 to 50 cm^{-1}, there is the branch R(1_1), essentially composed of lines associated with transitions $|0a;1m_i1\rangle \rightarrow |0s;2m_f1\rangle$ and $|0s;1m_i1\rangle \rightarrow |0a;2m_f1\rangle$ of equal intensities. Finally, the less intense lines constitute the branch R(1_0).

3.5.2.2. *Near-infrared (NIR) spectrum*

The integrated absorption coefficient (bar lines) involving vibration–inversion–orientation states is written as:

$$I^0(\omega) = \frac{8\pi^2 \mathcal{N} \omega}{3hc} \sum_{i,f,\alpha_i,\alpha_f}^{\alpha_i \neq \alpha_f} (\rho_a^i - \rho_a^f) \left| \left\langle 0\alpha_i; j_i m_i k_i \left| \left(\frac{d\vec{\mu}_0^A}{dq_2} \right)_{q_2=0} q_2 \right| 1\alpha_f; j_f m_f k_f \right\rangle \right|^2 \times \delta(\omega - \omega_{0\alpha_i;j_im_ik_i \rightarrow 1\alpha_f;j_fm_fk_f} - \Delta\omega_2) \quad [3.40]$$

where the population $\rho_a^f \cong 0$, as it involves high energy states of the excited vibrational level $v_2 = 1$. Given the complexity of the figure showing the vibration–inversion–orientation level schemes and the allowed transitions, it is not presented here.

Figure 3.6 presents the calculated bar spectrum and the experimental spectrum obtained with a resolution of 0.5 cm^{-1} by Abouaf *et al.* [ABO 77] at temperature $T = 8$ K.

The branch R(0_0) at ~ 975 cm^{-1} consists of two lines associated with transitions $|0a;000\rangle \rightarrow |1s;100\rangle$ (less intense) and $|0a;000\rangle \rightarrow |1s;1\pm10\rangle$ (more intense) due to degeneracy lifting in m.

The branch Q(1_1) (Q for the transitions $\Delta j = 0$) is split by inversion into two branches separated by ~28 cm^{-1} and associated with transitions $|0a;1m_i1\rangle \rightarrow |1s;1m_f1\rangle$ and $|0s;1m_i1\rangle \rightarrow |1a;1m_f1\rangle$.

The branch R(1_1) is split into two massifs corresponding to transitions $|0a;1m_i1\rangle \rightarrow |1s;2m_f1\rangle$ and $|0s;1m_i1\rangle \rightarrow |1a;2m_f1\rangle$, where $m_i = 0, \pm1$ and $m_f = 0, \pm1, \pm2$.

The remaining less intense lines P(1_0) (P for the transitions $\Delta j = -1$) and R(1_0) are, respectively, associated with transitions $|0s;1m_i0\rangle \rightarrow |1a;000\rangle$ and $|0s;1m_i0\rangle \rightarrow |1a;2m_f0\rangle$.

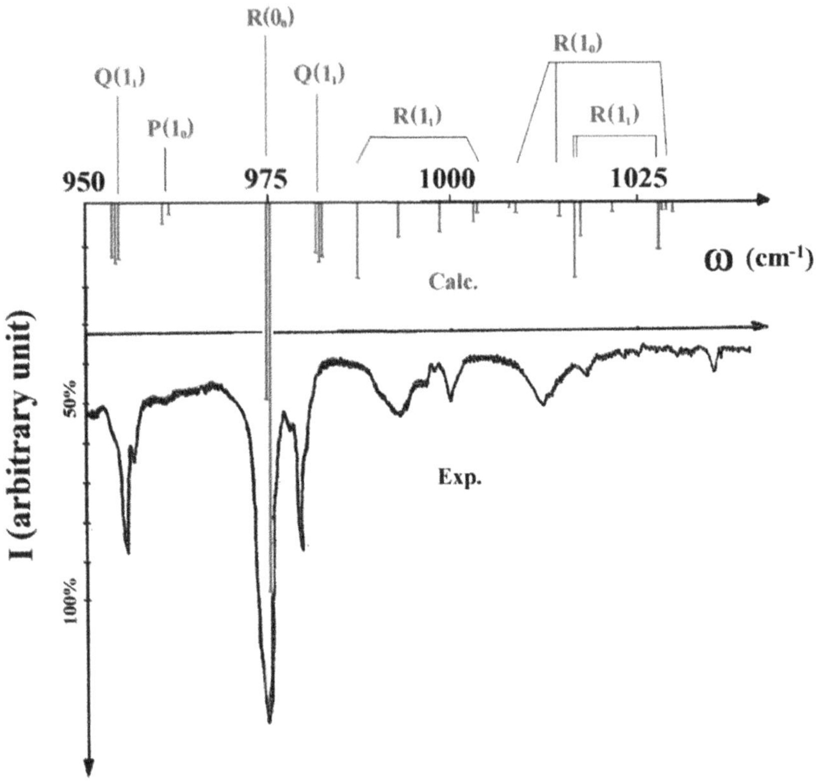

Figure 3.6. *Vibration–inversion–orientation bar spectrum of NH_3 in an argon matrix calculated at temperature T = 10 K and compared to the experimental spectrum obtained by Abouaf* et al. *[ABO 77]. The intensity of lines is expressed as the percentage of the most intense line. For a color version of this figure, see www.iste.co.uk/dahoo/infrared2.zip*

Finally, the comparison of the obtained bar spectra to the experimental spectra shows the good positioning of the calculated lines. This means that, at this stage, the chosen vibration–inversion–orientation basis is quite convenient for the study of the effects of dynamic coupling $\Delta\tilde{H}$.

3.5.3. *Spectral profile*

When the dynamic "orientation–bath" coupling is introduced, the line shifts $\Delta^{(1)}$ and $\Delta^{(2)}$ and their homogeneous width at mid-height $\Gamma^{(2)}$ are calculated.

FIR		NIR		Line	Δ		Γ	
ω (cm^{-1})	I/I_0	ω (cm^{-1})	I/I_0		$\Delta_{Resonant}$ (cm^{-1})	$\Delta_{Localized}$ (cm^{-1})	$\Gamma_{Resonant}$ (cm^{-1})	$\Gamma_{Localized}$ (cm^{-1})
20.3	100	975.3	100	$R(0_0)$	- 1.26	+ 0.65	0.68	0.04
19.4	48.7	974.7	51.4	"	- 1.48	- 0.58	0.86	0.04
48.0	7.9	1030.0	3.3	$R(1_0)$	- 0.42	+ 0.13	3.23	0.08
47.1	6.8	1029.0	2.9	"	- 0.69	- 1.10	3.07	0.08
42.3	6.8	1021.7	3.1	"	- 1.74	- 1.58	2.17	0.08
33.0	6.4	1014.8	3.9	"	+ 0.68	+ 3.29	1.45	0.08
		982.1	14.0	$Q(1_1)$	- 0.82	- 0.30	1.07	0.04
		981.8	15.2	"	0	0	1.00	0.04
		981.4	13.3	"	+ 0.82	+ 0.30	1.07	0.04
		954.6	14.0	"	+ 0.82	+ 0.30	1.07	0.04
		954.1	15.2	"	0	0	1.00	0.04
		953.8	13.3	"	- 0.82	- 0.30	1.07	0.04
48.9	6.4	1028.3	2.7	$R(1_1)$	+ 0.59	+ 0.40	1.51	0.06
48.5	11.9	1027.0	5.0	"	+ 0.09	- 1.38	2.05	0.06
46.7	18.2	1027.9	7.9	"	+ 0.05	+ 0.40	1.49	0.06
38.1	17.4	1017.5	9.3	"	- 0.65	- 0.68	1.55	0.06
35.4	33.9	1016.6	19.3	"	- 0.75	- 2.65	2.74	0.06
46.9	6.4	1003.4	2.7	"	- 0.59	- 0.40	1.51	0.06
46.6	11.9	1003.1	5.0	"	- 0.09	+ 1.38	2.05	0.06
43.6	18.2	998.3	7.9	"	- 0.05	- 0.40	1.49	0.06
36.1	17.4	992.5	9.3	"	+ 0.65	+ 0.68	1.55	0.06
32.4	33.9	987.2	19.3	"	+ 0.75	+ 2.65	2.74	0.06

Table 3.7. *Characteristics (position in frequency and intensity) of the lines in bar spectra and their shift and width at mid-height*

The line shifts are due to both elastic and inelastic collisions between the orientational motion of the molecule and one or two phonons of the lattice, including localized phonons (translation of the molecule).

The first-order processes with two phonons ($l_2 = 1$, expressions [3.35]) are the first to contribute to the line shift. The second-order processes with one or two phonons ($l_1 = 2$ and $l_2 = 2$, expressions [3.35]) have a substantial contribution to the shift.

Linewidths are due to the second order with one or two phonons ($l_1 = 2$ and $l_2 = 2$, expressions [3.35]), essentially resonant phonons (very weak participation of the translational motion of the molecule).

Table 3.7 presents the characteristics (position in frequency and intensity ≥ 5%) of the lines in far- and near-infrared ranges, as well as their shift and width at mid-height. The contributions brought by the coupling of the orientation–vibrations of the lattice (resonant phonons) and orientation–vibration of the molecule's center of mass (localized phonons) are presented separately.

3.5.3.1. *Far-infrared spectral profile*

According to the hypothesis of the Lorentzian model, the expression of the spectral profile of the trapped molecule can finally be written as:

$$I(\omega) = \frac{8\pi^2 \mathcal{N}\omega}{3hc} \sum_{i,f,\alpha_i,\alpha_f}^{\alpha_i \neq \alpha_f} (\rho_a^i - \rho_a^f) \left| \left\langle 0\alpha_i; j_i m_i k_i \left| \vec{\mu}_0^{eA} \right| 0\alpha_f; j_f m_f k_f \right\rangle \right|^2 \times \frac{\Gamma^{(2)}}{(\omega - \omega_{0\alpha_i;j_i m_i k_i \to 0\alpha_f;j_f m_f k_f} - \Delta^{(1)} - \Delta^{(2)})^2 + (\Gamma^{(2)})^2} \quad [3.41]$$

Using the results of Table 3.7, the spectral profile is presented in Figure 3.7.

3.5.3.2. *Near-infrared spectral profile*

In the NIR range, the expression leading to the spectral profile is written as:

$$I(\omega) = \frac{8\pi^2 \mathcal{N}\omega}{3hc} \sum_{i,f,\alpha_i,\alpha_f}^{\alpha_i \neq \alpha_f} (\rho_a^i - \rho_a^f) \left| \left\langle 0\alpha_i; j_i m_i k_i \left| \left(\frac{d\vec{\mu}_0^A}{dq_2} \right)_{q_2=0} q_2 \right| 1\alpha_f; j_f m_f k_f \right\rangle \right|^2 \times \frac{\Gamma^{(2)}}{(\omega - \omega_{0\alpha_i;j_i m_i k_i \to 1\alpha_f;j_f m_f k_f} - \Delta\omega_2 - \Delta^{(1)} - \Delta^{(2)})^2 + (\Gamma^{(2)})^2} \quad [3.42]$$

Using the results in Table 3.7, the spectral profile is presented in Figure 3.8.

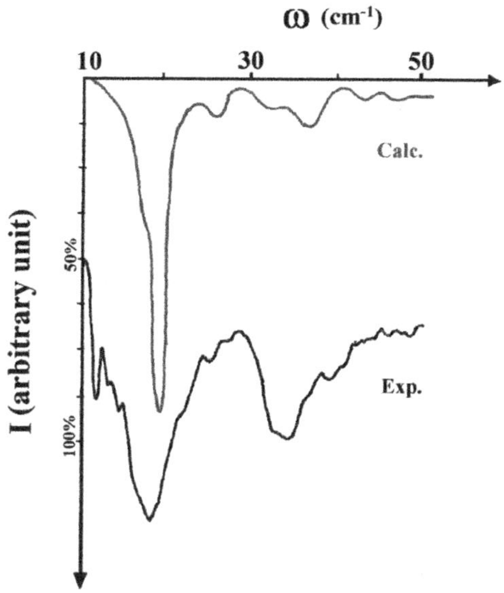

Figure 3.7. *Spectral profile of the inversion–orientation of NH_3 in an argon matrix calculated at T = 10 K and compared to the experimental spectrum. For a color version of this figure, see www.iste.co.uk/dahoo/infrared2.zip*

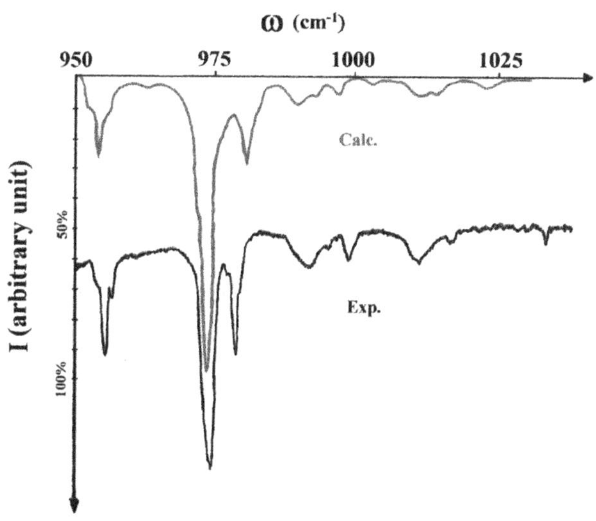

Figure 3.8. *Spectral profile of the vibration–inversion–orientation of NH_3 in an argon matrix calculated at T = 10 K and compared to the experimental spectrum. For a color version of this figure, see www.iste.co.uk/dahoo/infrared2.zip*

3.6. Appendices

3.6.1. *Appendix 1: Normal modes of vibrations of a Bravais lattice with face centered cubic (fcc) symmetry*

This appendix presents the analytical expressions of the normal modes of vibrations of the atoms of a perfect crystal of rare gas crystallizing in the fcc system depending on their instantaneous displacements of small amplitude, with respect to the fixed reference frame attached to the crystal and defined in Figure 3.2.

For this purpose, the molecule trapped in a simple substitution site and its 12 first neighbors are considered as a super-molecule that has the symmetry of the crystal; the trapped molecule appears as an isotropic body (point defect) whose mass and force constant differ from those of the substituted rare gas atom.

For a perfect crystal belonging to the octahedral symmetry group O_h, the total representation Γ_{fcc} can be decomposed, using the table of characters of this group in its irreducible representations [MAR 65]:

$$\Gamma_{\text{fcc}} = A_{1g} + A_{2g} + 2E_g + 2T_{1g} + 2T_{2g} + A_{2u} + E_u + 4T_{1u} + 2T_{2u}$$

where A, E and T are, respectively, irreducible representations with one, two or three dimensions.

The expressions of the normal components attached to the irreducible representations depending on the Cartesian components of the instantaneous displacement vectors $\vec{u}(0)$ of the trapped molecule and $\vec{u}(j)$ (j = 1,...,12) of the atoms (see Figure 3.2) of the crystal are given below, considering the mass of the atoms equal to the unity.

Mode A_{1g}

$$A_{1g} = \frac{1}{\sqrt{24}}\begin{pmatrix} u_X(1) - u_X(3) + u_X(5) - u_X(6) - u_X(7) + u_X(8) + u_X(9) - u_X(11) \\ +u_Y(2) - u_Y(4) + u_Y(5) + u_Y(6) - u_Y(7) - u_Y(8) + u_Y(10) - u_Y(12) \\ +u_Z(1) + u_Z(2) + u_Z(3) + u_Z(4) - u_Z(9) - u_Z(10) - u_Z(11) - u_Z(12) \end{pmatrix}$$

Mode A_{2g}

$$A_{2g} = \frac{1}{\sqrt{24}}\begin{pmatrix} -u_X(1) + u_X(3) + u_X(5) - u_X(6) - u_X(7) + u_X(8) - u_X(9) + u_X(11) \\ +u_Y(2) - u_Y(4) - u_Y(5) - u_Y(6) + u_Y(7) + u_Y(8) + u_Y(10) - u_Y(12) \\ +u_Z(1) - u_Z(2) + u_Z(3) - u_Z(4) - u_Z(9) + u_Z(10) - u_Z(11) + u_Z(12) \end{pmatrix}$$

Mode E_g^1

$$\alpha_{E_g^1} = \frac{1}{\sqrt{16}}\begin{pmatrix} -u_X(1)+u_X(3)+u_X(5)-u_X(6)-u_X(7)+u_X(8)-u_X(9)+u_X(11) \\ -u_Y(2)+u_Y(4)+u_Y(5)+u_Y(6)-u_Y(7)-u_Y(8)-u_Y(10)+u_Y(12) \end{pmatrix}$$

$$\beta_{E_g^1} = \frac{1}{\sqrt{16}}\begin{pmatrix} -u_X(5)+u_X(6)+u_X(7)-u_X(8)-u_Y(5)-u_Y(6)+u_Y(7)+u_Y(8) \\ +u_Z(1)+u_Z(2)+u_Z(3)+u_Z(4)-u_Z(9)-u_Z(10)-u_Z(11)-u_Z(12) \end{pmatrix}$$

Mode E_g^2

$$\alpha_{E_g^2} = \frac{1}{\sqrt{16}}\begin{pmatrix} -u_X(1)+u_X(3)-u_X(5)+u_X(6)+u_X(7)-u_X(8)-u_X(9)+u_X(11) \\ +u_Y(2)-u_Y(4)+u_Y(5)+u_Y(6)-u_Y(7)-u_Y(8)+u_Y(10)-u_Y(12) \end{pmatrix}$$

$$\beta_{E_g^2} = \frac{1}{\sqrt{16}}\begin{pmatrix} -u_X(5)+u_X(6)+u_X(7)-u_X(8)+u_Y(5)+u_Y(6)-u_Y(7)-u_Y(8) \\ +u_Z(1)-u_Z(2)+u_Z(3)-u_Z(4)-u_Z(9)+u_Z(10)-u_Z(11)+u_Z(12) \end{pmatrix}$$

Mode T_{1g}^1

$$\alpha_{T_{1g}^1} = \frac{1}{\sqrt{8}}\left(u_X(1)+u_X(3)-u_X(9)-u_X(11)-u_Z(1)+u_Z(3)-u_Z(9)+u_Z(11)\right)$$

$$\beta_{T_{1g}^1} = \frac{1}{\sqrt{8}}\left(u_X(2)+u_X(4)-u_X(10)-u_X(12)-u_Z(5)+u_Z(6)+u_Z(7)-u_Z(8)\right)$$

Mode T_{1g}^2

$$\alpha_{T_{1g}^2} = \frac{1}{\sqrt{8}}\left(u_Y(1)+u_Y(3)-u_Y(9)-u_Y(11)-u_Z(5)-u_Z(6)+u_Z(7)+u_Z(8)\right)$$

$$\beta_{T_{1g}^2} = \frac{1}{\sqrt{8}}\left(u_Y(2)+u_Y(4)-u_Y(10)-u_Y(12)-u_Z(2)+u_Z(4)-u_Z(10)+u_Z(12)\right)$$

Mode T_{1g}^3

$$\alpha_{T_{1g}^3} = \frac{1}{\sqrt{8}}\left(u_X(2)-u_X(4)+u_X(10)-u_X(12)-u_Y(1)+u_Y(3)-u_Y(9)+u_Y(11)\right)$$

$$\beta_{T_{1g}^3} = \frac{1}{\sqrt{8}}\left(-u_X(5)-u_X(6)+u_X(7)+u_X(8)+u_Y(5)-u_Y(6)-u_Y(7)+u_Y(8)\right)$$

Mode T_{2g}^{1}

$$\alpha_{T_{2g}^{1}} = \frac{1}{\sqrt{8}}\left(u_X(1)+u_X(3)-u_X(9)-u_X(11)+u_Z(1)-u_Z(3)+u_Z(9)-u_Z(11)\right)$$

$$\beta_{T_{2g}^{1}} = \frac{1}{\sqrt{8}}\left(u_X(2)+u_X(4)-u_X(10)-u_X(12)+u_Z(5)-u_Z(6)-u_Z(7)+u_Z(8)\right)$$

Mode T_{2g}^{2}

$$\alpha_{T_{2g}^{2}} = \frac{1}{\sqrt{8}}\left(u_Y(1)+u_Y(3)-u_Y(9)-u_Y(11)+u_Z(5)+u_Z(6)-u_Z(7)-u_Z(8)\right)$$

$$\beta_{T_{2g}^{2}} = \frac{1}{\sqrt{8}}\left(u_Y(2)+u_Y(4)-u_Y(10)-u_Y(12)+u_Z(2)-u_Z(4)+u_Z(10)-u_Z(12)\right)$$

Mode T_{2g}^{3}

$$\alpha_{T_{2g}^{3}} = \frac{1}{\sqrt{8}}\left(-u_X(2)+u_X(4)-u_X(10)+u_X(12)-u_Y(1)+u_Y(3)-u_Y(9)+u_Y(11)\right)$$

$$\beta_{T_{2g}^{3}} = \frac{1}{\sqrt{8}}\left(-u_X(5)-u_X(6)+u_X(7)+u_X(8)-u_Y(5)+u_Y(6)+u_Y(7)-u_Y(8)\right)$$

Mode A_{2u}

$$A_{2u} = \frac{1}{\sqrt{12}}\begin{pmatrix} -u_X(2)+u_X(4)+u_X(10)-u_X(12)-u_Y(1)+u_Y(3)+u_Y(9)-u_Y(11) \\ -u_Z(5)+u_Z(6)-u_Z(7)+u_Z(8) \end{pmatrix}$$

Mode E_{u}^{1}

$$E_{u}^{1} = \frac{1}{\sqrt{8}}\left(u_X(2)-u_X(4)-u_X(10)+u_X(12)-u_Y(1)+u_Y(3)+u_Y(9)-u_Y(11)\right)$$

Mode E_{u}^{2}

$$E_{u}^{2} = \frac{1}{\sqrt{8}}\left(-u_Y(1)+u_Y(3)+u_Y(9)-u_Y(11)+u_Z(5)-u_Z(6)+u_Z(7)-u_Z(8)\right)$$

Mode T_{1u}^{1}

$$\alpha_{T_{1u}^1} = u_X(0)$$

$$\beta_{T_{1u}^1} = \frac{1}{\sqrt{8}}\left(u_X(1)+u_X(3)+u_X(5)+u_X(6)+u_X(7)+u_X(8)+u_X(9)+u_X(11)\right)$$

$$\gamma_{T_{1u}^1} = \frac{1}{\sqrt{8}}\left(-u_Y(5)+u_Y(6)-u_Y(7)+u_Y(8)-u_Z(1)+u_Z(3)+u_Z(9)-u_Z(11)\right)$$

$$\delta_{T_{1u}^1} = \frac{1}{\sqrt{4}}\left(u_X(2)+u_X(4)+u_X(10)+u_X(12)\right)$$

Mode T_{1u}^{2}

$$\alpha_{T_{1u}^2} = u_Y(0)$$

$$\beta_{T_{1u}^2} = \frac{1}{\sqrt{8}}\left(u_Y(2)+u_Y(4)+u_Y(5)+u_Y(6)+u_Y(7)+u_Y(8)+u_Y(10)+u_Y(12)\right)$$

$$\gamma_{T_{1u}^2} = \frac{1}{\sqrt{8}}\left(-u_X(5)+u_X(6)-u_X(7)+u_X(8)-u_Z(2)+u_Z(4)+u_Z(10)-u_Z(12)\right)$$

$$\delta_{T_{1u}^2} = \frac{1}{\sqrt{4}}\left(u_Y(1)+u_Y(3)+u_Y(9)+u_Y(11)\right)$$

Mode T_{1u}^{3}

$$\alpha_{T_{1u}^3} = u_Z(0)$$

$$\beta_{T_{1u}^3} = \frac{1}{\sqrt{8}}\left(u_Z(1)+u_Z(2)+u_Z(3)+u_Z(4)+u_Z(9)+u_Z(10)+u_Z(11)+u_Z(12)\right)$$

$$\gamma_{T_{1u}^3} = \frac{1}{\sqrt{8}}\left(-u_X(1)+u_X(3)+u_X(9)-u_X(11)-u_Y(2)+u_Y(4)+u_Y(10)-u_Y(12)\right)$$

$$\delta_{T_{1u}^3} = \frac{1}{\sqrt{4}}\left(u_Z(5)+u_Z(6)+u_Z(7)+u_Z(8)\right)$$

Mode T_{2u}^{1}

$$\alpha_{T_{2u}^1} = \frac{1}{\sqrt{8}}\left(-u_X(1)+u_X(3)+u_X(9)-u_X(11)+u_Y(2)-u_Y(4)-u_Y(10)+u_Y(12)\right)$$

$$\beta_{T_{2u}^1} = \frac{1}{\sqrt{8}}\left(u_Z(1)-u_Z(2)+u_Z(3)-u_Z(4)+u_Z(9)-u_Z(10)+u_Z(11)-u_Z(12)\right)$$

Mode T_{2u}^{2}

$$\alpha_{T_{2u}^{2}} = \frac{1}{\sqrt{8}}\left(u_X(1)+u_X(3)-u_X(5)-u_X(6)-u_X(7)-u_X(8)+u_X(9)+u_X(11)\right)$$

$$\beta_{T_{2u}^{2}} = \frac{1}{\sqrt{8}}\left(u_Y(5)-u_Y(6)+u_Y(7)-u_Y(8)-u_Z(1)+u_Z(3)+u_Z(9)-u_Z(11)\right)$$

Mode T_{2u}^{3}

$$\alpha_{T_{2u}^{3}} = \frac{1}{\sqrt{8}}\left(u_Y(2)+u_Y(4)-u_Y(5)-u_Y(6)-u_Y(7)-u_Y(8)+u_Y(10)+u_Y(12)\right)$$

$$\beta_{T_{2u}^{3}} = \frac{1}{\sqrt{8}}\left(u_X(5)-u_X(6)+u_X(7)-u_X(8)-u_Z(2)+u_Z(4)+u_Z(10)-u_Z(12)\right)$$

It is important to note that the instantaneous displacement vector $\vec{u}(0)$ of the molecule's center of mass is only involved in mode T_{1u}.

3.6.2. *Appendix 2: Adjustment of the weakly perturbed rotational potential energy on the basis of the rotation matrix elements*

When the rotational motion of the molecule is not or only weakly perturbed by the environment, the potential energy surface V_{MC} can be numerically adjusted on the basis of the rotation matrix elements as follows:

$$V_{MC}(\varphi,\theta,\chi) = \sum_{l,p,q} A_{p,q}^{l} \times D_{p,q}^{l}(\varphi,\theta,\chi)$$

In order to determine the expression of the expansion coefficients $A_{p,q}^{l}$, the previous equation is multiplied on the right by $D_{p',q'}^{l'}(\varphi,\theta,\chi)$ and then integrated with respect to Euler angles $0 \le \varphi \le 2\pi$, $0 \le \theta \le \pi$, $0 \le \chi \le 2\pi$:

$$\iiint V_{MC} D_{p',q'}^{l'} \sin\theta d\theta d\varphi d\chi = \sum_{l,p,q} A_{p,q}^{l} \times \iiint D_{p,q}^{l} D_{p',q'}^{l'} \sin\theta d\theta d\varphi d\chi$$

Knowing that $D_{p,q}^{l} = \left(-1\right)^{p-q} D_{-p,-q}^{l\,*}$, the previous equation is written as:

$$\iiint V_{MC} D_{p',q'}^{l'} \sin\theta d\theta d\varphi d\chi = \sum_{l,p,q} (-1)^{p-q} A_{p,q}^{l} \times \iiint D_{-p,-q}^{l\,*} D_{p',q'}^{l'} \sin\theta d\theta d\varphi d\chi$$

but

$$\iiint D^{l}_{-p,-q}{}^{*} D^{l'}_{p',q'} \sin\theta d\theta d\varphi d\chi = \frac{8\pi^2}{(2l'+1)} \delta_{l,l'} \delta_{-p,p'} \delta_{-q,q'}$$

Then

$$\iiint V_{\mathrm{MC}} D^{l'}_{p',q'} \sin\theta d\theta d\varphi d\chi = (-1)^{q'-p'} \frac{8\pi^2}{(2l'+1)} A^{l'}_{-p',-q'}$$

Finally:

$$A^{l}_{p,q} = (-1)^{q-p} \frac{(2l+1)}{8\pi^2} \int_0^{2\pi} d\chi \int_0^{2\pi} d\varphi \int_0^{\pi} d\theta \sin\theta V_{\mathrm{MC}}(\varphi,\theta,\chi) D^{l}_{-p,-q}(\varphi,\theta,\chi)$$

3.6.3. *Appendix 3: Expansion coefficients of the coupling between the orientation of the molecule and lattice vibrations (phonons)*

The use of the two center expansion presented in Chapter 1 and the expressions of the normal vibrational modes of the doped crystal as a function of coordinates $\{u_\alpha(j)\}$ makes it possible to determine the coefficients X of expressions [3.35]. The coefficients of one phonon coupling are presented below. The coefficients of two phonon coupling (quite long expressions) can be found in [LAK 87a], which is freely available on open archive sites (HAL, for example).

One phonon coupling:

$$\Delta\tilde{H}_{\mathrm{1p}}\left(\vec{\Omega},\vec{u}\right) = \sum_{l_1=1}^{2} \sum_{p_1=-l_1}^{+l_1} X_{l_1,p_1}\left(\{R_j^{\mathrm{e}}\};\{u_\alpha(j)\}\right) \times D^{l_1}_{-p_1,0}(\varphi,\theta,\chi)$$

– For $l_1 = 1$:

$$X_{1,0} = -4\left(\frac{dA^{1}_{0,0}}{dR_j}\right)_{R_j^{\mathrm{e}}} \left[\alpha_{T^3_{1\mathrm{u}}} - \frac{1}{\sqrt{8}}\beta_{T^3_{1\mathrm{u}}} - \frac{1}{\sqrt{8}}\gamma_{T^3_{1\mathrm{u}}}\right]$$

$$X_{1,-1} = 2\sqrt{2}\left(\frac{dA^{1}_{0,0}}{dR_j}\right)_{R_j^{\mathrm{e}}} \left[\left(\alpha_{T^1_{1\mathrm{u}}} - \frac{1}{\sqrt{8}}\beta_{T^1_{1\mathrm{u}}} - \frac{1}{\sqrt{8}}\gamma_{T^1_{1\mathrm{u}}}\right) + i\times\left(\alpha_{T^2_{1\mathrm{u}}} - \frac{1}{\sqrt{8}}\beta_{T^2_{1\mathrm{u}}} - \frac{1}{\sqrt{8}}\gamma_{T^2_{1\mathrm{u}}}\right)\right]$$

$$X_{1,+1} = -2\sqrt{2}\left(\frac{dA^{1}_{0,0}}{dR_j}\right)_{R_j^{\mathrm{e}}} \left[\left(\alpha_{T^1_{1\mathrm{u}}} - \frac{1}{\sqrt{8}}\beta_{T^1_{1\mathrm{u}}} - \frac{1}{\sqrt{8}}\gamma_{T^1_{1\mathrm{u}}}\right) - i\times\left(\alpha_{T^2_{1\mathrm{u}}} - \frac{1}{\sqrt{8}}\beta_{T^2_{1\mathrm{u}}} - \frac{1}{\sqrt{8}}\gamma_{T^2_{1\mathrm{u}}}\right)\right]$$

– For $l_1 = 2$:

$$X_{2,0} = -\frac{\sqrt{3}}{2}\left(\frac{dA_{0,0}^2}{dR_j}\right)_{R_j^e}\left[\alpha_{E_g^1} - \beta_{E_g^1}\right]$$

$$X_{2,-1} = -\sqrt{\frac{3}{2}}\left(\frac{dA_{0,0}^2}{dR_j}\right)_{R_j^e}\left[\alpha_{T_{2g}^1} + i\times\beta_{T_{2g}^2}\right]$$

$$X_{2,+1} = \sqrt{\frac{3}{2}}\left(\frac{dA_{0,0}^2}{dR_j}\right)_{R_j^e}\left[\alpha_{T_{2g}^1} - i\times\beta_{T_{2g}^2}\right]$$

$$X_{2,-2} = -\sqrt{\frac{3}{2}}\left(\frac{dA_{0,0}^2}{dR_j}\right)_{R_j^e}\left[\frac{\sqrt{2}}{2}\left(\alpha_{E_g^2} - \beta_{E_g^2}\right) - i\times\beta_{T_{2g}^3}\right]$$

$$X_{2,+2} = -\sqrt{\frac{3}{2}}\left(\frac{dA_{0,0}^2}{dR_j}\right)_{R_j^e}\left[\frac{\sqrt{2}}{2}\left(\alpha_{E_g^2} - \beta_{E_g^2}\right) + i\times\beta_{T_{2g}^3}\right]$$

It is important to note that coefficients A are those defined in expression [3.20], but this time with their dependency on R_j.

4

Nanocages of Hydrate Clathrates

Nowadays, thanks to the techniques for the observation of the universe, the existence of gas hydrate clathrates in many cold media is an established fact in the astrophysics community.

In planetary systems, such as ours, for example, the gas hydrate clathrates (CO_2, CH_4, etc.) are present in many bodies, starting with our planet Earth, in marine sediments, permafrost and the underground lakes of the Antarctic. They could also exist below the surface of giant planets, on Titan (satellite of Saturn) or in the polar regions of the planet Mars.

In the cold regions of the interstellar medium, the observations revealed the presence of ice grains containing atoms, radicals, simple molecules, such as H_2, CO_2, CH_4 and NH_3, as well as more complex molecules.

Infrared spectroscopy is an investigation tool that can be used to obtain a large amount of information about the evolution of molecules in their medium, and hence understand the physico-chemical phenomena involved.

4.1. Introduction

The construction of the spectral profile (the shape and width of the spectral lines of a molecule trapped in a condensed medium) can be achieved, given the possibility of comparing theoretical results with high-resolution spectroscopy measurements.

For the CH_4 molecule trapped in a hydrate clathrate matrix, Dartois and his coworkers conducted a detailed experimental study using the Fourier transform infrared spectroscopy (FTIR) in the region of vibrational frequencies 3,000–6,000 cm^{-1}, containing the ν_3 antisymmetric stretching mode, the $(3\nu_4, 2\nu_3)$ modes and several combination $(\nu_1+\nu_4, \nu_3+\nu_4, \nu_2+\nu_3, ...)$ modes [DAR 08, DAR 10].

This chapter focuses on the development of theoretical models that can be used to determine the infrared absorption spectra of NH_3 or CH_4 molecules isolated in the nanocages of clathrates, at very low temperatures.

In these hydrate clathrates, molecules can be trapped in small and/or large cages of type I and/or II structures.

In this book, the fourth volume of this series, the theoretical models described in Volumes 1 and 2 are applied to the symmetric top molecule NH_3 and to the spherical top molecule CH_4 trapped in a nanocage. The perturbed motions of the molecule are studied using an atom–atom potential, to describe the interaction between the molecule trapped in a substitution site and its environment when a rare gas matrix is involved. When the molecule is trapped in a nanocage, such as the one present in clathrates or fullerene, the results of quantum mechanics calculations are used to place effective charges on the sites of molecules, or on the nanocage atoms. Then, the extended inclusion model developed by Lakhlifi–Dahoo is applied, allowing the determination of the probable trapping cage structures, the equilibrium position and the study of the motions of the molecule trapped in its site, using the atom–atom potential model to describe the molecule–nanocage interaction.

4.2. The extended substitution model

The substitution model used for the IR spectroscopy of trapped species in a rare gas matrix was extended to nanocages trapping the studied molecule, such as HCl, CO, CO_2, N_2O, O_3, NH_3, ND_3, CH_4 and CH_3F [LAK 15], using the methods developed in the group theory. The theoretical model can be used to deal with, for example, the degeneracy lifting of the energy levels of degenerate modes.

In order to present the extended model, and illustrate the application of the group theory to the study of the effect of an environment on the electronic energy levels of a chemical species, a model material is considered. It belongs to the class of compounds, characterized by a transition between two spin states (Spin Transition (ST) or Spin CrossOver (SCO)) when they are subjected to an external physical constraint (pressure, temperature, electric or magnetic field or light–matter interaction). These compounds can be used as pressure, temperature, light or electromagnetic field sensors. Similar to the molecules trapped in a nanocage of a solid medium, these complexes are characterized by an entity that is trapped in an environment. These entities are metallic cations of metals in the transition group (M^{n+}), with the d orbitals on a third electron shell featuring four to seven electrons, such as $3d_4$ to $3d_7$ ($n = 3$, $l = 2$ and $m = \pm 2, \pm 1, 0$), incorporated into a matrix composed of organic materials. The trapping geometric configuration may be tetrahedral or octahedral [LOU 07, LOU 08, CHI 12, CHI 14a, CHI 14b, CHI 15, JUR 16a, JUR 16b, JUR 16c, LIN 16, ALL 17, HAR 17, LIN17, LIN 18, LIN 20a].

The local electronic structure of the resulting ST complexes at the molecular scale is calculated using the ligand field theory. For example, in the octahedral environment, a degeneracy lifting of d orbitals in two groups is evidenced (Figure 4.1).

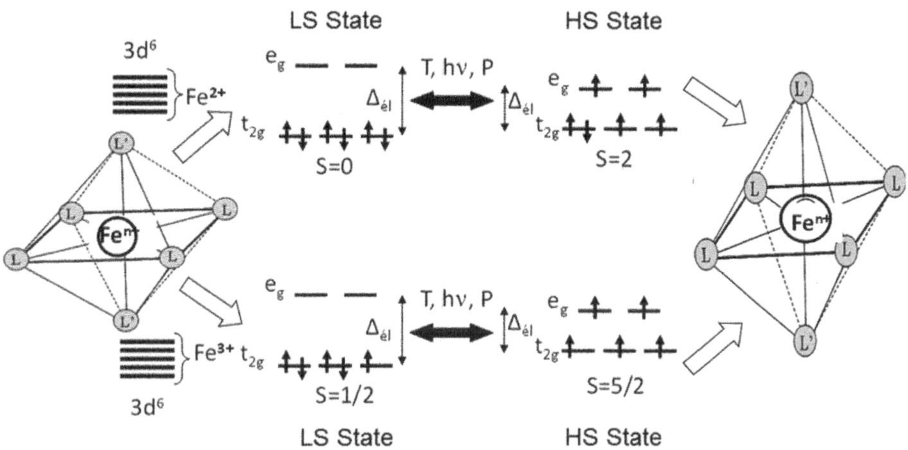

Figure 4.1. *LS state and HS state of Fe(II) and Fe(III) ions. For a color version of this figure, see www.iste.co.uk/dahoo/infrared2.zip*

As described in Volume 3, this degeneracy lifting can be interpreted using the approach proposed by Bethe [BET 29], which interprets the degeneracy lifting of atomic orbitals of atoms included in crystallographic sites within the group theory, by considering spin variables. It uses the notion of double symmetry groups by building a higher-order group by the tensor product of the point symmetry group, with a second-order group built on the space coordinate inversion operator.

In SCO compounds, the atomic levels of iron cations or metal cations of a transition element of the periodic table are characterized by splitting the electronic level into two subgroups in the crystal field of the surrounding medium, which can have an octahedral or tetrahedral symmetry. The Hamiltonian of an electron in the central field of an atom is invariant under all of the rotational symmetry operations and under the inversion operation with respect to the origin, which means that the symmetry group of the electron in an atom is the orthogonal group of dimension 3, denoted by O(3) or K_h, which is obtained by the direct product of the group of pure rotations and of the inversion group C_i.

Each level of the free ion belongs to one of the irreducible representations of the complete rotation group, characterized by the integral values of the orbital angular momentum L (s for L = 0, p for L = 1, d for L = 2, etc.). For the cations of transition metals, the levels d corresponding to L = 2 are degenerate, where the degeneracy is given by 2L + 1 = 5. When the cation is placed in a crystal field of octahedral symmetry, the atomic potential is no longer of spherical symmetry. The splitting of the degenerate levels of d orbital can be determined by applying methods of the group theory, in order to determine the new irreducible representations into which the d orbitals are separated.

The character $\chi(\varphi)$ of the state of the orbital angular momentum L is given by: $\chi(\varphi) = \sin(L + 1/2)\varphi / \sin(\varphi/2)$. Consequently, the characters of the five classes of transformations, consisting of the identity element E and four pure rotations, $8C_3(\varphi = \pi/3)$, $3C_4^2(\varphi = \pi)$, $6C_2(\varphi = \pi)$ and $6C_4(\varphi = \pi/2)$, of the crystal group O, are given by 5, -1, 1, 1 and -1, respectively. These are the characters of group O in the representation $D^{(2)}$ ($D^{(l)}$, $l = 2$) of dimension 5, ($2l+1 = 5$), of the rotation group. The resolution of this representation in the irreducible representations of group O [HAM 62] leads to the direct sum $D^{(2)} = E \oplus T_2$, where E is the irreducible representation of dimension 2 of group O, and T is the irreducible representation of dimension 3 of group O. If the crystal field is of symmetry $O_h = C_i \otimes O$, then it can be written that $D^{(2)} = E_g \oplus T_{2g}$, where g stands for gerade (even), since the d orbitals are invariant under inversion.

It is important to note that the d_{xy}, d_{yz} and d_{xz} orbitals of the irreducible representation t_{2g}, which point between ligands, have a lower energy than the $d_{x^2-y^2}$ and d_{z^2} orbitals of the irreducible representation e_g, which points towards the electrons. The gap between these two groups (a measure of the force of the ligand field), which depends on the distance between the metallic cation and the ligand, is denoted by Δ_O (O stands for octahedral). The distribution of the electrons d of the metallic ion in the orbital levels follows Pauli's principle and Hund's rule. In a strong field, the pairing energy of electrons (cost in energy when two electrons are in the same orbital) is not sufficient for Hund's rule to be obeyed, and only the t_{2g} orbitals are filled (low spin (LS) state). In a weak field, the two groups of t_{2g} and e_g orbitals can be filled (high spin (HS) state) (Figure 4.1). Their physical properties (magnetic, optical, thermal, electric, mechanical) depend on the change of the spin state of the metallic cation. These complexes can be classified, considering the cooperativity at the solid scale. These are bistable molecular materials with properties of commutation between the LS state and the HS state of multiphysics type (thermal, piezo-electric, magnetic and photonic). Hence, the LS and HS configurations can be represented by taking into account the manner in which the d orbitals are subdivided, and their filling according to Hund's rule. The method used allows the interpretation of the degeneracy lifting of the levels by spatial symmetry reduction.

In the case of a nanocage, the present static electric field is determined from the effective charges placed on the molecules that form the nanocage (section 4.2.2 of Volume 1 [DAH 17]). The effective charges are obtained from *ab initio* calculations of quantum chemistry or adjusted to the experimental results on the shift or broadening of lines. The long distance electrostatic interactions between the trapped molecules and the host molecules determined from these charges, make it possible to determine the perturbation undergone by the electronic potential driving the nuclear motions of the trapped molecule. In the case of rapid vibrational modes in comparison to slower translational modes and restricted rotation or libration, the process that was applied was described in Volumes 1 and 2 of this series [DAH 17, DAH 19a]. The vibration is decoupled from the slow modes of translation and rotation. The interaction energy expanded up to the second order is expressed in terms of normal coordinates, according to the hypothesis that the symmetry of the molecule is preserved. The molecule imposes its symmetry to the perturbing potential. A self-consistent process then allows the determination of the vibration energy at harmonic scale and the deduction of the shift and degeneracy [DAH 97, DAH 99a, CHA 99, DAH 99b, LAK 00, DAH 06].

In the case of a simple substitution site of octahedral symmetry O, or of a double substitution site of symmetry type D^2, the method applied to SCO for rotations can be applied. The stages leading to the calculation of the energy levels of the molecule are given in section 4.5 of Volume 1 [DAH 17]. For example, when the molecule is trapped in a clathrate cage, the short-range interaction with the water molecules modifies its degrees of freedom of translation and rotation, as the molecule is no longer free to move. Since the space of rotation of the molecule is no longer isotropic, it can have restricted rotational motions or librational motions.

In the case of the restricted rotational motion in a nanocage of clathrate or fullerene, it is possible to apply the methods of the group theory (as described below) for the electrons associated with the orbital l, when this orbital must evolve in a field of ligands.

For a diatomic molecule in a nanocage, the renormalized Hamiltonian of the orientational motion of the rigid molecule trapped in the equilibrium position of its center of mass, the formal expression of $\tilde{H}_{\text{orient}}$ (Chapter 4 of Volume 1 [DAH 17]) is written as:

$$\tilde{H}_{\text{orient}} = T_{\text{rot}} + V_{\text{MC}}^{\text{e}}(\varphi_0, \theta_0) \qquad [4.1]$$

where T_{rot} is the kinetic rotational operator of the free, rigid molecule (expression [2.82] of Volume 1 [DAH 17]) and $V^{\text{e}}_{\text{MC}}(\varphi_0,\theta_0)$ (expression [4.1]) is the perturbed orientation potential energy surface of the molecule in its inclusion site.

When a molecule is trapped in a nanocage, the interaction potential is expressed as a sum of terms, expanded on the basis of spherical harmonics. If a diatomic molecule can have a more or less perturbed rotational motion in a nanocage, then the potential energy $V^{\text{e}}_{\text{MC}}(\varphi_0,\theta_0)$ can be analytically expanded or numerically adjusted based on the spherical harmonics $Y_{l_0}^{p_0}(\theta_0,\varphi_0)$ in the form:

$$V^{\text{e}}_{\text{MC}}(\varphi_0,\theta_0) = \sum_{l_0=1}^{l_{\max}} \sum_{p_0=-l_0}^{+l_0} A_{p_0}^{l_0} \times Y_{l_0}^{p_0}(\theta_0,\varphi_0) \qquad [4.2]$$

In this expression, $A_{p_0}^{l_0}$ are the complex expansion coefficients, l_0 is a natural number associated with the nutation coordinate θ_0 and p_0 $(= -l_0,\ldots,0,\ldots,+l_0)$ is associated with the precession coordinate φ_0. The values taken by number l_0 are such that expression [4.2] is the best representation of the potential energy surface $V^{\text{e}}_{\text{MC}}(\varphi_0,\theta_0)$.

Given that the topology of the electrostatic potential inside the cage defines a specific position of the trapped molecule characterized by angles φ_0 and θ_0, the approximate local symmetry around this equilibrium position must be determined based on the dominant coefficients in the expansion of the potential of equation [4.2].

The irreducible representations of the symmetry must then be found. Since the space can be built from five rotation types, C^1, C^2, C^3, C^4 and C^6, a similar approach to that presented in [HAM 62] must be used, employing the methods developed in Volume 3 of this series to deal with rotation and methane, or as described for SCOs. Finding the effective symmetry of the potential energy surface $V^{\text{e}}_{\text{MC}}(\varphi_0,\theta_0)$ is required to determine the eigenenergies and the eigenvectors.

Solving Schrödinger's equation associated with Hamiltonian $\tilde{H}_{\text{orient}}$:

$$\tilde{H}_{\text{orient}} \left| jm \right\rangle = \left[T_{\text{rot}} + V^{\text{e}}_{\text{MC}}(\varphi_0,\theta_0) \right] \left| jm \right\rangle = E_{jm} \left| jm \right\rangle \qquad [4.3]$$

allows the determination of the eigenenergies E_{jm} and eigenvectors $|jm\rangle$ of the perturbed orientational motion, where j and m are the quantum numbers associated, respectively, with the angular variables of nutation θ_0 and precession φ_0. The techniques used to solve equation [4.3] depend on the shape of the potential energy surface $V_{\mathrm{MC}}^{\mathrm{c}}(\varphi_0,\theta_0)$ of the rigid molecule around its equilibrium orientation in its inclusion site, in particular on the amplitude of the perturbation of the rotational motion.

For the other types of molecules, except for linear triatomic molecules, the potential energy surfaces $V_{\mathrm{M}}^{\mathrm{sI-c}}(\varphi,\theta,\chi)$ can be numerically adjusted on the basis of rotation matrix elements $D_{p,q}^{l}(\varphi,\theta,\chi)$, in the form:

$$V_{\mathrm{M}}^{\mathrm{sI-c}}(\varphi,\theta,\chi)=\sum_{l=0}^{l_{\max}}\sum_{\substack{p=-l\\q=-l}}^{+l} A_{p,q}^{l}\times D_{p,q}^{l}(\varphi,\theta,\chi) \qquad [4.4]$$

where $A_{p,q}^{l}$ are the expansion coefficients that can be real or complex (see Appendix 2 (section 2.6.2) of Chapter 2). They are calculated using the Gauss–Legendre quadrature method. In expression [4.4], l is a natural number associated with the nutation angle θ, and p and q are integers associated, respectively, with the precession angle φ and the spinning angle χ.

Similar to the linear diatomic and triatomic molecules, a search for the effective symmetry of the potential energy surface $V_{\mathrm{M}}^{\mathrm{sI-c}}(\varphi,\theta,\chi)$ (Figures 4.5 and 4.8) is required to determine the eigenenergies and eigenvectors. Considering the position of the molecule in the cage (φ,θ,χ), it is necessary to determine the predominant coefficients in the expansion in terms of the rotation matrix $D_{p,q}^{l}(\varphi,\theta,\chi)$. Then, the type of symmetry attached to the local environment of the trapped molecule can be explored. The methods presented in [GIL 026], on the applications of Lie algebra for finding local symmetry, can also be used.

4.3. Clathrate structures

Gas hydrate clathrates are formed under specific conditions of temperature and pressure, from water containing dissolved gas species. During solidification, water molecules get organized around an atom or a small molecule, giving rise to nanocages according to well-defined structures, essentially of type sI, sII and sH (the latter requires the presence of larger species). Their stability is ensured by the hydrogen bonds between water molecules, and van der Waals bonds between the trapped species and the surrounding water molecules, starting from a minimum filling rate.

The characteristics of such structures were presented in Chapter 3 of Volume 2 [DAH 19a]. An overview of the characteristics of sI and sII structures is given in Table 4.1, and the shape of their cages is presented in Figure 4.2. It is important to note that these cages are not fully symmetric.

Clathrate type	**sI structure**		**sII structure**	
Unit cell	Cubic		Face-centered cubic	
Cell parameter *a* (Å)	12.00		17.30	
Number of H_2O per unit cell	46		136	
Size of the cage	Small 5^{12}	Large $5^{12}6^{2}$	Small 5^{12}	Large $5^{12}6^{4}$
Number of cages per unit cell	2	6	16	8
Mean radius (Å)	3.95	4.33	3.91	4.73

Table 4.1. *Data related to the clathrate sI and sII structures [SLO 08]*

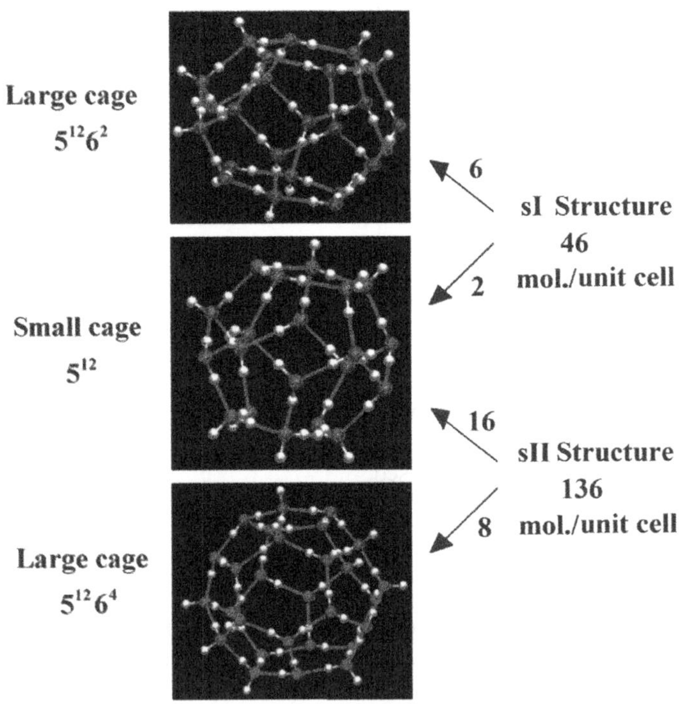

Figure 4.2. *Small and large cages forming the sI and sII structures, and the number of water molecules per unit cell. For a color version of this figure, see www.iste.co.uk/dahoo/infrared2.zip*

4.4. Inclusion of a CH_4 or NH_3 molecule in a clathrate nanocage

4.4.1. *Inclusion model*

Consider a CH_4 or NH_3 molecule, initially assumed to be rigid, trapped in a small or large cage of a clathrate matrix of sI or sII structure, which is also assumed to be rigid.

The positions of the hydrogen and oxygen atoms of the matrix are defined by their coordinates in a fixed reference frame attached to the crystal lattice, while those of the atoms constituting the trapped molecule are defined by their coordinates in the mobile reference frame $(G,\vec{x},\vec{y},\vec{z})$ attached to the equilibrium configuration of the molecule. Figure 4.3 presents, for example, the inclusion model of the CH_4 molecule in a small cage (a) and a large cage (b) of the sI structure of clathrate, and the fixed reference frame $(O,\vec{X},\vec{Y},\vec{Z})$ attached to this structure.

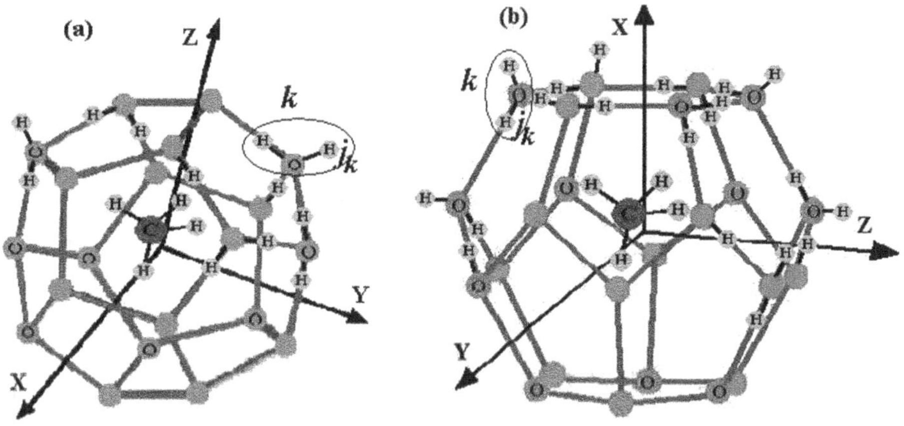

Figure 4.3. *Inclusion of a methane (or ammonia) molecule in a small cage (a) and a large cage (b) of the sI structure of clathrate.* $(O,\vec{X},\vec{Y},\vec{Z})$ *is the fixed (laboratory) reference frame attached to the crystal lattice of clathrate. For a color version of this figure, see www.iste.co.uk/dahoo/infrared2.zip*

Figure 4.4 presents the geometric characteristics of the CH_4 molecule in interaction with a partner molecule k of the nanocage. In the hydrate clathrate matrix, an atom is denoted by j_k, with $j = 1$, 2 or 3 representing the three atoms of the kth water molecule.

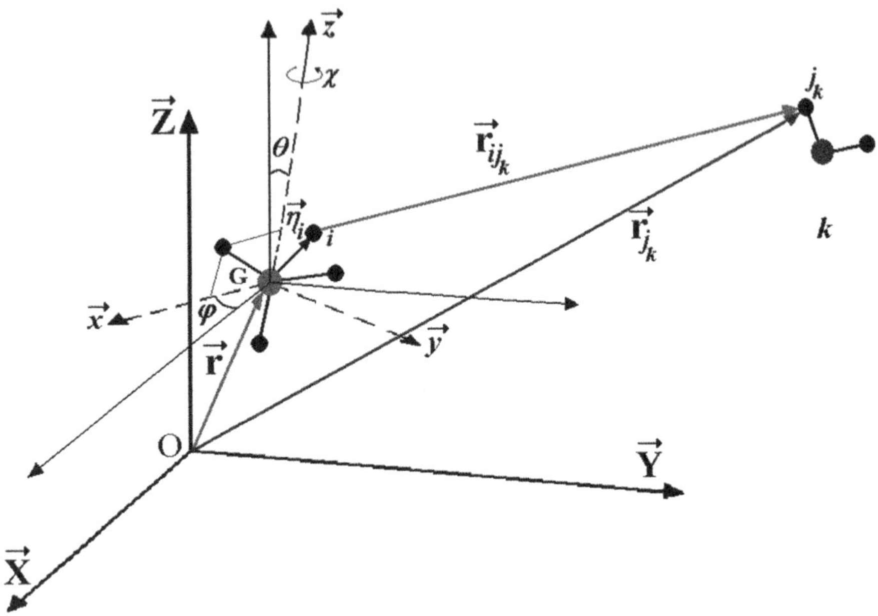

Figure 4.4. *Geometric characteristics of a CH_4–H_2O (or NH_3–H_2O) pair. The position of the center of mass G and the orientation of the molecule are defined with respect to the fixed reference frame* $(O,\vec{X},\vec{Y},\vec{Z})$ *attached to the clathrate matrix. For a color version of this figure, see www.iste.co.uk/dahoo/infrared2.zip*

The Cartesian coordinates $\vec{\eta}_i^e = (\eta_{i_x}^e, \eta_{i_y}^e, \eta_{i_z}^e)$ of the atoms of the rigid trapped NH_3 molecule in the reference frame $(G,\vec{x},\vec{y},\vec{z})$ attached to its equilibrium configuration, were presented in Table 3.3 of Chapter 3; those of the atoms of CH_4 molecule are presented in Table 4.2 of this chapter.

In the two molecules, the axes of the reference frame $(G,\vec{x},\vec{y},\vec{z})$ are chosen as follows:

– Gz coincides with a C_3 symmetry axis;

– Gx is the orthogonal to G_z, such that an H atom is located in the xGz plane;

– Gy is orthogonal to the xGz plane.

Atom \ Coordinates	$\eta^e_{i_x}$ (Å)	$\eta^e_{i_y}$ (Å)	$\eta^e_{i_z}$ (Å)
C (i = 1)	0	0	0
H (i = 2)	0	0	-1.0860
H (i = 3)	1.0239	0	0.3620
H (i = 4)	-0.5119	0.8867	0.3620
H (i = 5)	-0.5119	-0.8867	0.3620

Table 4.2. *Cartesian coordinates of the atoms of a CH_4 molecule in the reference frame $(G,\vec{x},\vec{y},\vec{z})$ attached to its equilibrium configuration*

4.4.2. *Interaction potential energy* – *equilibrium configuration*

The interaction energy between the trapped molecule and its environment is generally considered to result from binary interactions between this molecule and the atoms or molecules of the solid environment. The effects of the environment are often taken into account through effective parameters of this binary potential energy.

The binary potential energy is the sum of a dispersion–repulsion contribution modeled by an atom–atom potential energy of Lennard-Jones 6–12 type, and a charge–charge electric contribution, where the distance between the two molecules is of the order of their size.

The interaction potential energy $V_{\mathrm{MC}}^{\mathrm{s-c}}$ between the trapped molecule and the water molecules of the clathrate matrix of type "structure-cage s-c" (where s = sI or sII and c = s for small or l for large) is given by:

$$V_{\mathrm{MC}}^{\mathrm{s-c}} = \sum_{k=1}^{N_W} \sum_{j=1}^{3} \sum_{i=1}^{4\text{ ou }5} \left[4\varepsilon_{ij} \left\{ \left(\frac{\sigma_{ij}}{\left| \vec{\mathrm{r}}_{ij_k} \right|} \right)^{12} - \left(\frac{\sigma_{ij}}{\left| \vec{\mathrm{r}}_{ij_k} \right|} \right)^{6} \right\} + \frac{1}{4\pi\varepsilon_0} \frac{q_i q_j}{\left| \vec{\mathrm{r}}_{ij_k} \right|} \right] \qquad [4.5]$$

In this expression, $\vec{\mathrm{r}}_{ij_k}$ is the distance vector between the site i, of electric charge q_i, of the trapped molecule, on the one hand, and the site j, of electric charge q_j, of the water molecule k, on the other hand; while ε_{ij} and σ_{ij} are the site–site Lennard-Jones potential parameters determined by the Lorentz–Berthelot combination rules: $\varepsilon_{ij} = \sqrt{\varepsilon_{ii}\varepsilon_{jj}}$ and $2\sigma_{ij} = \sigma_{ii} + \sigma_{jj}$. The values of parameters ε_{ii} and σ_{ii} are given in Table 4.3 [LAK 84, WAL 95]. The effective distributions of electric charges of the ammonia molecule are $q_{\mathrm{H}} = +0.345\ e$ and $q_{\mathrm{N}} = -1.035\ e$ [PAR 07], those of the methane molecule are $q_{\mathrm{H}} = +0.143\ e$ and $q_{\mathrm{C}} = -0.572\ e$, and those of the water molecules of the clathrate matrix are $q_{\mathrm{H}} = +0.4238\ e$ and $q_{\mathrm{O}} = -0.8476\ e$ [ALA 09].

Pair *i–i*	**H–H**	**O–O**	**N–N**	**C–C**
ε_{ii} **(meV)**	0.7402	4.9470	3.2137	3.6947
σ_{ii} **(Å)**	2.810	3.030	3.310	3.210

Table 4.3. *Lennard-Jones parameters. Energy conversion factor 1 meV = 8.06554 cm^{-1}*

In expression [4.5], N_W is the number of water molecules in the clathrate matrix. The following factors are taken into account in the numerical calculations:

– a 4x4x4 matrix composed of 64 unit cells (or 2,944 water molecules), for the sI structure;

– a 3x3x3 matrix composed of 27 unit cells (or 3,672 water molecules), for the sII structure.

These matrices are quite large, which provides a good convergence of the "trapped molecule–clathrate matrix" interaction energies. Moreover, the coordinates of each hydrogen and oxygen atom of the matrix are listed in a data file for each type of structure.

The study of the motions of the molecule trapped in its cage requires the explicit expression of the dependency of the potential energy, in terms of the angular, vibrational and translational coordinates (coordinates of the center of mass) of the molecule. For this purpose, the distance vector $\vec{r}_{ij_k}$ is replaced in [4.5] by the expression: $\vec{r}_{j_k} - \vec{r} - \vec{\eta}_i$ (see Figure 3.3). In this expression:

– $\vec{r}\,(X,Y,Z)$ and $\vec{r}_{j_k}$ are, respectively, the position vector of the center of mass G of the trapped molecule and the position vector of atom *j* of the molecule *k* of the clathrate matrix, with respect to the fixed reference frame $(\mathrm{O},\vec{\mathrm{X}},\vec{\mathrm{Y}},\vec{\mathrm{Z}})$;

– $\vec{\eta}_i$ is the position vector of atom *i* of the molecule with respect to its reference frame $(\mathrm{G},\vec{x},\vec{y},\vec{z})$. It is a Taylor series expansion with respect to the normal vibration coordinates $\{q_v\}$ (see equation [3.7a] of Chapter 3) and is expressed with respect to the fixed reference frame $(\mathrm{G},\vec{\mathrm{X}},\vec{\mathrm{Y}},\vec{\mathrm{Z}})$ attached to the molecule, which allows the expression of the explicit dependence of angular coordinates $\vec{\Omega}(\varphi,\theta,\chi)$.

The equilibrium configuration $\left(\vec{r}^e(X^e, Y^e, Z^e), \vec{\Omega}^e(\varphi^e, \theta^e, \chi^e)\right)$ of a molecule, which is assumed to be rigid, trapped in a clathrate matrix, results from the minimization of the interaction energy V_{MC}^{s-c} with respect to the coordinates of the center of mass G and the angular coordinates of the molecule. Moreover, this allows the determination of the preferential structure-cage type of the molecule.

The results obtained for the CH_4 and NH_3 molecules are presented in Tables 4.4 and 4.5.

	sI – small	sI – large	sII – small	sII – large
X^e (Å)	0.03	- 0.02	0.00	- 0.60
Y^e (Å)	0.00	- 0.02	- 0.15	0.40
Z^e (Å)	- 0.10	- 0.27	0.00	- 0.40
φ^e (deg.)	180	85	135	105
θ^e (deg.)	110	40	80	60
χ^e (deg.)	0	100	30	15
V_{MCmin}^{s-c} (meV)	**- 266.6**	**- 263.4**	**- 238.1**	**- 229.3**

Table 4.4. *Equilibrium configurations and minimum energies of the CH_4 molecule trapped in a clathrate matrix of "s-c" type, s = sI or sII and c = s or l*

	sI – small	sI – large	sII – small	sII – large
X^e (Å)	0.16	- 0.32	0.12	1.00
Y^e (Å)	- 0.46	0.50	0.36	- 0.60
Z^e (Å)	- 0.24	- 0.98	0.00	0.60
φ^e (deg.)	108	325	120	160
θ^e (deg.)	92	40	90	120
χ^e (deg.)	0	95	40	40
V_{MCmin}^{s-c} (meV)	**- 468.7**	**- 450.4**	**- 307.1**	**- 237.0**

Table 4.5. *Equilibrium configurations and minimum energies of the NH_3 molecule trapped in clathrate matrix of "s-c" type, s = sI or sII and c = s or l*

The comparison of the values of minimum energy $V_{\mathrm{MC\,min}}^{\mathrm{s-c}}$ associated with each system considered as a "trapped molecule-structure-cage", clearly shows that the two molecules are preferentially trapped in small and large cages of the sI structure. In the case of CH_4, the results are consistent with that of Sloan and Koh [SLO 08]. Indeed, in Tables 4.4 and 4.5, the values of the potential energy $V_{\mathrm{MC\,min}}^{\mathrm{s-c}}$ of the first two columns associated with the sI structure are significantly lower than those of the last two columns associated with the sII structure.

Moreover, a look at the cages of the sI structure reveals that the center of mass of the CH_4 molecule is in an equilibrium position slightly off-center from the origin of the reference frame, where the cages are not fully symmetric. In the NH_3 molecule, in addition to the cage asymmetry, the significantly more off-center equilibrium position is due to the strong electrical polarity of this molecule. Indeed, the electrical contribution to the total energy $V_{\mathrm{MC}}^{\mathrm{s-c}}$ is about 8% and 4% for CH_4–sI–small cages and CH_4–sI–large cages, respectively, while it is of about 53% and 57% for NH_3–sI–small cages and NH_3–sI–large cages, respectively.

The following section of this chapter focuses on the calculations of the absorption spectra CH_4–sI and NH_3–sI systems. Moreover, since gas hydrate clathrate forms start with a minimum filling rate, both small and large cages at a theoretical cage per lattice rate of 2:6 of the sI structure, are considered.

4.5. System Hamiltonian and separation of movements

The Hamiltonian of the system consisting of the trapped molecule, which is assumed to be in its fundamental electronic state, and the clathrate matrix is given by:

$$H = H_{\mathrm{vib}} + T_{\mathrm{rot}} + \Delta T_{\mathrm{rot}}^{\mathrm{vib}} + T_{\mathrm{trans}} + H_{\mathrm{C}} + V_{\mathrm{MC}}^{\mathrm{s-c}} \qquad [4.6]$$

In this expression, H_{vib} is the Hamiltonian associated with the internal vibrational motions of the molecule in the gas phase and T_{rot} is its kinetic rotational operator:

$$\begin{aligned} H_{\mathrm{vib}} &= -\frac{1}{2}\sum_{v}\hbar\omega_{0_v}\left(\frac{\partial^2}{\partial q_v^2} - q_v^2\right) + \left\{\begin{array}{l} \sum\limits_{v',v'',v'''} k_{v'v''v'''} q_{v'} q_{v''} q_{v'''} \\ + \sum\limits_{v',v'',v''',v''''} k_{v'v''v'''v''''} q_{v'} q_{v''} q_{v'''} q_{v''''} + \dots \end{array}\right\} \\ T_{\mathrm{rot}} &= -B^e\left\{\frac{\partial^2}{\partial\theta^2} + \cot\theta\frac{\partial}{\partial\theta} + \frac{1}{\sin^2\theta}\frac{\partial^2}{\partial\varphi^2} + \left(\cot^2\theta + \frac{C^e}{B^e}\right)\frac{\partial^2}{\partial\chi^2} - 2\frac{\cos\theta}{\sin^2\theta}\frac{\partial^2}{\partial\varphi\partial\chi}\right\} \end{aligned} \qquad [4.7]$$

where $B^e = 9.941$ cm^{-1} and $C^e = 6.309$ cm^{-1} for NH_3, and $B^e = C^e = 5.252$ cm^{-1} for CH_4.

The significantly smaller term $\Delta T_{\text{rot}}^{\text{vib}}$ in expression [4.6] characterizes the vibration–rotation interaction due to the deformation of the molecule during its rotational motion in the gas phase. Its introduction is generally taken into account by the use of the rotational constants of the molecule, depending on its vibrational level.

Finally, T_{trans} is the kinetic translation operator of the center of mass G of the isolated and rigid molecule:

$$T_{\text{trans}} = \sum_{\alpha} \frac{P_\alpha^2}{2M_0} \qquad [4.8]$$

where P_α is the conjugate moment of translation along direction α (X,Y,Z) of the fixed reference system $(\text{O},\vec{\text{X}},\vec{\text{Y}},\vec{\text{Z}})$ and M_0 is the mass of the molecule. Finally, H_{C} is the Hamiltonian associated with all of the degrees of freedom of the water molecules of the clathrate matrix, in the absence (which is hypothetical, since the hydrate clathrates are formed in the presence of the gas species) of the trapped molecule, including their internal vibration, orientational and translational (vibration of their respective center of mass) motions.

However, given that clathrates are structures that are formed under significant high pressures and at low temperatures, due to strong hydrogen bonds between the hydrogen atoms and the oxygen atoms of different water molecules, it is reasonable to first deal with the spectroscopic phenomena of the molecule trapped in a crystal assumed to be rigid. Consequently, the Hamiltonian H_{C} can be ignored for the rest of this chapter.

The Born–Oppenheimer approximation (adiabatic approximation) can be used to separate the high-frequency internal vibration modes $\{q_v\}$ of the molecule from its low-frequency modes: translational $\vec{r}(X,Y,Z)$ and orientational $\vec{\Omega}(\varphi,\theta,\chi)$ motions with respect to the fixed reference frame $(\text{O},\vec{\text{X}},\vec{\text{Y}},\vec{\text{Z}})$ attached to the matrix. After a detailed study, the interaction potential energy $V_{\text{MC}}^{\text{s-c}}$ between the molecule and the clathrate matrix can be explicitly written as:

$$V_{\text{MC}}^{\text{s-c}} = V_{\text{MC}\min}^{\text{s-c}} + V_{\text{M}}^{\text{s-c}}(\vec{u},\vec{\Omega}) + W_{\text{M}}^{\text{s-c}}(\{q_v\}) + \Delta V_{\text{M}}^{\text{s-c}}(\vec{u},\vec{\Omega},\{q_v\}) \qquad [4.9]$$

where $\vec{u} = \vec{r} - \vec{r}^e$ is the instantaneous displacement vector of the center of mass of the molecule, in the vicinity of its equilibrium position vector $\vec{r}^e$. As already noted, in this expression, $V^{\text{s-c}}_{\text{MCmin}}$ is the value of the energy associated with the equilibrium configuration $(\vec{r}^e, \vec{\Omega}^e)$ of the rigid molecule in its cage.

The second term $V^{\text{s-c}}_{\text{M}}(\vec{u}, \vec{\Omega})$ of expression [4.9] characterizes the orientational and translational (motion of the center of mass) motions of the rigid molecule (low-frequency modes).

The third term $W^{\text{s-c}}_{\text{M}}(\{q_v\})$ characterizes the static perturbation of the internal vibration motions of the molecule (high-frequency modes) in its equilibrium configuration $(\vec{r}^e, \vec{\Omega}^e)$. It is due to the interaction of the trapped molecule with the nearest-neighbor atoms, due to the surrounding crystal field.

Experiments show that this term is generally a small perturbation driving a displacement and/or a degeneracy lifting of the vibrational levels, which can be written as a Taylor series expansion:

$$W^{\text{s-c}}_{\text{M}}(\{q_v\}) = \sum_{v} \beta_v q_v + \sum_{v} \sum_{v' \geq v} \beta_{vv'} q_v q_{v'} + \ldots \qquad [4.10]$$

where $\beta_v = \left(\frac{\partial V^{\text{s-c}}_{\text{M}}}{\partial q_v} \right)_0$ and $\beta_{vv'} = \left(\frac{\partial^2 V^{\text{s-c}}_{\text{M}}}{\partial q_v \partial q_{v'}} \right)_0$ are, respectively, the first derivative (force) and the second derivative (force constant) of the interaction potential energy $V^{\text{s-c}}_{\text{MC}}$, with respect to the dimensionless normal vibration coordinates of the trapped molecule. These terms are numerically calculated for "NH_3–sI–small cages" and "NH_3–sI–large cages", on the one hand, and "CH_4–sI–small cages" and "CH_4–sI–large cages", on the other hand.

Finally, the last term $\Delta V^{\text{s-c}}_{\text{M}}(\vec{u}, \vec{\Omega}, \{q_v\})$ in expression [4.10] characterizes the dynamic coupling between slow motions (orientational–translational) and rapid motions (vibrations) of the molecule in its cage. This coupling explains the potential energy transfers of vibrational modes towards translational and orientational modes (vibrational relaxation phenomenon).

However, an important step in the modeling of the infrared spectrum is the construction of the bar spectrum, i.e. the determination of the integrated intensity and frequency position of the spectral lines associated with the vibration–orientation transitions of the optically active molecule. The term $V_{\mathrm{M}}^{s-c}(\vec{u},\vec{\Omega})$ in expression [4.5] is then written as:

$$V_{\mathrm{M}}^{s-c}(\vec{u},\vec{\Omega}) = V_{\mathrm{M}}^{s-c}(\vec{\mathrm{r}}^{e},\vec{\Omega}) + V_{\mathrm{M}}^{s-c}(\vec{u},\vec{\Omega}^{e}) + \Delta V_{\mathrm{M}}^{s-c}(\vec{u},\vec{\Omega}) \quad [4.11]$$

where the first and second terms characterize, respectively, the orientational motion of the molecule trapped in its equilibrium position $\vec{\mathrm{r}}^{e}$ and its translational motion (oscillations of its center of mass) in the vicinity of its equilibrium position, for its equilibrium orientation $\vec{\Omega}^{e}$. The last term in expression [4.11] represents the dynamic orientation–translation coupling, responsible for potential energy transfers between the orientational motion and the translational motion (phase relaxation phenomenon), and can therefore induce shifts and broadening of the spectral lines.

Finally, this separation of motions, or renormalization of the Hamiltonian of the system, can be used to build the bar spectrum of the optically active molecule (or optically active vibration–orientation system), by solving the corresponding Schrödinger's eigenvalue equation.

According to the Born–Oppenheimer approximation, the Hamiltonian of the optically active system can be written as the sum of a vibrational Hamiltonian and of an orientational Hamiltonian:

$$\begin{aligned}\tilde{H}_{\mathrm{a}} &= \left[H_{\mathrm{vib}} + W_{\mathrm{M}}^{s-c}(\{q_{v}\})\right] + \left[T_{\mathrm{rot}} + V_{\mathrm{M}}^{s-c}(\vec{\mathrm{r}}^{e},\vec{\Omega})\right] \\ &= \tilde{H}_{\mathrm{vib}} + \tilde{H}_{\mathrm{orient}}\end{aligned} \quad [4.12]$$

that can be dealt with separately. The eigenfunctions $\left|\mathrm{v}_{v},\tilde{\Psi}_{\mathrm{orient}}\right\rangle$ of the Hamiltonian $\tilde{H}_{\mathrm{a}}$ can be written as the tensor product $\left|\mathrm{v}_{v}\right\rangle \otimes \left|\tilde{\Psi}_{\mathrm{orient}}\right\rangle_{\mathrm{v}_{v}}$ of the respective eigenfunctions of $\tilde{H}_{\mathrm{vib}}$ and $\tilde{H}_{\mathrm{orient}}$.

It is important to recall that the rotational and centrifugal distortion constants of the molecule in the gas phase depend on the vibrational state of the molecule. Schrödinger's equation associated with the orientational Hamiltonian $\tilde{H}_{\mathrm{orient}}$ must therefore be solved for each vibrational state involved in a transition.

4.6. Translational motion

When the (CH_4 or NH_3) molecule is trapped in a nanocage, its translational motion (motion of the center of mass) is no longer free. The molecule can only have translational oscillations around its equilibrium position $\vec{r}^e(X^e,Y^e,Z^e)$. The analysis of the potential energy surface $V_{\mathrm{M}}^{\mathrm{s-c}}(\vec{u},\vec{\Omega}^e)$ (expression [4.11]) shows that this motion is rather harmonic. A series expansion in successive powers of dynamic coordinates $u_X = X - X^e$, $u_Y = Y - Y^e$ and $u_Z = Z - Z^e$ can be formulated. The Hamiltonian of the motion is then written as:

$$\tilde{H}_{\mathrm{trans}} = -\frac{\hbar^2}{2M_0}\left\{\frac{\partial^2}{\partial u_X^2}+\frac{\partial^2}{\partial u_Y^2}+\frac{\partial^2}{\partial u_Z^2}\right\}+\sum_{l,m,n} K_{lmn} \times u_X^l \times u_Y^m \times u_Z^n \quad [4.13]$$

where K_{lmn} are the expansion coefficients characterizing the forces, the harmonic force constants (quadratic order) and the anharmonic force constants (higher orders) of the translational vibration motions of the center of mass of the molecule; l, m and n are natural numbers.

It is important to note that the coefficient K_{000} is excluded from this expression, as it corresponds to the minimum value $V_{\mathrm{MC\,min}}^{\mathrm{s-c}}$ of the energy (Table 4.4 for CH_4 and Table 4.5 for NH_3).

In fact, this series expansion cannot be performed analytically. However, the potential energy surface $V_{\mathrm{M}}^{\mathrm{s-c}}(\vec{u},\vec{\Omega}^e)$ can be numerically built and adjusted to successive powers, as in expression [4.13].

Given the more or less harmonic character of the translational motion of the molecule in the small and large cages of the sI structure, the eigenenergies $\tilde{E}_{\mathrm{n_x n_y n_z}}$ and the eigenvectors $|\mathrm{n_x n_y n_z}\rangle$ are determined by solving Schrödinger's eigenvalue equation associated with $\tilde{H}_{\mathrm{trans}}$. This can be performed on the basis of harmonic oscillators, the calculation of harmonic frequencies, and then the introduction of the anharmonic terms of the potential energy by the perturbation calculation.

4.6.1. *CH_4 – nanocages of the sI structure*

4.6.1.1. *CH_4–sI–small cage*

In the small cage of the sI structure, the equilibrium configuration of the CH_4 molecule is obtained for the position ($X^e = 0.03$ Å, $Y^e = 0$ Å, $Z^e = -0.10$ Å) of its center of mass, and the orientation ($\varphi^e = 180°, \theta^e = 110°, \chi^e = 0$), with an energy

$V_{\text{MC min}}^{\text{sI-s}} = -266.6$ meV (Table 4.4). The molecule is at a distance $\text{r}^e \approx 0.10$ Å from the origin of the reference frame $(\text{O},\vec{\text{X}},\vec{\text{Y}},\vec{\text{Z}})$ of the nanocage.

When the molecule is maintained at its equilibrium orientation, the potential energy surface of its translational motion $V_{\text{M}}^{\text{sI-s}}(\vec{u},\vec{\Omega}^e)$ is nearly equal to that of a three-dimensional oscillator, which is more or less harmonically characterized by the motions of quite high amplitude around its equilibrium position. $V_{\text{M}}^{\text{sI-s}}(\vec{u},\vec{\Omega}^e)$ is then adjusted to a series of successive powers, as indicated in expression [4.13]. The resolution of Schrödinger's eigenvalue equation associated with the corresponding Hamiltonian $\tilde{H}_{\text{trans}}$ yields the following frequencies: $\omega_X = 133\ \text{cm}^{-1}$, $\omega_Y = 108\ \text{cm}^{-1}$ and $\omega_Z = 120\ \text{cm}^{-1}$.

4.6.1.2. *CH_4–sI–large cage*

In the large cage, the equilibrium configuration of the molecule is obtained for the position ($X^e = -0.02$ Å, $Y^e = -0.02$ Å, $Z^e = -0.27$ Å) and the orientation ($\varphi^e = 85°, \theta^e = 40°, \chi^e = 100°$), with an energy $V_{\text{MC min}}^{\text{sI-l}} = -263.4$ meV (Table 4.4). The molecule is at the distance $\text{r}^e \approx 0.27$ Å from the origin of the reference frame $(\text{O},\vec{\text{X}},\vec{\text{Y}},\vec{\text{Z}})$ of the nanocage.

When the molecule is maintained at its equilibrium orientation, the potential energy surface of its translational motion $V_{\text{M}}^{\text{sI-l}}(\vec{u},\vec{\Omega}^e)$ characterizes two oscillators that are rather anharmonic and of quite high amplitudes, the first one along the X axis, perpendicular to the two hexagons of the cage and the second one in the YZ plane. The adjustment of $V_{\text{M}}^{\text{sI-l}}(\vec{u},\vec{\Omega}^e)$ and the resolution of Schrödinger's eigenvalue equation associated with the corresponding $\tilde{H}_{\text{trans}}$, yield the following frequencies: $\omega_X = 63\ \text{cm}^{-1}$, $\omega_Y = 52\ \text{cm}^{-1}$ and $\omega_Z = 47\ \text{cm}^{-1}$.

4.6.2. *NH_3 – nanocages of the sI structure*

4.6.2.1. *NH_3–sI–small cage*

In the small cage of the sI structure, the equilibrium configuration of the NH_3 molecule is obtained for the position ($X^e = 0.16$ Å, $Y^e = -0.48$ Å, $Z^e = -0.28$ Å) of its center of mass, and the orientation ($\varphi^e = 108°, \theta^e = 92°, \chi^e = 0$), with an energy $V_{\text{MC min}}^{\text{sI-s}} = -468.7$ meV (Table 4.5). The molecule is at the distance $\text{r}^e \approx 0.58$ Å from the origin of the reference frame $(\text{O},\vec{\text{X}},\vec{\text{Y}},\vec{\text{Z}})$ of the nanocage.

When the molecule is maintained at its equilibrium orientation, the potential energy surface of its translational motion $V_{\mathrm{M}}^{\mathrm{sI-s}}(\vec{u},\vec{\Omega}^e)$ characterizes a three-dimensional harmonic oscillator of small amplitude around the equilibrium position. The adjustment of $V_{\mathrm{M}}^{\mathrm{sI-s}}(\vec{u},\vec{\Omega}^e)$ and the resolution of Schrödinger's equation associated with the corresponding Hamiltonian $\tilde{H}_{\mathrm{trans}}$, yield the following frequencies: $\omega_X = 125$ cm^{-1}, $\omega_Y = 95$ cm^{-1} and $\omega_Z = 113$ cm^{-1}.

4.6.2.2. *NH_3–sI–large cage*

In the large cage, the equilibrium configuration of the molecule is obtained for the position ($X^e = -0.32$ Å, $Y^e = 0.50$ Å, $Z^e = -0.98$ Å) and the orientation ($\varphi^e = 325°, \theta^e = 40°, \chi^e = 95°$), with an energy $V_{\mathrm{MC\,min}}^{\mathrm{sI-l}} = -450.4$ meV (Table 4.5). The molecule is at the distance $\mathrm{r}^e \approx 1.15$ Å, strongly off-center, with respect to the origin of the reference frame $(\mathrm{O},\vec{\mathrm{X}},\vec{\mathrm{Y}},\vec{\mathrm{Z}})$ of the nanocage.

When the molecule is maintained at its equilibrium orientation, the potential energy surface of its translational motion $V_{\mathrm{M}}^{\mathrm{sI-l}}(\vec{u},\vec{\Omega}^e)$ characterizes two oscillators that are rather anharmonic and of quite high amplitudes, the first one along the X axis, perpendicular to the two hexagons of the cage and the second one in the YZ plane. The adjustment of $V_{\mathrm{M}}^{\mathrm{sI-l}}(\vec{u},\vec{\Omega}^e)$ and the resolution of Schrödinger's equation associated with the corresponding $\tilde{H}_{\mathrm{trans}}$, yield the following frequencies: $\omega_X = 120$ cm^{-1}, $\omega_Y = 84$ cm^{-1} and $\omega_Z = 88$ cm^{-1}.

4.7. Vibrational motions

When the molecule is trapped in the clathrate matrix, it is immersed in an asymmetric environment, the intermolecular forces generate displacements and/or the degeneracy lifting of the frequencies of the vibration modes, with respect to its frequencies in the gas state ω_{gaz}. When expressed in a relative value, the modifications do not generally exceed 1–2%.

These displacements include a static contribution due to the inclusion of the molecule in the matrix assumed to be rigid and a dynamic contribution responsible for the energy transfer, due to the coupling between the motions of the crystal lattice and the vibration modes of the molecule.

In the approximation of the rigid matrix, the values of the coefficients β_{ν} and $\beta_{\nu\nu'}$ of the expansion of the term $W_{\mathrm{M}}^{\mathrm{s-c}}(\{q_{\nu}\})$ (expression [4.10]) are numerically determined. As previously mentioned, since this term is generally small (perturbation), it is introduced using the method of first-order stationary perturbations.

For a given vibrational mode ν, the frequency shift of the molecule is given by:

$$\Delta\omega_{\nu} = \hbar^{-1}\left[\langle 0...1_{\nu}...0|W_{\mathrm{M}}(\{q_{\nu}\})|0...1_{\nu}...0\rangle - \langle 0...0...0|W_{\mathrm{M}}(\{q_{\nu}\})|0...0...0\rangle\right] \quad [4.14]$$

where $|0...0...0\rangle$ and $|0...1_{\nu}...0\rangle$ are the eigenvectors of the vibration Hamiltonian H_{vib} of the (isolated) molecule in the gas phase, and are associated, respectively, with the fundamental state and the first excited state of ν mode, with the other modes being in their fundamental state. For the NH_3 molecule, these eigenvectors were determined and presented in Appendix 4.7 (section 4.7.4) of Volume 3 [DAH 21].

If the vibration mode ν is degenerate, a degeneracy lifting can occur and distinct shifts are obtained for the same "molecule – s-c" system.

As already noted, in small and large cages of the sI structure, the CH_4 molecule accepts a position that is slightly off-center with respect to the center of the cages and can have translational motions of quite high amplitudes. In this case, the shifts of vibrational frequencies towards low frequencies (red), with respect to those of the gas state of the molecule, are to be expected.

On the contrary, the NH_3 molecule has a significantly more off-center position (than the CH_4 molecule) with respect to the center of the cages, and is submitted to a strong crystal field. Its translational motions have small amplitudes. Vibrational frequency shifts towards high frequencies (blue), with respect to those of the gas state of the molecule, are to be expected in this case.

4.7.1. *CH_4 – nanocages of the sI structure*

The methane molecule has four vibrational modes: ν_1 of non-degenerate symmetric stretch and ν_2 of doubly degenerate symmetric bending, both of which are inactive in the infrared; ν_3 of antisymmetric stretch and ν_4 of antisymmetric bending, both being triply degenerate and active in the infrared.

As far as the CH_4 molecule is concerned, the rest of this chapter focuses on the calculation of the infrared spectrum in the region of the ν_3 mode, whose detailed spectroscopic study was experimentally conducted by Dartois *et al.* [DAR 08].

The frequency shift and the degeneracy lifting of the ν_3 mode (active in IR) obtained for "CH_4–sI–small cage" and "CH_4–sI–large cage" are presented in Table 4.6.

Vibration mode	ω_ν^{gas} (exp.)	sI–small cage		sI–large cage	
		$\Delta\omega_\nu^{sI\text{-}s}$	$\omega_\nu^{sI\text{-}s}$	$\Delta\omega_\nu^{sI\text{-}l}$	$\omega_\nu^{sI\text{-}l}$
ν_{3a}	3,022.5	- 6.5	3,016.0	- 18.7	3,003.8
ν_{3b}	3,022.5	- 2.3	3,020.2	- 10.7	3,011.8
ν_{3c}	3,022.5	- 8.7	3,013.8	- 16.1	3,006.4

Table 4.6. *The frequencies and frequency shifts (cm^{-1}) of the vibrational mode ν_3 of the antisymmetric stretch of the CH_4 molecule trapped in the clathrate matrix of "s-c" type, s = sI and c = s or l. ω_ν^{gas} (exp.), are the experimental frequency of CH_4 measured in the gas phase*

4.7.2. *NH_3 – nanocages of the sI structure*

For the ammonia molecule, the vibration modes were presented in Chapter 4 of Volume 3 [DAH 21]. The frequency shifts obtained for "NH_3–sI–small cage" and "NH_3–sI–large cage" are presented in Table 4.7.

Vibration mode	ω_ν^{gas} (exp.)	sI–small cage		sI–large cage	
		$\Delta\omega_\nu^{sI\text{-}s}$	$\omega_\nu^{sI\text{-}s}$	$\Delta\omega_\nu^{sI\text{-}l}$	$\omega_\nu^{sI\text{-}l}$
ν_1	3,337	24.3	3,361.3	15.2	3,352.2
ν_2	950	17.4	967.4	29.4	979.4
ν_{3a}	3,448	11.1	3,459.1	6.9	3,454.9
ν_{3b}	3,448	21.2	3,469.2	11.4	3,459.4
ν_{4a}	1,627	10.9	1,637.9	3.8	1,630.8
ν_{4b}	1,627	5.5	1,632.5	4.4	1,631.4

Table 4.7. *The frequencies and frequency shifts (cm^{-1}) of the vibrational modes of the NH_3 molecule trapped in the clathrate matrix of "s-c" type, s = sI and c = s or l. ω_ν^{gas} (exp.), are the experimental frequencies of NH_3 measured in the gas phase*

As noted in Table 4.7, the frequency shifts due to the inclusion of the molecule in the two cages are positive: shifts towards high frequencies or blue shifts. This reflects a quite strong confinement of the molecule in its environment.

4.7.2.1. NH_3 *vibration–inversion motion*

In the clathrate matrix and both in the small and large cages of the sI structure, the molecule is immersed in a non-symmetric crystal field. Moreover, due to the strong polarity of this molecule, the molecule–matrix potential energy surface only has one minimum in each cage (Table 4.5), corresponding to the equilibrium configuration of the molecule. The potential energy function (double well) associated with the vibration–inversion mode of the isolated molecule only has one well; therefore, the inversion motion is not allowed and the vibrational level splitting no longer exists.

4.8. Orientational motion

Determining the orientational level scheme of the included molecule assumed to be rigid and maintained in its equilibrium position $\vec{r}^e(X^e, Y^e, Z^e)$ (position of its center of mass), requires the resolution of Schrödinger's eigenvalue equation associated with the Hamiltonian $\tilde{H}_{\text{orient}}$ (expression [4.12]), which yields the orientational eigenenergies $\tilde{E}_{\text{orient}}$ and eigenfunctions $\left|\tilde{\Psi}_{\text{orient}}\right\rangle$. An in-depth study of the orientational potential energy surface $V_{\text{M}}^{\text{s-c}}(\vec{r}^e, \vec{\Omega})$ of the molecule trapped in its cage is therefore necessary.

The resolution model to be used depends on the nature of this orientational motion: free rotation, more or less free rotation, or higher or lower amplitude libration around an equilibrium orientation.

4.8.1. CH_4 *– nanocages of the sI structure*

Detailed analysis of orientational potential energy surfaces $V_{\text{M}}^{\text{sI-s}}(\varphi, \theta, \chi)$ and $V_{\text{M}}^{\text{sI-l}}(\varphi, \theta, \chi)$ of the CH_4 molecule trapped, respectively, in the small and large cages of the sI structure, shows that the orientational motion has a more or less perturbed rotational character along energy wells, whose respective barrier height is 17 and 5 meV, or 137 and 40 cm^{-1}. Figure 4.5 presents, for example, energy contour maps $V_{\text{M}}^{\text{sI-s}}(\varphi, \theta)$ and $V_{\text{M}}^{\text{sI-l}}(\varphi, \theta)$ after minimization, for each pair (φ, θ), with respect to the spinning angle χ.

Consequently, the potential energy surfaces $V_{\mathrm{M}}^{\mathrm{sl-c}}(\varphi,\theta,\chi)$ can be numerically adjusted based on the rotation matrix elements $D_{p,q}^{l}(\varphi,\theta,\chi)$ in the form:

$$V_{\mathrm{M}}^{\mathrm{sl-c}}(\varphi,\theta,\chi)=\sum_{l=0}^{l_{\max}}\sum_{\substack{p=-l\\q=-l}}^{+l}A_{p,q}^{l}\times D_{p,q}^{l}(\varphi,\theta,\chi) \qquad [4.15]$$

where $A_{p,q}^{l}$ are the expansion coefficients that can be real or complex (see Appendix 2 (section 3.6.2) of Chapter 3). They are calculated using the Gauss–Legendre quadrature method. In expression [4.15], l is an natural number associated with the nutation angle θ, and p and q are integers associated, respectively, with the precession angle φ and the spinning angle χ.

(a)

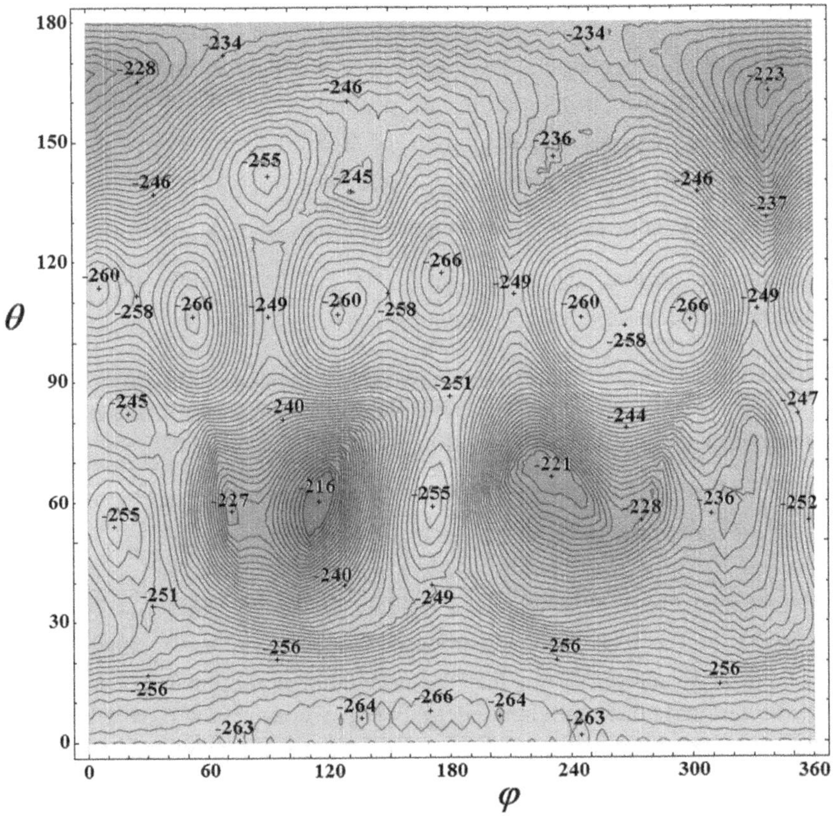

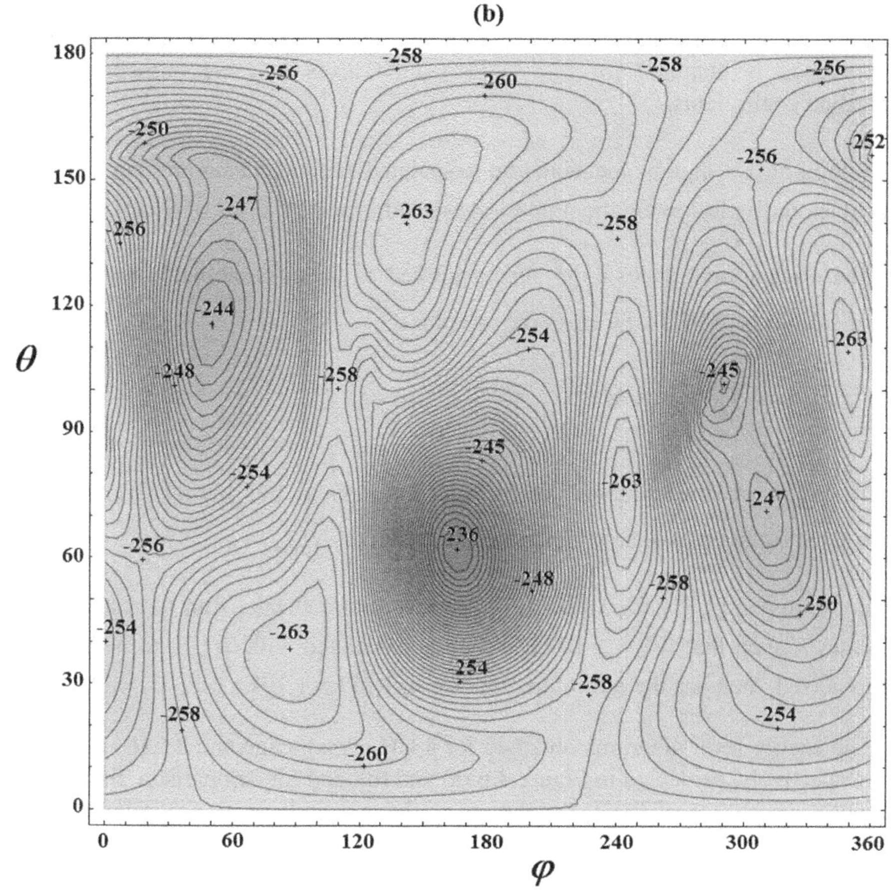

Figure 4.5. *Contour maps of the potential energy surfaces* $V_{\mathrm{M}}^{\mathrm{sI-c}}(\varphi,\theta)$ *(meV) of* CH_4 *trapped in the sI structure: (a) small cage and (b) large cage. For a color version of this figure, see www.iste.co.uk/dahoo/infrared2.zip*

However, since the Gz axis of the CH_4 molecule coincides with a C_3 symmetry axis, an integer q associated with the spinning angle χ can only take the values $q = 0, \pm 3, \pm 6$ and so on, and such that $-l \leq q \leq +l$.

Finally, the maximum value l_{max} that l can take depends on the proper convergence of the adjustment. For CH_4–sI small and large cages, the proper convergence is obtained for l_{max} = 10, which leads to large numbers for the expansion coefficients.

Schrödinger's eigenvalue equation associated with the Hamiltonian $\tilde{H}_{\text{orient}}$ is solved similarly, following the process presented in section 3.4.2 of Chapter 3.

The orientational states $\left|j\gamma\right\rangle$ are expressed as linear combinations of the free rotational states $\left|JMK\right\rangle$ of CH_4 in the gas phase:

$$\left|j\gamma\right\rangle = \sum_{J,M,K} \Lambda_{JMK}^{j\gamma} \left|JMK\right\rangle \qquad [4.16]$$

where $\Lambda_{JMK}^{j\gamma}$ are the complex coefficients determined by solving Schrödinger's eigenvalue equation associated with the Hamiltonian $\tilde{H}_{\text{orient}}$.

In expansion [4.15], the value of l_{max} indicates that a state j can be contaminated by states j' such that $j' \leq j + 10$. The matrix associated with $\tilde{H}_{\text{orient}}$ to be built and diagonalized, then has the minimum dimension 1,771 x 1,771.

The energy level diagrams obtained for a CH_4–small cage and a CH_4–large cage are, respectively, presented in Figure 4.6 (a) and (b), and for comparison, the diagram of energy levels (c) of the free rotation of the molecule is also presented in this figure.

Without going into further detail, Figure 4.6 clearly shows that the orientational states of the molecule are completely modified with respect to those of the free motion. Indeed, the calculations show that each state is a mixture of several free rotational states. The figure also shows that the thin lines (in red) represent the non-degenerate states that can be characterized by $\gamma = 0$, and whose linear combination contains only the free rotational states of quantum numbers $K = 0, \pm 3$, ± 6 and so on. The thick lines (in green) represent the doubly degenerate states that can be characterized by $\gamma = \pm$, and whose linear combination contains the free rotational states of quantum numbers $K = \pm 1, \pm 2, \pm 4, \pm 5$ and so on.

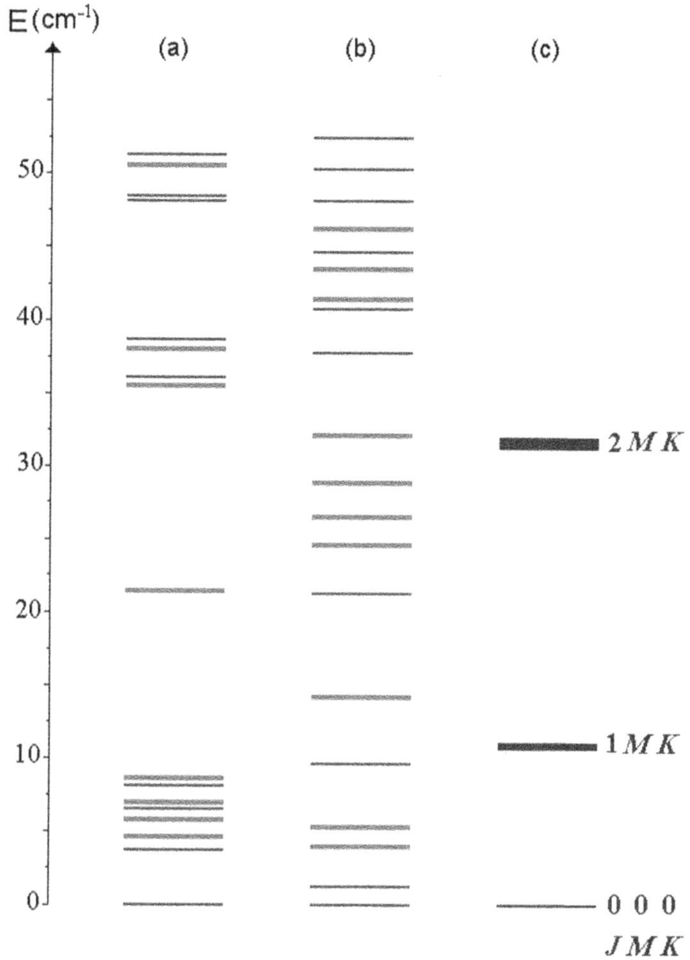

Figure 4.6. *Orientational level diagrams of the CH_4 molecule trapped in the clathrate matrix of the sI structure: (a) small cage, (b) large cage, and for comparison, (c) the free rotational level diagram of the molecule in the gas phase. The thin lines (red) represent non-degenerate levels and the thick lines represent doubly degenerate levels. For a color version of this figure, see www.iste.co.uk/dahoo/infrared2.zip*

4.8.2. *NH_3 – nanocages of the sI structure*

4.8.2.1. *NH_3–sI–small cage*

The detailed analysis of the orientational potential energy surface $V_{\mathrm{M}}^{\mathrm{sI-s}}(\varphi,\theta,\chi)$ of the NH_3 molecule trapped in the small cage of the sI structure shows that the orientational motion has a three-dimensional librational (oscillatory) character of small amplitudes around the equilibrium orientation of the molecule (see Table 4.5).

Three new coordinates $u_\theta = \cos\theta$, $u_\varphi = \varphi - \varphi^e$ and $u_\chi = \chi - \chi^e$ (u_φ and u_χ expressed in radian) varying around 0 are introduced. The potential energy $V_{\mathrm{M}}^{\mathrm{sl-s}}(\varphi,\theta,\chi)$ can be expanded as a series of successive powers of u_φ, u_θ and u_χ in the form:

$$V_{\mathrm{M}}^{\mathrm{sl-s}}(u_\varphi,u_\theta,u_\chi) = \sum_{\delta,\gamma,\lambda} K_{\delta\gamma\lambda} \times u_\varphi^\delta \times u_\theta^\gamma \times u_\chi^\lambda \qquad [4.17]$$

where:

– $K_{\delta\gamma\lambda}$ are the expansion coefficients characterizing the forces, harmonic force constants (quadratic order) and anharmonic force constants (higher orders) of the librational motion of the molecule;

– δ, γ and λ are the natural numbers different from zero. Indeed, the coefficient K_{000} is the minimum value $V_{\mathrm{MC\,min}}^{\mathrm{sl-s}}$ of the energy.

Solving Schrödinger's equation $\tilde{H}_{\mathrm{orient}}\,|jmk\rangle = \tilde{E}_{jmk}\,|jmk\rangle$ associated with the orientational Hamiltonian $\tilde{H}_{\mathrm{orient}} = T_{\mathrm{rot}} + V_{\mathrm{M}}^{\mathrm{sl-s}}(u_\varphi,u_\theta,u_\chi)$, requires the expansion of the rotational kinetic energy operator T_{rot} (equation [4.49] of Volume 3 [DAH 21]) in terms of new coordinates $(u_\varphi,u_\theta,u_\chi)$. The quantum numbers j, m and k are associated, respectively, with librators in θ, φ and χ.

After several rearrangements and fourth-order Taylor series expansions, the following expression is obtained:

$$T_{\mathrm{rot}} \approx -B^e \left\{ \begin{aligned} &\frac{\partial^2}{\partial u_\theta^2} - u_\theta^2 \frac{\partial^2}{\partial u_\theta^2} - 2u_\theta \frac{\partial}{\partial u_\theta} \\ &+\left(1+u_\theta^2+u_\theta^4\right)\frac{\partial^2}{\partial u_\varphi^2} + \left(\frac{C^e}{B^e} + u_\theta^2 + u_\theta^4\right)\frac{\partial^2}{\partial u_\chi^2} \\ &-2u_\theta\left(1+u_\theta^2\right)\frac{\partial^2}{\partial u_\varphi \partial u_\chi} + \ldots \end{aligned} \right\} \qquad [4.18]$$

The expansion of $V_{\mathrm{M}}^{\mathrm{sl-s}}(u_\varphi,u_\theta,u_\chi)$ shows the quasi-harmonic nature of the motion (particularly in θ and φ), the quadratic order coefficients being much greater than those of higher orders. Table 4.8 only presents the expansion coefficients obtained up to the order $\delta+\gamma+\lambda=6$.

K_{020}	K_{200}	K_{002}	K_{400}	K_{600}	K_{220}	K_{420}	K_{040}	K_{060}
288.3	144.7	118.3	9.1	-10.6	-88.6	-16.0	77.2	67.2
K_{240}	K_{004}	K_{006}	K_{202}	K_{204}	K_{022}	K_{042}	K_{024}	K_{222}
-42.9	-159.2	73.9	39.0	70.7	-200.4	-43.2	130.9	156.6

Table 4.8. *Expansion coefficients $K_{\delta\gamma\lambda}$ (meV) of the potential energy surface, in terms of successive powers of coordinates u_θ, u_φ and u_χ of the NH_3 molecule trapped in a small cage of the sI structure. The energy conversion factor is 1 meV = 8.06554 cm^{-1}*

Hence, Schrödinger's equation is solved on the basis of harmonic (librators) oscillators. The orientational Hamiltonian is then written as:

$$\tilde{H}_{\text{orient}} \approx K_{000} + h_0(u_\theta) + h_0(u_\varphi) + h_0(u_\chi) + h_p(u_\theta, u_\varphi, u_\chi) \qquad [4.19]$$

where:

$$h_0(u_\theta) = -B^e \frac{\partial^2}{\partial u_\theta^2} + K_{020} u_\theta^2 \qquad [4.20a]$$

$$h_0(u_\varphi) = -B^e \frac{\partial^2}{\partial u_\varphi^2} + K_{200} u_\varphi^2 \qquad [4.20b]$$

$$h_0(u_\chi) = -C^e \frac{\partial^2}{\partial u_\chi^2} + K_{002} u_\chi^2 \qquad [4.20c]$$

are the Hamiltonians associated with the three independent harmonic librators $(u_\theta, u_\varphi, u_\chi)$, and

$$h_p(u_\theta, u_\varphi, u_\chi) = -B^{\acute{e}} \left\{ \begin{aligned} & -u_\theta^2 \frac{\partial^2}{\partial u_\theta^2} - 2u_\theta \frac{\partial}{\partial u_\theta} + u_\theta^2\left(1+u_\theta^2\right)\left(\frac{\partial^2}{\partial u_\varphi^2} + \frac{\partial^2}{\partial u_\chi^2}\right) \\ & -2u_\theta\left(1+u_\theta^2\right)\frac{\partial^2}{\partial u_\varphi \partial u_\chi} + \ldots \end{aligned} \right\} + \sum_{\delta,\gamma,\lambda}^{*} K_{\delta\gamma\lambda} \times u_\varphi^\delta \times u_\theta^\gamma \times u_\chi^\lambda \qquad [4.21]$$

is the perturbation Hamiltonian that will be processed using the stationary perturbations theory. It is important to note that in expression [4.21], the symbol * indicates that K_{020}, K_{200} and K_{002} are excluded.

The eigenenergy and the eigenvector of an orientational state can be written as:

$$\begin{aligned}&\tilde{E}_{jmk} = E_j^{(0)} + E_m^{(0)} + E_k^{(0)} + \Delta E_{jmk}\\&|jmk\rangle = |j\rangle \otimes |m\rangle \otimes |k\rangle\end{aligned} \qquad [4.22]$$

The corresponding eigenenergies $E_j^{(0)}$, $E_m^{(0)}$ and $E_k^{(0)}$ and eigenvectors are:

$$\begin{aligned}&E_j^{(0)} = \hbar\omega_\theta\left(j+\tfrac{1}{2}\right) \quad \text{with} \quad \hbar\omega_\theta = 2\sqrt{B^e K_{020}} \text{ and } \beta_\theta = \left(\frac{K_{020}}{B^e}\right)^{1/4}\\&|j\rangle = N_j \exp\left(-\beta_\theta^2 u_\theta^2/2\right) H_j\left(\beta_\theta u_\theta\right)\\&N_j = \left(\frac{\beta_\theta^2}{\pi}\right)^{1/4}\left(2^j j!\right)^{-1/2}\end{aligned} \qquad [4.23a]$$

$$\begin{aligned}&E_m^{(0)} = \hbar\omega_\varphi\left(m+\tfrac{1}{2}\right) \quad \text{with} \quad \hbar\omega_\varphi = 2\sqrt{B^e K_{200}} \text{ and } \beta_\varphi = \left(\frac{K_{200}}{B^e}\right)^{1/4}\\&|m\rangle = N_m \exp\left(-\beta_\varphi^2 u_\varphi^2/2\right) H_m\left(\beta_\varphi u_\varphi\right)\\&N_m = \left(\frac{\beta_\varphi^2}{\pi}\right)^{1/4}\left(2^m m!\right)^{-1/2}\end{aligned} \qquad [4.23b]$$

$$\begin{aligned}&E_k^{(0)} = \hbar\omega_\chi\left(k+\tfrac{1}{2}\right) \quad \text{with} \quad \hbar\omega_\chi = 2\sqrt{C^e K_{002}} \text{ and } \beta_\chi = \left(\frac{K_{002}}{C^e}\right)^{1/4}\\&|k\rangle = N_k \exp\left(-\beta_\chi^2 u_\chi^2/2\right) H_k\left(\beta_\chi u_\chi\right)\\&N_k = \left(\frac{\beta_\chi^2}{\pi}\right)^{1/4}\left(2^k k!\right)^{-1/2}\end{aligned} \qquad [4.23c]$$

where N are normalization factors and H represents the Hermite polynomials. The quantum numbers j, m and k are natural numbers, independent of each other.

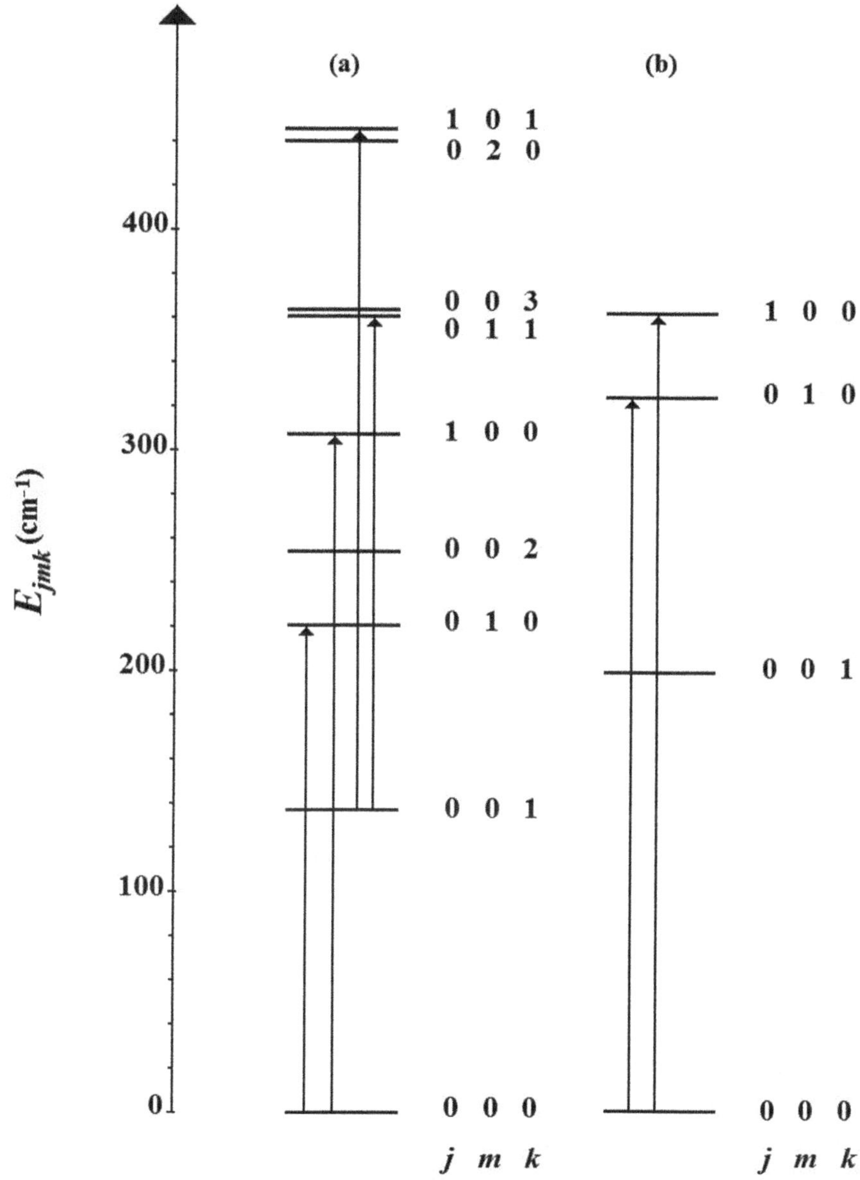

Figure 4.7. *Diagrams of the orientational levels of the NH_3 molecule trapped in the clathrate matrix of the sI structure: (a) small cage and (b) large cage. The possible orientational transitions are also presented*

Solving Schrödinger's equation associated with $h_0(u_\theta)$, $h_0(u_\varphi)$ and $h_0(u_\chi)$ yields the pairs of values (frequency ω and coefficient β): $(\omega_\theta = 304.1\ \text{cm}^{-1}, \beta_\theta = 3.911)$, $(\omega_\varphi = 215.4\ \text{cm}^{-1}, \beta_\varphi = 3.292)$ and $(\omega_\chi = 155.2\ \text{cm}^{-1}, \beta_\chi = 3.507)$.

The corrective values to eigenenergies are then obtained by using the perturbational treatment of $h_p(u_\theta, u_\varphi, u_\chi)$ at the first order in energy: $\Delta E_{jmk} = \langle jmk | h_p(u_\theta, u_\varphi, u_\chi) | jmk \rangle$. These corrective values are included in the energy level diagram in Figure 4.7(a).

The non-zero orientational matrix elements used in the calculation of the analytical expression of ΔE_{jmk}, as well as the latter, were partly provided in Appendix 3.7 (section 3.7.2) of Volume 2 [DAH 19a]. The terms that depend on u_χ can be similarly calculated.

It is important to note that at the first order of perturbation, the corrections due to the odd power terms in u_θ, u_φ and u_χ $(\delta + \gamma + \lambda = \text{odd})$ are zero. This explains the absence of the corresponding expansion coefficients from Table 4.8.

4.8.2.2. *NH_3–sI–large cage*

The detailed analysis of the orientational potential energy surface $V_{\text{M}}^{\text{sI-l}}(\varphi, \theta, \chi)$ of the molecule trapped in the large cage of the sI structure shows that the orientational motion has a more or less harmonic librational character, of small amplitudes around the equilibrium orientation of the molecule (see Table 4.5). Figure 4.8 presents, for example, the energy contour maps of the potential energy surface $V_{\text{M}}^{\text{sI-l}}(\varphi, \theta)$, minimized at each point with respect to the spinning angle χ.

Three new coordinates are introduced $u_\theta = \sin(\theta - \theta^e)$, $u_\varphi = \varphi - \varphi^e$ and $u_\chi = \chi - \chi^e$ (u_φ and u_χ expressed in radian) varying around 0. The potential energy $V_{\text{M}}^{\text{sI-l}}(\varphi, \theta, \chi)$ and the kinetic energy operator T_{rot} can be dealt with similarly to the NH_3–small cage. The expansion coefficients $K_{\delta\gamma\lambda}$ are presented in Table 4.9. It is important to note that the coefficient K_{000} corresponds to the minimum value $V_{\text{MC min}}^{\text{sI-l}}$ of the energy.

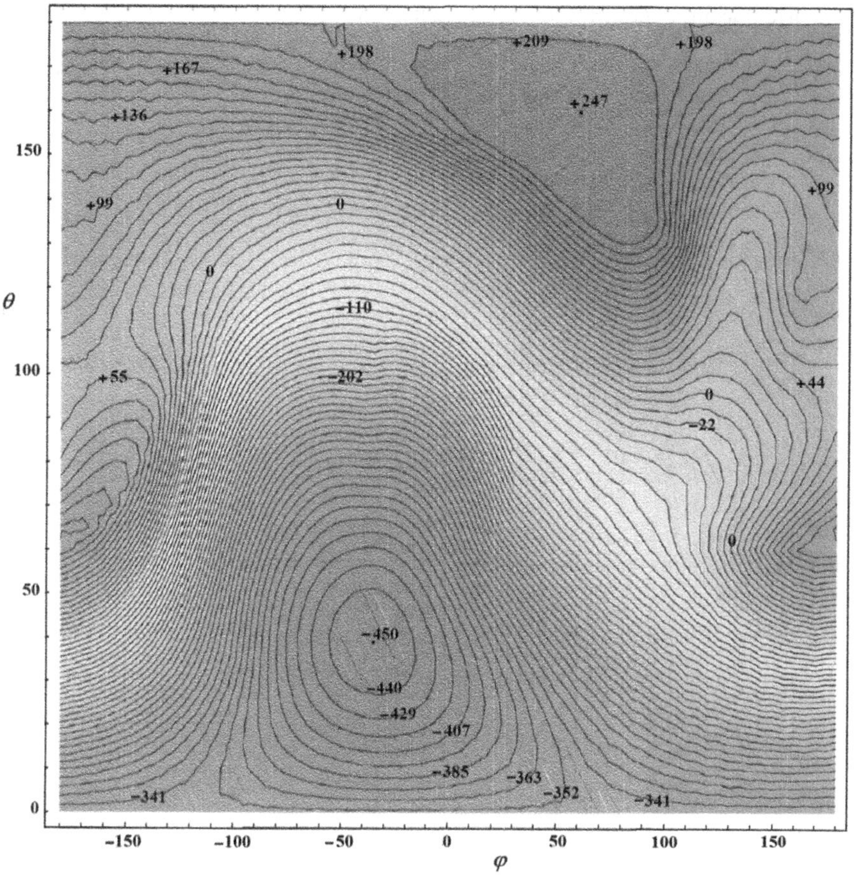

Figure 4.8. *Contour maps of the potential energy surface* $V_{\mathrm{M}}^{\mathrm{sI-1}}(\varphi,\theta)$ *(meV) of* NH_3 *trapped in a large cage of the sI structure. For a color version of this figure, see www.iste.co.uk/dahoo/infrared2.zip*

K_{020}	K_{200}	K_{002}	K_{400}	K_{600}	K_{220}	K_{420}	K_{040}	K_{060}
253.3	109.6	32.7	21.0	-3.0	-11.5	-19.7	171.6	90.6
K_{240}	K_{004}	K_{202}	K_{204}	K_{402}	K_{022}	K_{042}	K_{024}	K_{222}
-119.0	70.5	-116.3	40.1	-12.6	-38.8	-36.7	6.0	34.2

Table 4.9. *Expansion coefficients* $K_{\delta\gamma\lambda}$ *(meV) of the potential energy surface, in terms of the successive powers of coordinates* u_θ, u_φ *and* u_χ *of the* NH_3 *molecule trapped in a large cage of the sI structure*

The orientational Hamiltonian $\tilde{H}_{\text{orient}}$ takes the form:

$$\tilde{H}_{\text{orient}} \approx K_{000} + h_0(u_\theta) + h_0(u_\varphi) + h_0(u_\chi) + h_p(u_\theta, u_\varphi, u_\chi) \qquad [4.24]$$

where:

$$h_0(u_\theta) = -B^e \frac{\partial^2}{\partial u_\theta^2} + K_{020} u_\theta^2 \qquad [4.25a]$$

$$h_0(u_\varphi) = -B^{effect} \frac{\partial^2}{\partial u_\varphi^2} + K_{200} u_\varphi^2 \quad \text{with} \quad B^{effect} = 2.420 B^e \qquad [4.25b]$$

$$h_0(u_\chi) = -C^{effect} \frac{\partial^2}{\partial u_\chi^2} + K_{002} u_\chi^2 \quad \text{with} \quad C^{effect} = \left(C^e + 0.704 B^e\right) \qquad [4.25c]$$

are the Hamiltonians associated with the three independent harmonic librators $(u_\theta, u_\varphi, u_\chi)$, and

$$h_p(u_\theta, u_\varphi, u_\chi) = -B^{\acute{e}} \left\{ \begin{array}{l} u_\theta \left(0,704 + 1,011 u_\theta + 1,200 u_\theta^2 + 0,820 u_\theta^3\right) \frac{\partial}{\partial u_\theta} \\ + u_\theta \left(-5,768 + 12,731 u_\theta - 25,037 u_\theta^2 + 47,147 u_\theta^3\right) \frac{\partial^2}{\partial u_\varphi^2} \\ + u_\theta \left(2,861 + 5,306 u_\theta + 8,321 u_\theta^2 + 11,795 u_\theta^3\right) \frac{\partial^2}{\partial u_\chi^2} \\ -3,708 \begin{pmatrix} 1 - 3,223 u_\theta + 6,760 u_\theta^2 - 13,568 u_\theta^3 \\ + 25,532 u_\theta^4 \end{pmatrix} \frac{\partial^2}{\partial u_\varphi \partial u_\chi} + \ldots \end{array} \right\} + \sum_{\delta,\gamma,\lambda}^{*} K_{\delta\gamma\lambda} \times u_\varphi^\delta \times u_\theta^\gamma \times u_\chi^\lambda \qquad [4.26]$$

is the perturbation Hamiltonian, which will be dealt with using the stationary perturbations theory. It is important to note that in expression [4.26], the symbol * indicates that K_{020}, K_{200} and K_{002} are excluded.

Solving Schrödinger's equation associated with $h_0(u_\theta)$, $h_0(u_\varphi)$ and $h_0(u_\chi)$, yields the pairs of values (frequency ω and coefficient β): $(\omega_\theta = 285.0 \text{ cm}^{-1}, \beta_\theta = 3.786)$, $(\omega_\varphi = 291.7 \text{ cm}^{-1}, \beta_\varphi = 2.462)$ and $(\omega_\chi = 118.5 \text{ cm}^{-1}, \beta_\chi = 2.110)$. The expressions of eigenenergies and eigenvectors are of the same type as those of expressions [4.23].

The corrective values to eigenenergies are then calculated by using the perturbational processing of $h_p(u_\theta, u_\varphi, u_\chi)$ at the first order in energy: $\Delta E_{jmk} = \langle jmk | h_p(u_\theta, u_\varphi, u_\chi) \, | jmk \rangle$. These corrective values are included in the level diagram shown in Figure 4.7(b).

The non-zero orientational matrix elements used in the calculation of the analytical expression of ΔE_{jmk}, as well as the latter, were partly provided in Appendix 3.7 (section 3.7.2) of Volume 2 [DAH 19a]. The terms that depend on u_χ can be similarly calculated.

It is important to note that at the first order of perturbation, the corrections due to the odd power terms in u_θ, u_φ and u_χ ($\delta + \gamma + \lambda = \text{odd}$) are zero. This explains the absence of the corresponding expansion coefficients from Table 4.9.

4.9. Bar spectra

The calculation of the bar spectrum is an important step of the model developed, as it provides information about the quality of the bases chosen for the treatment of the various motions of the "trapped molecule–nanocage" system, and gives an idea on the coupling between the optical system and the bath. Indeed, to the extent that the bar spectrum can be used to reconstruct the majority of the features of the experimental spectrum, it can be said that the coupling is, and remains, a small perturbation during the evolution of the total system.

The bar spectrum is calculated, in each case, after solving Schrödinger's eigenvalue equation associated with the orientational Hamiltonian $\tilde{H}_{\text{orient}}$, in order to determine the energy levels.

As already noted, the CH_4 and NH_3 molecules stabilize the sI structure of clathrate in the small and large cages. The number $\mathcal{N}$ of optically active molecules per unit volume can be written as the sum $\mathcal{N}^{(s)} + \mathcal{N}^{(l)}$ of the number of optically active molecules trapped, per unit volume, in small and large cages, respectively.

4.9.1. *Far infrared*

NOTE. As the spherical top CH_4 molecule does not have a permanent dipole moment ($\mu^e = 0$), it does not absorb in the far infrared range (no rotational spectrum). Therefore, the rest of this section focuses on the NH_3 molecule.

In the far infrared, the integrated absorption coefficient of a number $\mathcal{N}^{(c)}$ of optically active NH_3 molecules that do not interact, trapped per unit volume in cages c (c = s or l), at temperature T, is given by:

$$I^{0(c)}(\omega) = \frac{8\pi^2 \mathcal{N}^{(c)} \omega}{3hc} \sum_{i,f} \frac{e^{-\beta E^{(c)}_{j_i m_i k_i}} - e^{-\beta E^{(c)}_{j_f m_f k_f}}}{Z^{(c)}} \times \left| {}^{(c)}\left\langle j_i m_i k_i \middle| \vec{\mu}^{e}_{A} \middle| j_f m_f k_f \right\rangle^{(c)} \right|^2 \times \delta\left(\omega - \omega^{(c)}_{j_i m_i k_i \to j_f m_f k_f}\right) \quad [4.27]$$

where $\omega^{(c)}_{j_i m_i k_i \to j_f m_f k_f} = \hbar^{-1}\left(E^{(c)}_{j_f m_f k_f} - E^{(c)}_{j_i m_i k_i}\right)$ is the frequency associated with the transition between the orientational states $\left| j_i m_i k_i \right\rangle^{(c)}$ and $\left| j_f m_f k_f \right\rangle^{(c)}$ of respective energy $E^{(c)}_{j_i m_i k_i}$ and $E^{(c)}_{j_f m_f k_f}$ (see Figure 4.6), $\vec{\mu}^{e}_{A}$ is the permanent dipole moment vector of the molecule, expressed in the reference frame $(G, \vec{X}, \vec{Y}, \vec{Z})$ attached to it (see Appendix 2 (section 4.10.2)) and $Z^{(c)}$ is the partition function, the distribution of optically active molecules over all of the orientational eigenstates.

Figure 4.7 presents the diagrams of the orientational (librational) levels obtained for NH_3 trapped in the small (a) and large (b) cages of sI clathrate, as well as the possible transitions induced by the dipole moment in the approximation of its expansion up to the first order, in terms of coordinates u_θ, u_φ and u_χ.

Cage (c)	$\left\| j_i m_i k_i \right\rangle^{(c)} \to \left\| j_f m_f k_f \right\rangle^{(c)}$	$\omega^{(c)}_{j_i m_i k_i \to j_f m_f k_f}$ **(cm^{-1})**	$I^{0(c)}$ **(%)**
Small (s)	$\left\|000\right\rangle^{(s)} \to \left\|010\right\rangle^{(s)}$	219.7	27.1
	$\left\|000\right\rangle^{(s)} \to \left\|100\right\rangle^{(s)}$	306.4	26.5
Large (l)	$\left\|000\right\rangle^{(l)} \to \left\|010\right\rangle^{(l)}$	323.8	86.6
	$\left\|000\right\rangle^{(l)} \to \left\|100\right\rangle^{(l)}$	363.0	100

Table 4.10. *Integrated intensities of the far infrared spectrum of NH_3 molecules trapped in the sI structure of clathrate at T = 10 K*

Assuming a theoretical filling rate of the molecules trapped in the sI structure, hence $\mathcal{N}^{(s)} = 25\% \mathcal{N}$ and $\mathcal{N}^{(l)} = 75\% \mathcal{N}$, the integrated intensities of the far infrared spectrum (librational spectrum) at temperature T = 10 K are presented in

Table 4.10. It is worth noting that at this temperature, only the fundamental librational level $|000\rangle^{(c)}$ (c = s or l) is populated. The first excited level $|001\rangle^{(p)}$ of energy 135.9 cm^{-1} (see Figure 4.7(a)) starts to have a population $\geq 2\%$, from temperature $T = 50$ K.

Hence, the far infrared spectrum or librational spectrum consists of four lines well isolated from one another (see Table 4.10).

4.9.2. *Near infrared*

In near infrared, the integrated absorption coefficient associated with the vibrational mode ν of $\mathcal{N}^{(c)}$ optically active trapped molecules is:

$$I_{\nu}^{0\,(c)}(\omega) = \frac{8\pi^2 \mathcal{N}^{(c)}\omega}{3hc} \left|\langle 0_\nu | q_\nu | 1_\nu \rangle\right|^2 \times \sum_{i,f} \frac{e^{-\beta E^{(c)}_{0,j_i m_i k_i}} - e^{-\beta E^{(c)}_{1,j_f m_f k_f}}}{Z^{(c)}} \times$$

$$\left| {}^{(c)}_{0_\nu}\left\langle j_i m_i k_i \left| \left(\frac{\partial \vec{\mu}_A}{\partial q_\nu}\right)_0^{(c)} \right| j_f m_f k_f \right\rangle^{(c)}_{1_\nu} \right|^2 \times \delta(\omega - \omega^{(c)}_{0,j_i m_i k_i \to 1,j_f m_f k_f}) \qquad [4.28]$$

where $\omega^{(c)}_{0,j_i m_i k_i \to 1,j_f m_f k_f} = \omega^{(c)}_\nu + \hbar^{-1}(E^{(c)}_{j_f m_f k_f} - E^{(c)}_{j_i m_i k_i})$ is the frequency associated with the transition between the vibration–orientation states $|0_\nu\, j_i m_i k_i\rangle^{(c)}$ and $|1_\nu\, j_f m_f k_f\rangle^{(c)}$. Moreover, in the harmonic approximation of normal vibrational modes of the trapped molecule, the transition element of the dimensionless vibrational coordinate of mode ν is considered $\langle 0_\nu | q_\nu | 1_\nu \rangle = 1/\sqrt{2}$, and the expressions of the components of the $\left(\frac{\partial \vec{\mu}_A}{\partial q_\nu}\right)$ vector, with respect to the reference frame $(G, \vec{X}, \vec{Y}, \vec{Z})$ attached to the molecule, are presented in Appendix 4.10 (section 4.10.2) and in section 4.6.5 of Volume 3 [DAH 21].

In order to determine the integrated intensities of the near infrared spectrum, as done previously, a theoretical filling rate of the molecules trapped in the sI structure is assumed, namely $\mathcal{N}^{(s)} = 25\%\mathcal{N}$ and $\mathcal{N}^{(l)} = 75\%\mathcal{N}$.

4.9.2.1. CH_4 – nanocages of the sI structure

NOTE.– For the CH_4 molecule, the quantum numbers (*jmk*) characterizing the librational states are replaced by ($j\gamma$) in expression [4.28], characterizing the rotational states (see section 4.8.1).

Figure 4.9 presents the calculated bar spectrum, as well as the experimental spectrum obtained by Dartois and his coworkers at the temperature T = 7 K [DAR 08, DAR 10], in the frequency range of the triply degenerate vibrational mode ν_3 of CH_4 trapped in the clathrate nanocages of the sI structure.

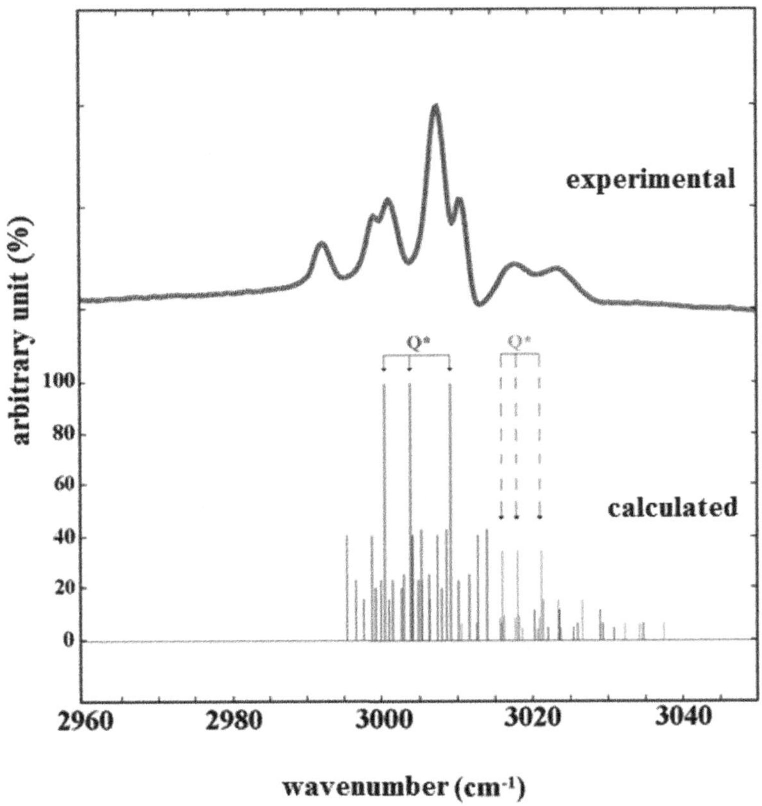

Figure 4.9. *Bar spectrum calculated in the vibrational mode ν_3 range of CH_4 trapped in small cages (green lines) and large cages (red lines) of the sI structure. For a color version of this figure, see www.iste.co.uk/dahoo/infrared2.zip*

Two sets of spectral lines can be noted in the calculated bar spectrum: between 2,995 and 3,015 cm^{-1}, on the one hand, and between 3,015 and 3,040 cm^{-1}, on the other hand. These sets each consist of three blocks of lines, whose existence is due to the degeneracy lifting of the vibrational mode ν_3 of CH_4.

The first set, which is compact, consists of spectral lines (red lines) that cover a range of about 20 cm^{-1} and are due to the absorption of 75% of CH_4 molecules trapped in the large cages. Three lines of strongest intensity, of pure vibration (branch Q*) can be distinguished, each of which is surrounded by high- and low-frequency vibration–rotation lines (branches R* and P*, respectively).

The second set, far less compact than the previous one, consists of weaker spectral lines (green lines) covering a range of about 23 cm^{-1} and are due to the absorption of 25% of CH_4 molecules trapped in the small cages. There are still three pure vibrational lines (branches Q*) surrounded by several vibration–rotation lines. This set also includes a part of vibration–rotation lines (branches R*) due to the absorption of molecules trapped in the large cages.

4.9.2.2. *NH_3 – nanocages of the sI structure*

Considering the same previous filling rate, the integrated intensities of the near infrared spectrum of the ammonia molecule, at temperature $T = 10$ K, are presented in Table 4.11. It is important to note that at this temperature, only the fundamental level $|0_\nu 000\rangle^{(c)}$ is populated.

Mode ν (cage)	ω_ν (cm^{-1})	I_ν^0 (%)
ν_1 (large) (small)	3,352.2 3,361.3	100 33.4
ν_2 (small) (large)	967.4 979.4	32.9 100
ν_3 (large) (small) (large) (small)	3,454.9 3,459.1 3,459.4 3,469.2	100 33.4 37.8 12.6
ν_4 (large) (large) (small) (small)	1,630.8 1,631.4 1,632.5 1,637.9	60.1 100 33.4 20.1

Table 4.11. *Integrated intensities of the near infrared spectra of the NH_3 molecule trapped in the sI structure of clathrate at $T = 10$ K. As the intensities associated with the vibration–orientation transitions are weak (less than 4%), they are not included in this table*

In each frequency range associated with the vibrational modes of the molecule, only the pure vibrational spectral lines (branches Q*) are presented. The calculations show that the intensities associated with the vibration–orientation (vibration–libration) transitions are less than 4% of those of the pure vibrational transitions. These intensities are therefore ignored and are not included in Table 4.11.

4.10. Appendices

4.10.1. *Appendix 1: Expressions of the orientational transition elements in the harmonic librators approximation*

The second-order series expansion of trigonometric functions $\cos\varphi$, $\sin\varphi$, $\cos\theta$, $\sin\theta$, $\cos\chi$ and $\sin\chi$ in terms of variables u_φ, u_θ and u_χ introduced in the study of the orientational motions of the molecule trapped in the small and large cages (section 4.8.2.1), can be used to obtain the expressions of the orientational transition elements, as follows.

4.10.1.1. *NH_3–small cage*

$$\cos\varphi \approx -0.309 - 0.951\times u_\varphi + 0.155\times u_\varphi^2, \quad \sin\varphi \approx 0.951 - 0.309\times u_\varphi - 0.475\times u_\varphi^2$$

$$\cos\theta = u_\theta, \quad \sin\theta \approx 1 - 0.5\times u_\theta^2$$

$$\cos\chi \approx 1 - 0.5\times u_\chi^2, \quad \sin\chi \approx u_\chi$$

The energy level diagram in Figure 4.7(a) shows the presence of the energy levels of quantum numbers j, m and k with possible values of 1 and 2; therefore, the following expressions are presented up to the second order, in terms of coordinates u:

$$\left|\langle j_i m_i k_i \left|\cos\varphi\sin\theta\right| j_f m_f k_f\rangle\right|^2 \approx \left|\left\langle j_i m_i k_i \left| \begin{matrix} -0.309 - 0.951\times u_\varphi \\ +0.155\times u_\theta^2 + 0.155\times u_\varphi^2 \end{matrix} \right| j_f m_f k_f \right\rangle\right|^2$$

$$\left|\langle j_i m_i k_i \left|\sin\varphi\sin\theta\right| j_f m_f k_f\rangle\right|^2 \approx \left|\left\langle j_i m_i k_i \left| \begin{matrix} 0.951 - 0.309\times u_\varphi \\ -0.475\times u_\theta^2 - 0.475\times u_\varphi^2 \end{matrix} \right| j_f m_f k_f \right\rangle\right|^2$$

$$\left|\langle j_i m_i k_i \left|\cos\theta\right| j_f m_f k_f\rangle\right|^2 \approx \left|\langle j_i m_i k_i \left|u_\theta\right| j_f m_f k_f\rangle\right|^2$$

$$\left|\left\langle j_i m_i k_i \left|\cos\varphi\cos\theta\cos\chi - \sin\varphi\sin\chi\right| j_f m_f k_f \right\rangle\right|^2 \approx$$

$$\left|\left\langle j_i m_i k_i \left| \begin{matrix} -0.309\times u_\theta - 0.951\times u_\chi \\ -0.951\times u_\theta \times u_\varphi + 0.309\times u_\varphi \times u_\chi \end{matrix} \right| j_f m_f k_f \right\rangle\right|^2$$

$$\left|\left\langle j_i m_i k_i \left|\sin\varphi\cos\theta\cos\chi + \cos\varphi\sin\chi\right| j_f m_f k_f \right\rangle\right|^2 \approx \left|\left\langle j_i m_i k_i \left|0.951\times u_\theta - 0.309\times u_\chi\right| j_f m_f k_f \right\rangle\right|^2$$

$$\left|\left\langle j_i m_i k_i \left|-\sin\theta\cos\chi\right| j_f m_f k_f \right\rangle\right|^2 \approx \left|\left\langle j_i m_i k_i \left|-1\right| j_f m_f k_f \right\rangle\right|^2$$

$$\left|\left\langle j_i m_i k_i \left|-\cos\varphi\cos\theta\sin\chi - \sin\varphi\cos\chi\right| j_f m_f k_f \right\rangle\right|^2 \approx \left|\left\langle j_i m_i k_i \left| \begin{matrix} -0.951 + 0.309\times u_\varphi \\ +0.309\times u_\theta \times u_\chi \\ +0.475\times u_\varphi^2 + 0.475\times u_\chi^2 \end{matrix} \right| j_f m_f k_f \right\rangle\right|^2$$

$$\left|\left\langle j_i m_i k_i \left|-\sin\varphi\cos\theta\sin\chi + \cos\varphi\cos\chi\right| j_f m_f k_f \right\rangle\right|^2 \approx \left|\left\langle j_i m_i k_i \left| \begin{matrix} -0.309 - 0.951\times u_\varphi \\ -0.951\times u_\theta \times u_\chi \\ +0.155\times u_\varphi^2 + 0.155\times u_\chi^2 \end{matrix} \right| j_f m_f k_f \right\rangle\right|^2$$

$$\left|\left\langle j_i m_i k_i \left|\sin\theta\sin\chi\right| j_f m_f k_f \right\rangle\right|^2 \approx \left|\left\langle j_i m_i k_i \left|u_\chi\right| j_f m_f k_f \right\rangle\right|^2$$

4.10.1.2. *NH_3—large cage*

$$\cos\varphi \approx 0.819 + 0.574\times u_\varphi - 0.410\times u_\varphi^2, \qquad \sin\varphi \approx -0.574 + 0.819\times u_\varphi + 0.287\times u_\varphi^2$$

$$\cos\theta \approx 0.766 - 0.643\times u_\theta - 0.383\times u_\theta^2, \qquad \sin\theta \approx 0.643 + 0.766\times u_\theta - 0.321\times u_\theta^2$$

$$\cos\chi \approx -0.087 - 0.996\times u_\chi + 0.044\times u_\chi^2, \qquad \sin\chi \approx 0.996 - 0.087\times u_\chi - 0.483\times u_\chi^2$$

It can be noted that the orientational energy level diagram in Figure 4.7(b) consists of few levels; therefore, the second orders in u are ignored in the following expressions:

$$\left|\left\langle j_i m_i k_i \left|\cos\varphi\sin\theta\right| j_f m_f k_f \right\rangle\right|^2 \approx \left|\left\langle j_i m_i k_i \left|0.527 + 0.627\times u_\theta - 0.369\times u_\varphi\right| j_f m_f k_f \right\rangle\right|^2$$

$$\left|\left\langle j_i m_i k_i \left|\sin\varphi\sin\theta\right| j_f m_f k_f \right\rangle\right|^2 \approx \left|\left\langle j_i m_i k_i \left|-0.369 - 0.440\times u_\theta + 0.527\times u_\varphi\right| j_f m_f k_f \right\rangle\right|^2$$

$$\left|\langle j_i m_i k_i |\cos\theta| j_f m_f k_f \rangle\right|^2 \approx \left|\langle j_i m_i k_i |0.766 - 0.643\times u_\theta| j_f m_f k_f \rangle\right|^2$$

$$\left|\langle j_i m_i k_i |\cos\varphi\cos\theta\cos\chi - \sin\varphi\sin\chi| j_f m_f k_f \rangle\right|^2 \approx \left|\left\langle j_i m_i k_i \middle| \begin{matrix} -0.627 + 0.046\times u_\theta \\ +0.778\times u_\varphi - 0.575\times u_\chi \end{matrix} \middle| j_f m_f k_f \right\rangle\right|^2$$

$$\left|\langle j_i m_i k_i |\sin\varphi\cos\theta\cos\chi + \cos\varphi\sin\chi| j_f m_f k_f \rangle\right|^2 \approx \left|\left\langle j_i m_i k_i \middle| \begin{matrix} 0.854 - 0.032\times u_\theta \\ +0.517\times u_\varphi + 0.367\times u_\chi \end{matrix} \middle| j_f m_f k_f \right\rangle\right|^2$$

$$\left|\langle j_i m_i k_i |-\sin\theta\cos\chi| j_f m_f k_f \rangle\right|^2 \approx \left|\langle j_i m_i k_i |0.056 + 0.067\times u_\theta + 0.640\times u_\chi| j_f m_f k_f \rangle\right|^2$$

$$\left|\langle j_i m_i k_i |-\cos\varphi\cos\theta\sin\chi - \sin\varphi\cos\chi| j_f m_f k_f \rangle\right|^2 \approx \left|\left\langle j_i m_i k_i \middle| \begin{matrix} -0.675 + 0.525\times u_\theta \\ -0.367\times u_\varphi - 0.517\times u_\chi \end{matrix} \middle| j_f m_f k_f \right\rangle\right|^2$$

$$\left|\langle j_i m_i k_i |-\sin\varphi\cos\theta\sin\chi + \cos\varphi\cos\chi| j_f m_f k_f \rangle\right|^2 \approx \left|\left\langle j_i m_i k_i \middle| \begin{matrix} 0.367 - 0.368\times u_\theta \\ +0.675\times u_\varphi - 0.854\times u_\chi \end{matrix} \middle| j_f m_f k_f \right\rangle\right|^2$$

$$\left|\langle j_i m_i k_i |\sin\theta\sin\chi| j_f m_f k_f \rangle\right|^2 \approx \left|\langle j_i m_i k_i |0.640 + 0.763\times u_\theta - 0.056\times u_\chi| j_f m_f k_f \rangle\right|^2$$

The selection rules associated with variables u in the harmonic approximation are:

$$\langle i|u|f\rangle = \frac{1}{\beta}\left(\frac{2^i \times i!}{2^f \times f!}\right)^{\frac{1}{2}} \left\{\frac{1}{2}\delta_{i,f+1} + f\delta_{i,f-1}\right\}$$

$$\langle i|u^2|f\rangle = \frac{1}{\beta^2}\left(\frac{2^i \times i!}{2^f \times f!}\right)^{\frac{1}{2}} \left\{\frac{1}{4}\delta_{i,f+2} + \frac{2f+1}{2}\delta_{i,f} + f(f-1)\delta_{i,f-2}\right\}$$

4.10.2. *Appendix 2: Dipole moment as a function of dimensionless normal coordinates*

At the first order of the expansion as a function of dimensionless normal coordinates, the dipole moment of a symmetric top molecule of type XY_3 is written as:

$$\vec{\mu} = \mu^e \vec{e}_z + \left(\left(\frac{\partial \mu_z}{\partial q_1} \right)_0 q_1 + \left(\frac{\partial \mu_z}{\partial q_2} \right)_0 q_2 \right) \vec{e}_z + \left(\left(\frac{\partial \mu_x}{\partial q_{3a}} \right)_0 q_{3a} + \left(\frac{\partial \mu_x}{\partial q_{4a}} \right)_0 q_{4a} \right) \vec{e}_x$$
$$+ \left(\left(\frac{\partial \mu_y}{\partial q_{3b}} \right)_0 q_{3b} + \left(\frac{\partial \mu_y}{\partial q_{4b}} \right)_0 q_{4b} \right) \vec{e}_y$$

For the NH_3 molecule:

$$\mu^e = 1.476 \text{ D}$$

$$\left(\frac{\partial \mu_z}{\partial q_1} \right)_0 = 7.83 \times 10^{-2} \text{ D}$$

$$\left(\frac{\partial \mu_z}{\partial q_2} \right)_0 = -41.17 \times 10^{-2} \text{ D}$$

$$\left(\frac{\partial \mu_x}{\partial q_{3a}} \right)_0 = -15.04 \times 10^{-2} \text{ D}$$

$$\left(\frac{\partial \mu_x}{\partial q_{4a}} \right)_0 = 16.06 \times 10^{-2} \text{ D}$$

$$\left(\frac{\partial \mu_y}{\partial q_{3b}} \right)_0 = 9.24 \times 10^{-2} \text{ D}$$

$$\left(\frac{\partial \mu_y}{\partial q_{4b}} \right)_0 = -20.71 \times 10^{-2} \text{ D}$$

For a spherical top molecule of type XY_4, the dipole moment is written as:

$$\vec{\mu} = \left(\sum_\nu \left(\frac{\partial \mu_x}{\partial q_\nu} \right)_0 q_\nu \right) \vec{e}_x + \left(\sum_\nu \left(\frac{\partial \mu_y}{\partial q_\nu} \right)_0 q_\nu \right) \vec{e}_y + \left(\sum_\nu \left(\frac{\partial \mu_z}{\partial q_\nu} \right)_0 q_\nu \right) \vec{e}_z$$

where $\nu = 3a,\ 3b,\ 3c,\ 4a,\ 4b,\ 4c$ are the only active modes in the infrared. Moreover, when the permanent dipole moment of such a molecule is zero, it has no spectral signature in the far infrared range (no rotational transitions).

For the CH_4 molecule, according to [MOU 98, CAS 13]:

q_v	q_{3a}	q_{3b}	q_{3c}	q_{4a}	q_{4b}	q_{4c}
$\left(\frac{\partial\mu_x}{\partial q_v}\right)_0 \times 10^{-2}$	3.0624	3.0624	-6.1248	-3.0842	-3.0842	6.1684
$\left(\frac{\partial\mu_y}{\partial q_v}\right)_0 \times 10^{-2}$	-5.3042	5.3042	0	5.3420	-5.3420	0
$\left(\frac{\partial\mu_z}{\partial q_v}\right)_0 \times 10^{-2}$	4.3309	4.3309	4.3309	-4.3617	-4.3617	-4.3617

Table 4.12. *First derivatives of the dipole moment of CH_4 as a function of dimensionless normal coordinates in the molecular frame*

The unitary transformation matrix or the transition matrix from the reference frame attached to the equilibrium configuration of the molecule to the laboratory reference frame [ROS 67] can be used to write the components of the dipole moment as follows:

$$\text{Permanent moment: } \vec{\mu}_A^e = \mu^e \begin{pmatrix} \cos\varphi\sin\theta \\ \sin\varphi\sin\theta \\ \cos\theta \end{pmatrix}$$

$$\text{Mode } \nu_1: \left(\frac{\partial\vec{\mu}_A}{\partial q_1}\right) = \left(\frac{\partial\mu_z}{\partial q_1}\right)\begin{pmatrix} \cos\varphi\sin\theta \\ \sin\varphi\sin\theta \\ \cos\theta \end{pmatrix}$$

$$\text{Mode } \nu_2: \left(\frac{\partial\vec{\mu}_A}{\partial q_2}\right) = \left(\frac{\partial\mu_z}{\partial q_2}\right)\begin{pmatrix} \cos\varphi\sin\theta \\ \sin\varphi\sin\theta \\ \cos\theta \end{pmatrix}$$

$$\text{Mode } \nu_{3_a}: \left(\frac{\partial\vec{\mu}_A}{\partial q_{3_a}}\right) = \left(\frac{\partial\mu_x}{\partial q_{3_a}}\right)\begin{pmatrix} \cos\varphi\cos\theta\cos\chi - \sin\varphi\sin\chi \\ \sin\varphi\cos\theta\cos\chi + \cos\varphi\sin\chi \\ -\sin\theta\cos\chi \end{pmatrix}$$

$$\text{Mode } \nu_{3_b}: \left(\frac{\partial \vec{\mu}_A}{\partial q_{3_b}}\right) = \left(\frac{\partial \mu_y}{\partial q_{3_b}}\right) \begin{pmatrix} -\cos\varphi\cos\theta\sin\chi - \sin\varphi\cos\chi \\ -\sin\varphi\cos\theta\sin\chi + \cos\varphi\cos\chi \\ \sin\theta\sin\chi \end{pmatrix}$$

$$\text{Mode } \nu_{3c}: \left(\frac{\partial \vec{\mu}_A}{\partial q_{3c}}\right) = \left(\frac{\partial \mu_z}{\partial q_{3c}}\right) \begin{pmatrix} \cos\varphi\sin\theta \\ \sin\varphi\sin\theta \\ \cos\theta \end{pmatrix}$$

$$\text{Mode } \nu_{4_a}: \left(\frac{\partial \vec{\mu}_A}{\partial q_{4_a}}\right) = \left(\frac{\partial \mu_x}{\partial q_{4_a}}\right) \begin{pmatrix} \cos\varphi\cos\theta\cos\chi - \sin\varphi\sin\chi \\ \sin\varphi\cos\theta\cos\chi + \cos\varphi\sin\chi \\ -\sin\theta\cos\chi \end{pmatrix}$$

$$\text{Mode } \nu_{4_b}: \left(\frac{\partial \vec{\mu}_A}{\partial q_{4_b}}\right) = \left(\frac{\partial \mu_y}{\partial q_{4_b}}\right) \begin{pmatrix} -\cos\varphi\cos\theta\sin\chi - \sin\varphi\cos\chi \\ -\sin\varphi\cos\theta\sin\chi + \cos\varphi\cos\chi \\ \sin\theta\sin\chi \end{pmatrix}$$

$$\text{Mode } \nu_{4c}: \left(\frac{\partial \vec{\mu}_A}{\partial q_{4c}}\right) = \left(\frac{\partial \mu_z}{\partial q_{4c}}\right) \begin{pmatrix} \cos\varphi\sin\theta \\ \sin\varphi\sin\theta \\ \cos\theta \end{pmatrix}$$

5

Fullerene Nanocage

Up until 1985, only two crystal phases of carbon were known, namely diamond and graphite. However, between 1960–1970, research conducted by theoretical scientists Osawa and Yoshida [OSA 70, YOS 71] led to the prediction of the existence of a stable molecule exclusively composed of carbon atoms, namely fullerene C_{60}.

The regular icosahedral structure of the fullerene C_{60} molecule was discovered experimentally as result of the cooperation between astrophysicists and spectroscopists.

In 1985, Harold Kroto attempted to identify the unusual infrared (IR) spectrum emitted by large aggregates composed of carbon, known to be generated by red giant stars. In order to explain these spectra, he advanced the hypothesis according to which hydrocarbon molecules were present in the interstellar space.

At the same time, Richard Smalley's group developed a laser vaporization technique for the synthesis of various species of aggregates, but more specifically of carbon aggregates. Their works suggested the possibility of creating unusual carbon molecules or aggregates.

Kroto and Smalley started their collaboration in view of the synthesis of hydrocarbon molecules from graphite and the study of their IR spectrum. They were surprised to notice the systematic presence of a molecule that could correspond to 60 carbon atoms. The systematic presence and the abundance of this molecule were indicators of its high stability.

After considerable effort, they formulated the hypothesis that this molecule was composed only of carbon atoms and had icosahedral geometry [KRO 85].

This marked the birth of the family of fullerenes, aggregates consisting only of carbon atoms!

These molecules have the remarkable property of being formed of closed cages shaped as a sphere (C_{60}, the smallest structure), an ellipsoid (C_{70}, C_{76}, C_{78}, C_{80}, etc.) or a tube (nanotubes).

It was in 2010 that astronomers at the University of Western Ontario [CAM 10] observed fullerene C60 molecules for the first time. Using the Spitzer Space Telescope, the molecules were detected in a cloud of cosmic dust around a star at 6,500 light-years.

The observation of possible molecular species trapped in these fullerene nanocages in the interstellar medium can therefore be expected. Information on the modifications induced in the spectral signatures of the species trapped in these nanocages is important for the characterization – from a physico-chemical point of view – of the observation medium of these fullerenes.

5.1. Introduction

During the years 1990–2000, several experimental and theoretical works were conducted on the trapping of atoms, ions and diatomic molecules in fullerenes. A non-exhaustive list of corresponding references is available in [LAK 11].

Then, Whitener *et al.* succeeded in chemically opening nanocages of fullerene C_{60} in order to trap molecules such as NH_3 and CH_4 [WHI 08, WHI 09, WHI 10]. One of the conclusions of these works indicates that methane can freely rotate in its cage.

Theoretically, calculations using *ab initio* methods were conducted for the study of the trapping mechanism of the ammonia molecule in a fullerene C_{60} nanocage [ERK 03, SLA 05, GAN 09]. According to the calculations of Erkoç and Türker [ERK 03], a nanocage can trap up to six NH_3 molecules. On the contrary, Ganji *et al.* [GAN 09] show that only one NH_3 molecule can be trapped and form a stable complex. Finally, Slanina *et al.* concluded that, at equilibrium, the NH_3 molecule remains oriented to pentagons.

In this chapter, the theoretical models developed in Volumes 1 and 2 [DAH 17, DAH 19a], and in Chapters 3 and 4 of this volume, are applied to the case of the symmetric top NH_3 molecule trapped in a fullerene C_{60} nanocage, in order to determine the infrared absorption spectra at very low temperature, particularly in the region of vibration–inversion ν_2 mode.

The extended Lakhlifi–Dahoo inclusion model leads to the determination of the equilibrium configuration of the trapped molecule and the study of its motions in its site using the atom–atom potential model to describe the interaction between the molecule and its environment.

Nevertheless, previous models are not applicable as such to the NH_3 molecule trapped in the fullerene nanocage. Indeed, an effect of setting the molecule into translational motion upon inversion and/or orientational motion must be taken into account. This is the inertial model.

5.2. Ammonia molecule trapped in a fullerene C_{60} nanocage

5.2.1. *Structure of the fullerene C_{60} nanocage*

The fullerene C_{60} nanocage is a super-molecule composed of 60 carbon atoms bound by covalent bonds, of which 60 are simple bonds of 1.455 Å length and 30 are double bonds of 1.398 Å length, forming a spherical structure of radius r = 3.556 Å similar to a soccer ball [HED 91]. It has 20 hexagons, 12 pentagons each surrounded by 6 hexagons and thus having no common side, with the 60 atoms occupying the vertices of the structure, as shown in Figure 5.1.

In this figure, the reference frame $(C_j, \vec{x}_j, \vec{y}_j, \vec{z}_j)$ attached to an atom j is such that the $\vec{z}_j$ axis is directed from the center of the nanocage towards this atom, and the plane $x_j C_j y_j$ is tangent to the surface, therefore perpendicular to the $\vec{z}_j$ axis. In the reference frame $(O,\vec{X},\vec{Y},\vec{Z})$ of the nanocage the position of each atom is defined by its Cartesian coordinates $\vec{r}_j = (X_j, Y_j, Z_j)$ and/or its spherical coordinates $\vec{r}_j = (r, \varphi_j, \theta_j)$.

The covalent bonds between the carbon atoms induce aspheric atom charge distributions leading to an anisotropy of the dipole polarizability tensor $\vec{\vec{\alpha}}_j$ of each atom j and whose non-zero components, expressed in the reference frame attached to this atom $(C_j, \vec{x}_j, \vec{y}_j, \vec{z}_j)$, are:

$$\alpha_{j_{xx}} = \alpha_{j_{yy}} = \alpha_{\perp}^{c} \text{ and } \alpha_{j_{zz}} = \alpha_{//}^{c},$$

where $\alpha_{\perp}^{c} = 1.44$ Å^3 and $\alpha_{//}^{c} = 0.41$ Å^3 are, respectively, the perpendicular and parallel components of polarizability.

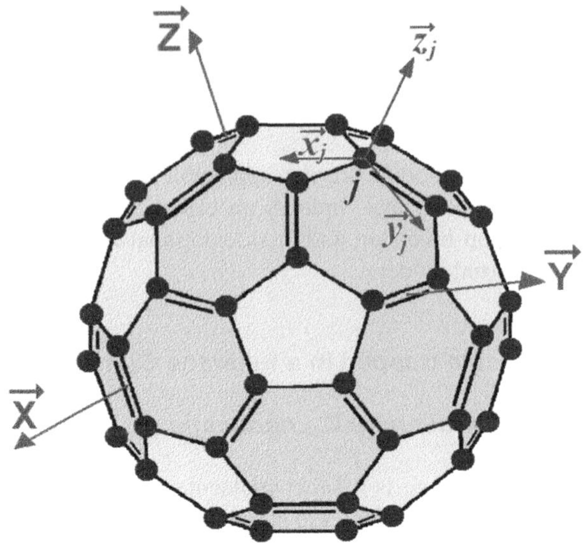

Figure 5.1. *Geometry of the fullerene C_{60} molecule.* $(O,\vec{X},\vec{Y},\vec{Z})$ *and* $(C_j,\vec{x}_j,\vec{y}_j,\vec{z}_j)$ *are, respectively, the fixed (laboratory) reference frame attached to the nanocage and the reference frame attached to the carbon atom C_j. For a color version of this figure, see www.iste.co.uk/dahoo/infrared2.zip*

5.2.2. *Inclusion of NH_3 in a fullerene C_{60} nanocage*

The infrared vibrational spectrum of the fullerene C_{60} molecule has a discrete set of isolated lines corresponding to vibrational modes, the slowest with a frequency of 273 cm^{-1}.

When the molecule is trapped in the nanocage, the renormalization of the Hamiltonian of the "molecule-nanocage" system leads to minimizing the dynamic coupling between the degrees of freedom of the molecule and those of the nanocage. The latter can then be assumed to be rigid.

Moreover, the mass ratio $\mathbf{M_{Fullerene}}/\mathbf{M_{Ammonia}} \sim 720/17$ and the ratio of the inertia moments $\mathbf{I_{Fullerene}}/\mathbf{I_{Ammonia}} \geq 2{,}300$, the fullerene C_{60} molecule (nanocage) can also be assumed to be fixed and its reference frame $(O,\vec{X},\vec{Y},\vec{Z})$ can be considered the absolute reference frame of the total system (laboratory reference frame).

The Cartesian coordinates $\vec{\eta}_i^e = (\eta_{i_x}^e, \eta_{i_y}^e, \eta_{i_z}^e)$ of the atoms of the rigid NH_3 molecule with respect to the reference frame $(G, \vec{x}, \vec{y}, \vec{z})$ attached to its equilibrium configuration were presented in Table 3.3 of Chapter 3.

On the contrary, the non-zero components of its dipole moment vector, of its dipole polarizability tensor and of its quadrupole moment tensor, also expressed in its reference frame, are, respectively:

- $\mu_{0_z} = \mu^e$,
- $\alpha_{0_{xx}} = \alpha_{0_{yy}} = \alpha_{\perp}^e$ and $\alpha_{0_{zz}} = \alpha_{//}^e$,
- $\Theta_{0_{xx}} = \Theta_{0_{yy}} = -\frac{1}{2}\Theta^e$ and $\Theta_{0_{zz}} = \Theta^e$.

where $\mu^e = 1.48$ D is the permanent dipole moment, $\alpha_{\perp}^e = 2.18$ Å^3 and $\alpha_{//}^e = 2.42$ Å^3 are the perpendicular and parallel components of the polarizability, and $\Theta^e = -2.93$ DÅ is the permanent quadrupole moment.

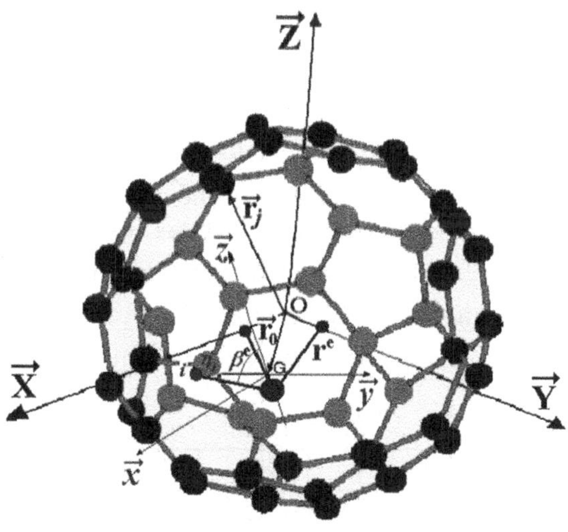

Figure 5.2. *Geometric characteristics of a symmetric top NH_3 molecule trapped in a fullerene C_{60} nanocage. $(G,\vec{x},\vec{y},\vec{z})$ is the reference frame attached to the equilibrium configuration of the molecule. For a color version of this figure, see www.iste.co.uk/dahoo/infrared2.zip*

Figure 5.2 presents the geometric characteristics of the inclusion of the NH_3 molecule in the fullerene C_{60} nanocage, where $\vec{r}_0$ is the position vector (displacement vector) of its center of mass, $r^e = 1.0156$ Å and $\beta^e = 67.96°$ are its internal coordinates (length of N–H bond and angle between this bond the $\vec{z}$ axis of the molecule (C_3 symmetry axis)) and $\vec{r}_j$ is the position vector of the carbon atom j of the nanocage.

5.2.3. *Interaction potential energy – equilibrium configuration*

The interaction potential energy V_{MF} between the ammonia molecule and the rigid fullerene nanocage is the sum, by molecule–carbon atom j pairs, of Lennard-Jones dispersion–repulsion and induction contributions, hence:

$$V_{MF} = \sum_{j=1}^{60}\left[\sum_{i=1}^{4} 4\varepsilon_{ij}\left\{\left(\frac{\sigma_{ij}}{|\vec{r}_{ij}|}\right)^{12} - \left(\frac{\sigma_{ij}}{|\vec{r}_{ij}|}\right)^{6}\right\} - \frac{1}{2}\vec{E}_{0\to j}\cdot\vec{\vec{\alpha}}_j^A\cdot\vec{E}_{0\to j}\right] \quad [5.1]$$

with

$$\vec{E}_{0\to j} = \frac{1}{4\pi\varepsilon_0}\left\{\vec{\nabla}\vec{\nabla}\left(\frac{1}{|\vec{r}_{0j}|}\right)\cdot\vec{\mu}_0^A - \frac{1}{3}\vec{\nabla}\vec{\nabla}\vec{\nabla}\left(\frac{1}{|\vec{r}_{0j}|}\right):\vec{\vec{\Theta}}_0^A\right\} \quad [5.2]$$

In these expressions, $\vec{r}_{ij}$ and $\vec{r}_{0j}$ are the distance vectors between the atom i of the trapped molecule and the carbon atom j of the nanocage, on the one hand, and between the center of mass of the molecule and this very atom, on the other hand, and $\vec{E}_{0\to j}$ is the electric field vector created by the multipole moments of the molecule at the place of the carbon atom. Moreover, ε_0 is the vacuum permittivity, and ε_{ij} and σ_{ij} are the Lennard-Jones atom–atom potential parameters whose values are given in Table 5.1.

Pair i–j	H–C	N–C
ε_{ij} (meV)	2.27	2.81
σ_{ij} (Å)	2.96	3.39

Table 5.1. *Lennard-Jones atom–atom parameters. Energy conversion factor 1 meV = 8.06554 cm^{-1}*

The dependency on vibrational coordinates $\{q_v\}$ of the molecule is contained in the position vector $\vec{\eta}_i$ of the atom i of the trapped molecule through the distance vector $\vec{\mathrm{r}}_{ij}$ written as:

$$\vec{\mathrm{r}}_{ij} = \vec{\mathrm{r}}_j - \vec{\mathrm{r}}_0 - \vec{\eta}_i \qquad [5.3]$$

and also in its multipole moments expressed with respect to the equilibrium configuration reference frame $(\mathrm{G},\vec{x},\vec{y},\vec{z})$, as presented in expressions [3.7] of Chapter 3.

As for the dependency on angular coordinates $\vec{\Omega}(\varphi,\theta,\chi)$, it is contained in these same quantities when they are expressed with respect to the laboratory reference frame attached to the molecule $(\mathrm{G},\vec{\mathrm{X}},\vec{\mathrm{Y}},\vec{\mathrm{Z}})$ (see Appendix 1 (section 2.5.1) of Chapter 2).

Moreover, in expression [5.1], $\vec{\vec{\alpha}}_j^A$ is the polarizability tensor of carbon atom j expressed with respect to the laboratory reference frame attached to this atom $(\mathrm{C}_j,\vec{\mathrm{X}},\vec{\mathrm{Y}},\vec{\mathrm{Z}})$ using the matrix $\mathbf{M}(\varphi_j,\theta_j)$ of transition from the reference frame $(\mathrm{C}_j,\vec{\mathrm{X}},\vec{\mathrm{Y}},\vec{\mathrm{Z}})$ to the reference frame $(\mathrm{C}_j,\vec{x}_j,\vec{y}_j,\vec{z}_j)$.

Furthermore, the Born–Oppenheimer approximation (adiabatic approximation) according to which the high-frequency internal vibrational modes of the molecule are separated from its low-frequency external modes, orientation $\vec{\Omega} = (\varphi,\theta,\chi)$ and translation $\vec{\mathrm{r}}_0$, can be used to write the potential energy V_{MF} as:

$$V_{\mathrm{MF}}\left(\vec{\mathrm{r}}_0,\vec{\Omega},\{q_v\}\right) = V_{\mathrm{M}}\left(\vec{\mathrm{r}}_0,\vec{\Omega}\right) + W_{\mathrm{M}}\left(\{q_v\}\right) + \Delta V_{\mathrm{M}}\left(\vec{\mathrm{r}}_0,\vec{\Omega},\{q_v\}\right) \qquad [5.4]$$

where $V_{\mathrm{M}}\left(\vec{\mathrm{r}}_0,\vec{\Omega}\right)$ is the dependency on external degrees of freedom of the potential energy of the rigid molecule and $W_{\mathrm{M}}\left(\{q_v\}\right)$ characterizes the vibrational dependency of the molecule in its equilibrium configuration, position $\vec{\mathrm{r}}_0^{\,e}$ and orientation $\vec{\Omega}^e = (\varphi^e,\theta^e,\chi^e)$, and that can be written as a Taylor series expansion, as presented in expression [4.10] of Chapter 4. As this term is generally small (perturbation), it is introduced by the method of first-order stationary perturbations.

Finally, the part $\Delta V_{\mathrm{M}}\left(\vec{\mathrm{r}}_0,\vec{\Omega},\{q_v\}\right)$ in expression [5.4] characterizes the dynamic coupling between the external modes (orientation and translation) and the internal modes (vibrations) and that allows the vibrational energy transfer.

The search for the equilibrium configuration $\left(\vec{r}_0^e(X^e, Y^e, Z^e), \vec{\Omega}^e(\varphi^e, \theta^e, \chi^e)\right)$ of the rigid NH_3 molecule trapped in the fullerene C_{60} nanocage is achieved by minimizing the interaction energy $V_M\left(\vec{r}_0, \vec{\Omega}\right)$ with respect to the coordinates of the center of mass G and to the angular coordinates of the molecule.

The resulting minimum potential energy is then $V_{\text{Mmin}} = -181.02$ meV, the $\vec{z}$ axis of the molecule (C_3 symmetry axis) being oriented along a "carbon – O – carbon" direction joining a pair of carbon atoms symmetrically opposite with respect to the center of the nanocage and its center of mass being shifted by $r_0^e = 0.184$ Å along this same direction, but anti-parallel to the $\vec{z}$ axis, hence: $\vec{z} \cdot \vec{r}_0^e = -r_0^e$.

5.3. Potential energy surfaces – inertial model

5.3.1. *Orientation–translational motion*

When the $\vec{z}$ axis of the molecule is along a "pentagon – O – pentagon" direction joining two symmetrically opposite pentagons, its center of mass shifts by $r_0^e = 0.188$Å along this same direction and as in the case of "carbon – O – carbon" (anti-parallel), the corresponding energy is -179.16 meV.

And when the $\vec{z}$ axis is oriented along a "hexagon – O – hexagon" direction joining two symmetrically opposite hexagons, its center of mass shifts by $r_0^e = 0.180$Å along this same direction and similarly, the corresponding energy is -180.40 meV.

It is important to note that the induction and dispersion–repulsion contributions represent, respectively, ~ 35% and ~ 65% of the total energy.

Finally, the results of the calculations of the potential energy surface $V_M\left(\vec{r}_0, \vec{\Omega}\right)$ show that:

– the $\vec{z}$ axis of the molecule is always oriented along the direction through the center of the nanocage, its center of mass G shifting by r_0^e such that $\vec{z} \cdot \vec{r}_0^e = -r_0^e$;

– when the center of mass shifts by 0.01 Å around r_0^e along this same direction, the energy variation remains negligible, ≤ 0.5 meV;

– the spinning motion in χ of the molecule about its axis is quasi-free, with the height of the energy barrier being low, ≤ 1 meV.

Consequently, considering $r_0^e = 0.184$ Å as an average value of the displacement for all the position vectors $\vec{r}_0^{\,e}$ of the center of mass of the molecule and all its orientations $\vec{\Omega} = (\varphi, \theta, \chi)$, the trajectory of the center of mass can be approximately represented by the surface of a sphere of radius $r_0^e = 0.184$ Å about the center of the nanocage (such as the orbit of the Moon around the Earth, to which it always shows the same face). The potential energy surface $V_{\mathrm{M}}\left(\vec{r}_0(\vec{\Omega})\right)$ is then approximately flat, since the height of the energy barrier is $\leq (181.02 - 179.16) = 1.86$ meV. The orientation–translational motion can then be assumed as a free rotation–translational motion.

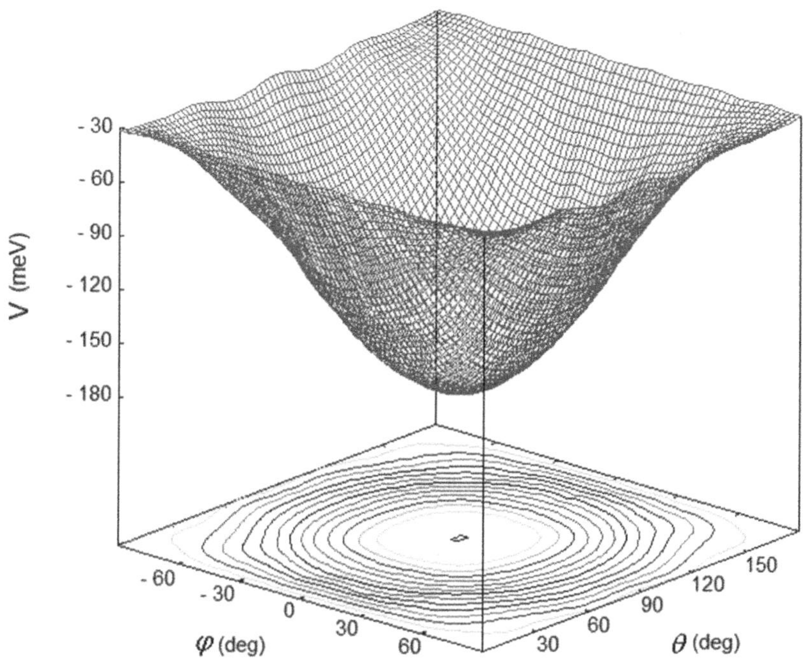

Figure 5.3. *Potential energy surface $V_{\mathrm{M}}(\varphi, \theta)$ of the symmetric top NH_3 molecule trapped in a fullerene C_{60} nanocage. For a color version of this figure, see www.iste.co.uk/dahoo/infrared2.zip*

Furthermore, when the center of mass of the molecule is maintained in its equilibrium position $\vec{r}_0^{\,e}$, its orientational motion (motion of its axis) in (φ, θ) has a librational character (oscillatory motions) of low amplitudes in the vicinity of its equilibrium orientation $\vec{\Omega}^e(\varphi^e, \theta^e, \chi^e)$. The resulting potential energy surface

$V_M(\varphi,\theta)$ is very close to that of a two-dimensional harmonic oscillator and is presented in Figure 5.3. The resulting associated librational frequencies are $\omega_\varphi \approx 155$ cm^{-1} and $\omega_\theta \approx 151$ cm^{-1}. It can then be considered that the axis of the molecule is constantly anti-parallel to the displacement vector $\vec{r}_0$ of its center of mass.

5.3.2. *Translational motion*

When the molecule is fixed at its equilibrium orientation $\vec{\Omega}^e(\varphi^e,\theta^e,\chi^e)$, the radial potential energy function $V_M(r_0)$ of the translational motion of its center of mass, in the vicinity of its equilibrium position $r_0^e = 0.184$ Å, is determined and presented in Figure 5.4. The resulting curve is similar to that of a quasi-harmonic oscillator of force constant $k = 2975.6$ meV.Å^{-2}. The frequency associated with this motion is then calculated: $\omega_{radial} \approx 218$ cm^{-1}.

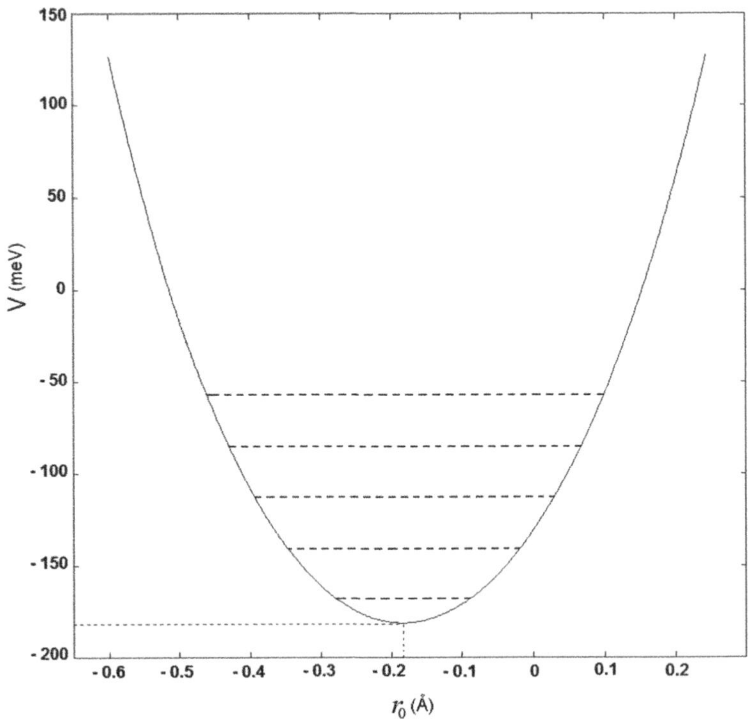

Figure 5.4. *Radial potential energy function $V_M(r_0)$ of the symmetric top NH_3 molecule trapped, at its equilibrium orientation, in a fullerene C_{60} nanocage*

5.3.3. *Vibration–inversion–translational motion*

As previously noted (see Chapter 4 of Volume 3), for the vibration–inversion mode ν_2 of the NH_3 molecule, the hydrogen H atoms can simultaneously pass from a given configuration to its opposite through the σ_h plane perpendicular to the symmetry axis of the molecule. This mode is generally described as the one-dimensional motion of a fictitious particle of mass μ_2 (see further on) moving in a double symmetric potential well $V_{\nu_2}(\theta_2)$ (see expression [3.18] of Chapter 3) [SWA 62].

In the fullerene nanocage, this inversion motion can only take place if the center of mass of the molecule moves simultaneously from a position $\left(\vec{r}_0^{\,e},\vec{\Omega}\right)$ to the opposite position $\left(-\vec{r}_0^{\,e},\vec{\Omega}\right)$ through the center of the cage where the molecule is in flat configuration (the atom N is in the plane formed by the three H atoms). The energy barrier to cross in this case is 50 meV.

Finally, it is clear that the vibration–inversion–translational motion $\left(r_0,\theta_2\right)$, on the one hand, and the orientation–translational motion $\left(\vec{r}_0^{\,e},\vec{\Omega}\right)$, on the other hand, are strongly coupled.

The expression of the interaction potential energy V_{MF} can be rewritten in the form:

$$\begin{aligned} V_{MF}\left(\vec{r}_0,\vec{\Omega},\{q_\nu\}\right) = V_M\left(\vec{r}_0^{\,e},\vec{\Omega}\right) + V_M\left(r_0,\theta_2\right) + W_M\left(\{q_\nu\}\right) \\ + V_M\left(r_0\right) + \Delta V_M\left(\vec{r}_0,\vec{\Omega},\{q_\nu\}\right) \end{aligned} \qquad [5.5]$$

Hence, the kinetic energy operators associated with these motions must be treated classically before their quantum treatment.

5.3.4. *Kinetic Lagrangian*

The kinetic Lagrangian associated with the vibration–inversion, rotational and translational motions considered to be classical motions is defined by:

$$L_K = \frac{1}{2}\mu_2 r^{e2}\dot{\theta}_2^2 + \frac{1}{2}I_B^e\left(\dot{\theta}^2 + \dot{\varphi}^2\sin^2\theta\right) + \frac{1}{2}I_C^e\left(\dot{\varphi}\cos\theta + \dot{\chi}\right)^2 + \frac{1}{2}M\dot{\vec{r}}_0^{\,2} \qquad [5.6]$$

where μ_2 is the reduced mass of the molecule associated with the vibration–inversion mode of coordinate θ_2, I_B^e and I_C^e are the main moments of inertia associated, respectively, with the rotational motion of the axis (C_3 symmetry axis) of the rigid molecule and of rotation about this axis, M is the mass of the molecule, and r^e is the length of the N–H bonds.

Moreover, from the results discussed above, the position vector $\vec{r}_0$ of the center of mass of the molecule can be written as:

$$\vec{r}_0 = \vec{r}_0^{\,e}(\varphi,\theta,\theta_2) + \vec{u}_0 \qquad [5.7]$$

where $\vec{u}_0$ is the dynamic displacement vector in the vicinity of the equilibrium position $\vec{r}_0^{\,e}(\varphi,\theta,\theta_2)$ that statically depends on angular and vibration–inversion coordinates. These components are:

$$\vec{r}_0^{\,e}(\varphi,\theta,\theta_2) = \left| \begin{array}{l} -r_0^e(\theta_2)\sin\theta\cos\varphi \\ -r_0^e(\theta_2)\sin\theta\sin\varphi \\ -r_0^e(\theta_2)\cos\theta \end{array} \right. \qquad [5.8]$$

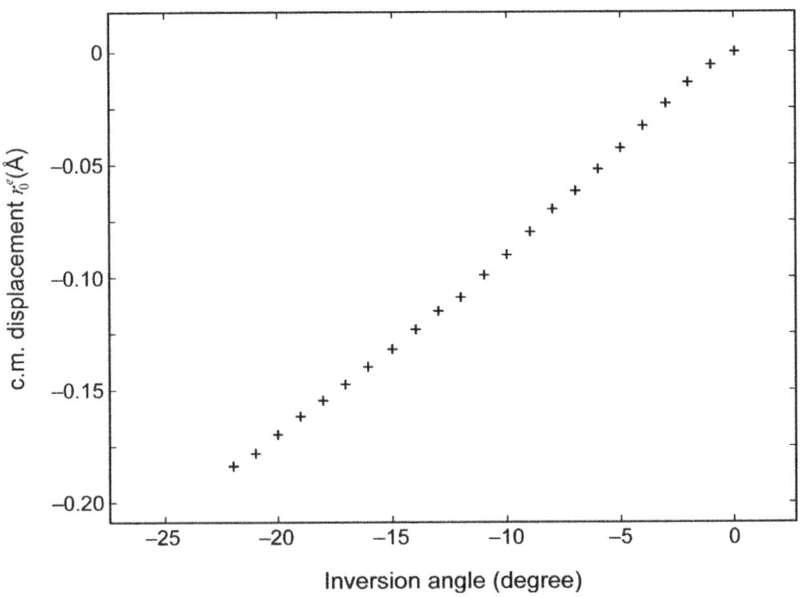

Figure 5.5. *Variation of the equilibrium position of the NH_3 molecule trapped in the fullerene C_{60} nanocage depending on the angular coordinate of inversion*

In this expression, $r_0^e(\theta_2)$ is the variation of the equilibrium position of the molecule, for a given orientation (φ, θ) (direction of the axis of the molecule), depending on the angular coordinate of inversion. The quasi-linear shape of such a variation for the "carbon-O-carbon" direction is presented in Figure 5.5.

It is important to recall that the rotational motion about the axis of the molecule (spinning) is considered to be free, which explains why the angular coordinate χ does not appear in expression [5.8].

The translational velocity vector of the center of mass of the molecule is the derivative with respect to time of the position vector $\vec{r}_0$. This yields:

$$\dot{\vec{r}}_0 = \left(\frac{\partial \vec{r}_0^{\,e}}{\partial \varphi} \dot{\varphi} + \frac{\partial \vec{r}_0^{\,e}}{\partial \theta} \dot{\theta} + \frac{\partial \vec{r}_0^{\,e}}{\partial \theta_2} \dot{\theta}_2 \right) + \dot{\vec{u}}_0 \qquad [5.9]$$

Without detailing the calculations, using expressions [5.8] and [5.9], the task is to determine $\dot{r}_0^2$ and then insert it in the expression of the kinetic Lagrangian [5.6]. The interested reader is invited to do these calculations.

Furthermore, given that this chapter is devoted to the calculation of bar spectra (frequencies and integrated intensities) of the NH_3 molecule trapped in the fullerene nanocage, in the adiabatic approximation, the dynamic coupling between the various degrees of freedom is ignored.

Hence, expression [5.6] of the kinetic Lagrangian of the molecule becomes:

$$L_K = \frac{1}{2} \mu_2^{\text{eff}} r^{e2} \dot{\theta}_2^2 + \frac{1}{2} I_B^{\text{eff}} \left(\dot{\theta}^2 + \dot{\varphi}^2 \sin^2 \theta \right) + \frac{1}{2} I_C^e \left(\dot{\varphi} \cos\theta + \dot{\chi} \right)^2 + \frac{1}{2} M \dot{u}_0^2 \qquad [5.10]$$

where μ_2^{eff} is the effective reduced mass of the molecule and I_B^{eff} its effective moment of inertia written as:

$$\mu_2^{\text{eff}} = \mu_2 + \frac{M}{r^{e2}} \left(\frac{\partial r_0^e}{\partial \theta_2} \right)^2 ; \qquad I_B^{\text{eff}} = I_B^e + M r_0^{e2} \qquad [5.11]$$

where $\frac{\partial r_0^e}{\partial \theta_2} = 0.479$ Å is the leading coefficient of the line obtained and presented in Figure 5.5.

5.4. Quantum treatment

The previously determined potential energy surfaces and the classical treatment of motions can be used to express the Hamiltonian of the optically active molecule, considered a quantum system, in the form:

$$H_{\mathrm{a}}^{\mathrm{eff}} = H_{\mathrm{vib}}^{\mathrm{eff}} + H_{\mathrm{orient}}^{\mathrm{eff}} \qquad [5.12]$$

where $H_{\mathrm{vib}}^{\mathrm{eff}}$ and $H_{\mathrm{orient}}^{\mathrm{eff}}$ are, respectively, the effective renormalized Hamiltonians associated, on the one hand, with vibrational modes $\{q_\nu\}$ and with vibration–inversion mode θ_2, and, on the other hand, with orientational motions of the molecule taking into account the $\left(\mathrm{r}_0^e, \theta_2\right)$ and $\left(\vec{\mathrm{r}}_0^{\,e}, \vec{\Omega}\right)$ couplings.

5.4.1. *Vibrational modes – frequency shifts*

Section 4.7 of Chapter 4 presents the frequency shift of a given vibrational mode ν of the molecule, due to its inclusion in a nanocage. This frequency shift is defined as follows:

$$\Delta\omega_\nu = \hbar^{-1}\left[\langle 0...1_\nu...0|W_{\mathrm{M}}\left(\{q_\nu\}\right)|0...1_\nu...0\rangle - \langle 0...0...0|W_{\mathrm{M}}\left(\{q_\nu\}\right)|0...0...0\rangle\right] \qquad [5.13]$$

where $|0...0...0\rangle$ and $|0...1_\nu...0\rangle$ are the eigenvectors of the vibrational Hamiltonian H_{vib} of the molecule in the gas phase (isolated) and are associated, respectively, with the fundamental state and with the first excited state of mode ν, the other modes being in their fundamental state. The eigenvectors of vibrations of the NH_3 molecule in the gas phase were determined and presented in Appendix D (section 4.7.4) of Chapter 4 of Volume 3 of this set.

Moreover, if the vibration mode ν is degenerate, a degeneracy lifting is possible, and there are distinct shifts of its components.

In the equilibrium configuration (position and orientation) of the NH_3 molecule in the fullerene C_{60} nanocage, the Taylor series expansion of the vibrational contribution $W_{\mathrm{M}}\left(\{q_\nu\}\right)$ is used to calculate the frequency shifts presented in Table 5.2.

This table shows that the shifts are positive. The four vibrational modes are therefore shifted towards high frequencies (blue shift). However, as the next section shows, mode ν_2 undergoes a shift towards low frequencies (red shift). Moreover, the degeneracy of modes ν_3 and ν_4 is lifted.

Vibrational modes	ω_ν^{gas} (exp.) (cm^{-1})	$\Delta\omega_\nu$ (cm^{-1})
ν_1	3,337	133.0
ν_2	950	47.0
ν_{3a}	3,448	171.0
ν_{3b}		69.6
ν_{4a}	1,627	47.1
ν_{4b}		34.0

Table 5.2. *Vibrational frequency shifts of NH_3 trapped in a C_{60} nanocage, ω_ν^{gas} are the frequencies in the gas phase*

Finally, the rather high values indicate a strong confinement of the molecule in the nanocage, compared to the results obtained for the NH_3 trapped in clathrate nanocages. Indeed, the radius value 3.556 Å of C_{60} is significantly smaller than that of the small cage of clathrate (3.95 Å). The molecule is therefore much more restricted in C_{60} than in clathrates.

5.4.2. *Vibration–inversion mode*

Figure 5.6 presents the potential energy functions $V_{\nu_2}(\theta_2)$ of the vibration–inversion motion of the molecule: (a) in the gas phase (with $\mu_2 = 2.563$ g.mole^{-1}), (b) trapped in the fullerene nanocage (with $\mu_2^{eff} = 6.344$ g.mole^{-1}) and numerically obtained. The energy levels, eigenvalues of the associated Hamiltonian and the fundamental frequencies are also presented.

In the gas phase, the vibration–inversion mode ν_2 has two components, one at 932.5 cm^{-1} and the other at 968.3 cm^{-1}. The experimental values are 931.7 cm^{-1} and 968.3 cm^{-1}.

In the fullerene nanocage, mode ν_2 is shifted by 47 cm^{-1} (see Table 5.2), and because of the 50 meV increase of the inversion barrier and the reduced mass from 2.563 to 6.344 g.mole^{-1}, the frequency of the vibration–inversion mode is calculated at (704.6 + 47) cm^{-1} or 751.6 cm^{-1}. Therefore, the mode is quasi-degenerate and shifted towards low frequencies (red shift).

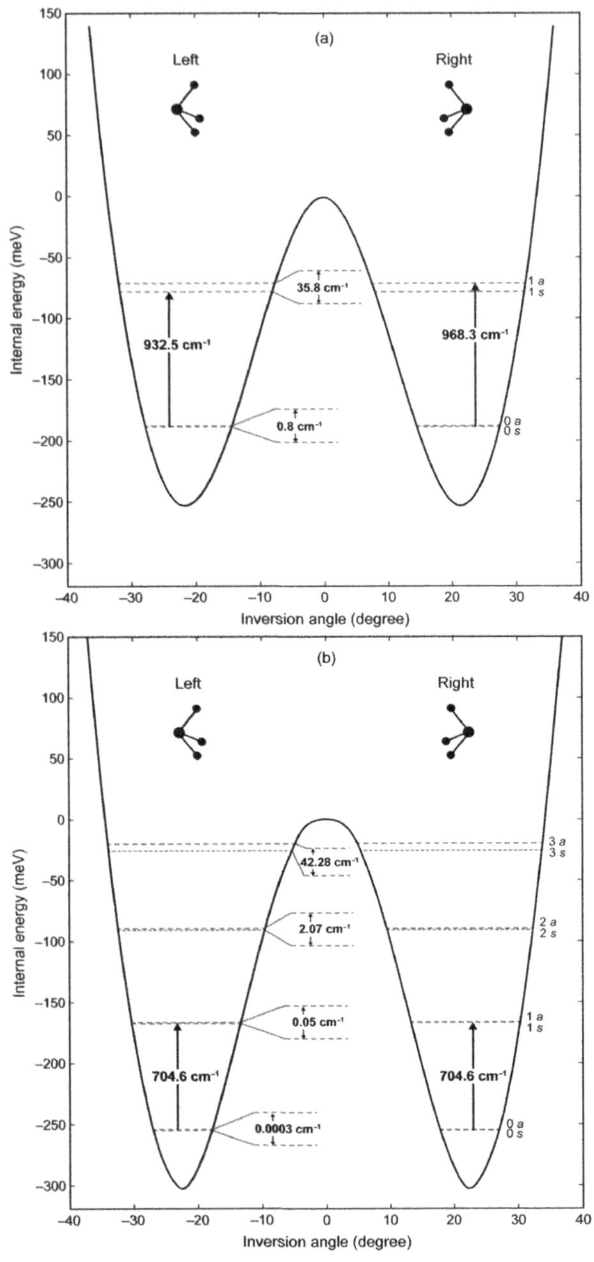

Figure 5.6. *Potential energy of the vibration–inversion mode of the NH_3 molecule as a function of the inversion angle between the N–H bond and the plane formed by the H atoms: (a) in the gas phase and (b) trapped in C_{60}*

5.4.3. *Orientational motions*

Due to the quantization of the orientational motions of the molecule in the nanocage, the orientational Hamiltonian $H_{\text{orient}}^{\text{eff}}$ can be written as:

$$H_{\text{orient}}^{\text{eff}} = T_{\text{rot}}^{\text{eff}} + V_{\text{M}}\left(\vec{r}_0^{\,e}, \vec{\Omega}\right) \qquad [5.14]$$

where $T_{\text{rot}}^{\text{eff}}$ is the effective rotational kinetic energy operator

$$T_{\text{rot}}^{\text{eff}} = -B^{\text{eff}}\left\{\frac{\partial^2}{\partial\theta^2} + \cot\theta\frac{\partial}{\partial\theta} + \frac{1}{\sin^2\theta}\frac{\partial^2}{\partial\varphi^2} + \left(\cot^2\theta + \frac{C^e}{B^{\text{eff}}}\right)\frac{\partial^2}{\partial\chi^2} - 2\frac{\cot\theta}{\sin\theta}\frac{\partial^2}{\partial\varphi\partial\chi}\right\} \qquad [5.15]$$

In this expression, $B^{\text{eff}} = \hbar^2/2I_B^{\text{eff}}$ is the effective rotational constant associated with the rotation–translational motion (see section 5.3.1) of the symmetry axis of NH_3 trapped in the C_{60} nanocage. The obtained value is $B^{\text{eff}} = 7.419\ \text{cm}^{-1}$ and it is smaller than the value $B^e = 9.941\ \text{cm}^{-1}$ of the rigid molecule in the gas phase. On the contrary, $C^e = \hbar^2/2I_C^e$ is the rotational constant about the symmetry axis of the molecule (spinning) in the gas phase, $C^e = 6.309\ \text{cm}^{-1}$.

As already mentioned, the rotation–translational motion is assumed to be free as the potential energy surface barrier $V_{\text{M}}\left(\vec{r}_0^{\,e}, \vec{\Omega}\right)$ is low $\leq 1.86\ \text{meV}$. Therefore, the Hamiltonian $H_{\text{orient}}^{\text{eff}}$ becomes $H_{\text{rot}}^{\text{eff}} = T_{\text{rot}}^{\text{eff}}$, free rotational Hamiltonian of a symmetric oblate top, but with a rotational constant $B^{\text{eff}} = 7.419\ \text{cm}^{-1}$ instead of $B^e = 9.941\ \text{cm}^{-1}$.

Solving Schrödinger's eigenvalue equation associated with this Hamiltonian leads to eigenenergies and eigenfunctions (or eigenvalues and eigenvectors):

$$\begin{aligned} E_{JK} &= B^{\text{eff}} J(J+1) + (C^e - B^{\text{eff}})K^2 \\ \left|JMK\right\rangle &= (-1)^{M-K}\sqrt{\frac{2J+1}{8\pi^2}}D_{-M,-K}^{J}(\varphi,\theta,\chi) \end{aligned} \qquad [5.16]$$

In expressions [5.16], J, K and M are the quantum numbers verifying the relations: $J \geq 0$, $-J \leq M \leq J$ and $-J \leq K \leq J$. The quantum number M is not present in the expression of the eigenenergy; each energy level of quantum number J is associated with $2J+1$ eigenfunctions, being $2J+1$ times degenerate. Moreover, since the quantum number K is squared in expressions [5.16], the energy levels are doubly degenerate, except for $K = 0$ (see Chapter 4 of Volume 3).

Finally, $D^{J}_{-M,-K}(\varphi,\theta,\chi)$ are rotation matrix elements (see Appendix 2 (section 2.5.2) of Chapter 2).

The diagram of rotational levels obtained for NH_3 trapped in the fullerene nanocage is completely different from that of NH_3 in the gas phase, as shown in Figure 5.7. Indeed, in the fullerene nanocage, the molecule behaves as a symmetric top molecule of type XH_3Y, the CH_3F molecule or the deuterated form of methane CH_3D in the gas phase, for example [SUG 65, TOW 75].

NOTE.– Our calculations do not take into account the corrective terms of centrifugal distortion as their impact on the energy levels is negligible.

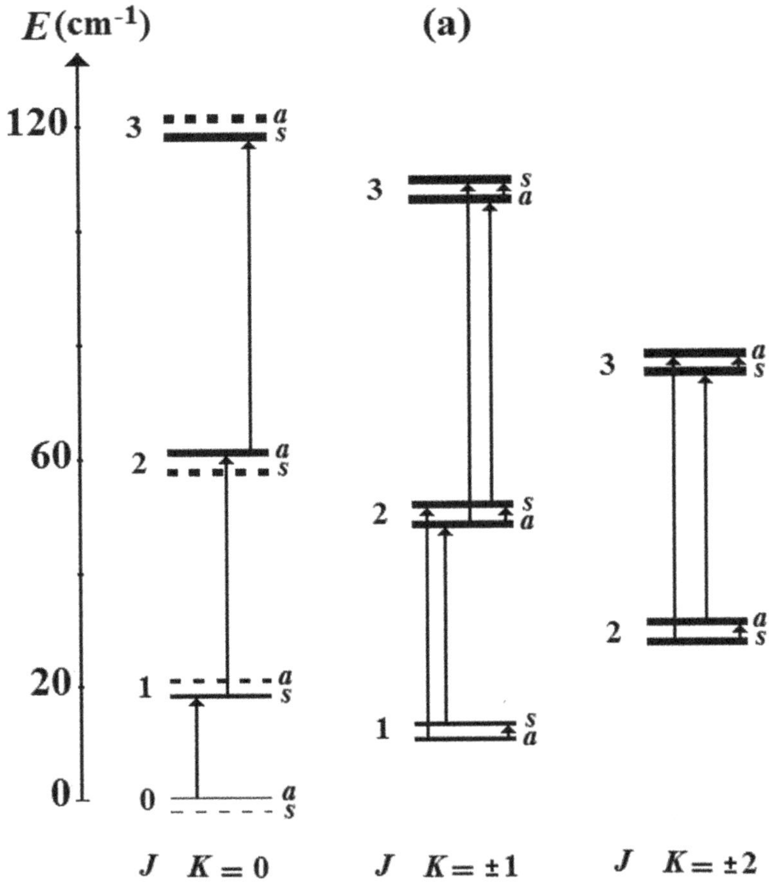

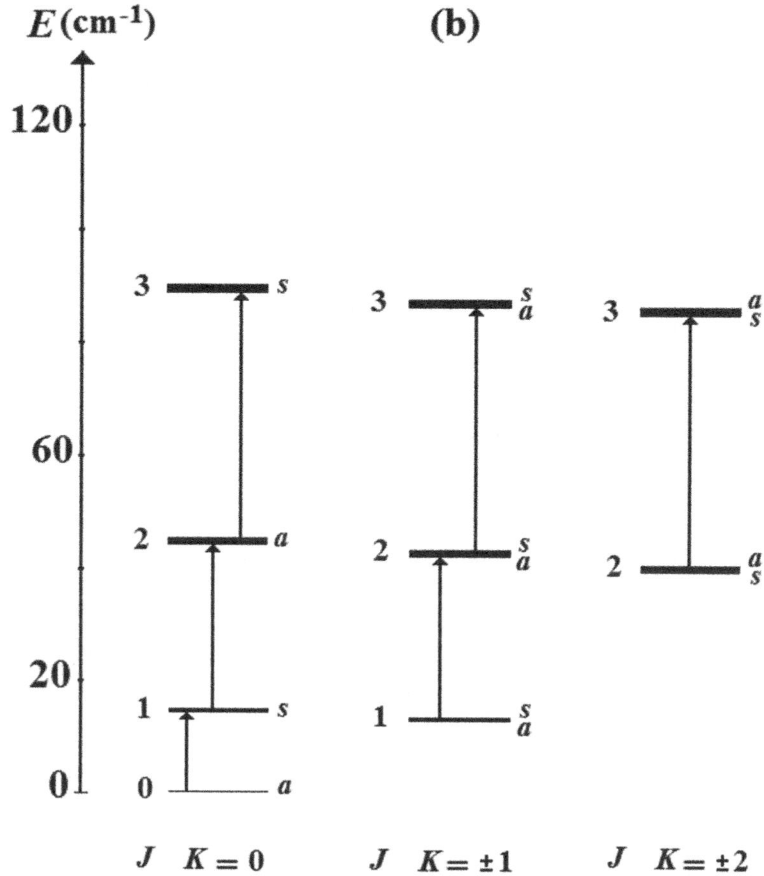

Figure 5.7. *Diagrams of rotational levels of the symmetric top NH_3 molecule: (a) in the gas phase (b) trapped in the fullerene C_{60} nanocage. The degeneracy in M of each level is schematically represented by line thickness*

5.5. Bar spectra

The general expressions of the infrared absorption coefficient of an optically active molecule are presented on several occasions. More specifically, those concerning the bar spectra can be found in section 4.9.1, expression [4.27] and section 4.9.2, expression [4.28] of Chapter 4.

5.5.1. *Far infrared and microwaves*

When the NH_3 molecule is in the gas phase, the transitions allowed by the selection rules are indicated in Figure 5.7(a).

In the far infrared range (rotational spectrum), the selection rules that allow rotational transitions are:

$\Delta J = +1$, $\Delta M = 0, \pm 1$ and $\Delta K = 0$ (branches $R(J_K)$) and the rules $s \leftrightarrow a$ due to the inversion motion of the fundamental vibrational level. The number of lines associated with values $K \neq 0$ of the quantum number K is then multiplied by 2.

In the microwave range (inversion spectrum), the selection rules are:

$$s \leftrightarrow a \text{ with } \Delta J = 0,\ \Delta M = 0, \pm 1 \text{ and } \Delta K = 0 .$$

On the contrary, when the molecule is trapped in the fullerene nanocage, the separation of the inversion doublet of fundamental vibrational level is so small that $s \leftrightarrow a$ transitions are indiscernible. In this case, only the rotational spectrum exists with the rules $\Delta J = +1$, $\Delta M = 0, \pm 1$ and $\Delta K = 0$ (branches $R(J_K)$), and whatever the value of K, the number of spectral lines remains simple (absence of multiplicity). The spectrum is similar to that of CH_3D.

5.5.2. *Near infrared*

In near infrared, the bar spectrum associated with the fundamental transition $|0_\nu\rangle \rightarrow |1_\nu\rangle$ of the vibrational mode ν (the other modes remain in their fundamental state) of NH_3 trapped in C_{60}, consists of rovibrational lines according to the rotational selection rules $\Delta J = 0$, $\Delta M = 0, \pm 1$ and $\Delta K = 0$ (lines $Q(J_K)$), $\Delta J = -1$, $\Delta M = 0, \pm 1$ and $\Delta K = 0$ (lines $P(J_K)$) and $\Delta J = +1$, $\Delta M = 0, \pm 1$ and $\Delta K = 0$ (lines $R(J_K)$).

As an illustration, Figure 5.8 show the rovibrational bar spectra associated with the vibration–inversion mode ν_2, calculated for 10 K, 30 K and 100 K temperatures. The same spectra can be built in the range of frequencies of the other vibrational modes of the molecule.

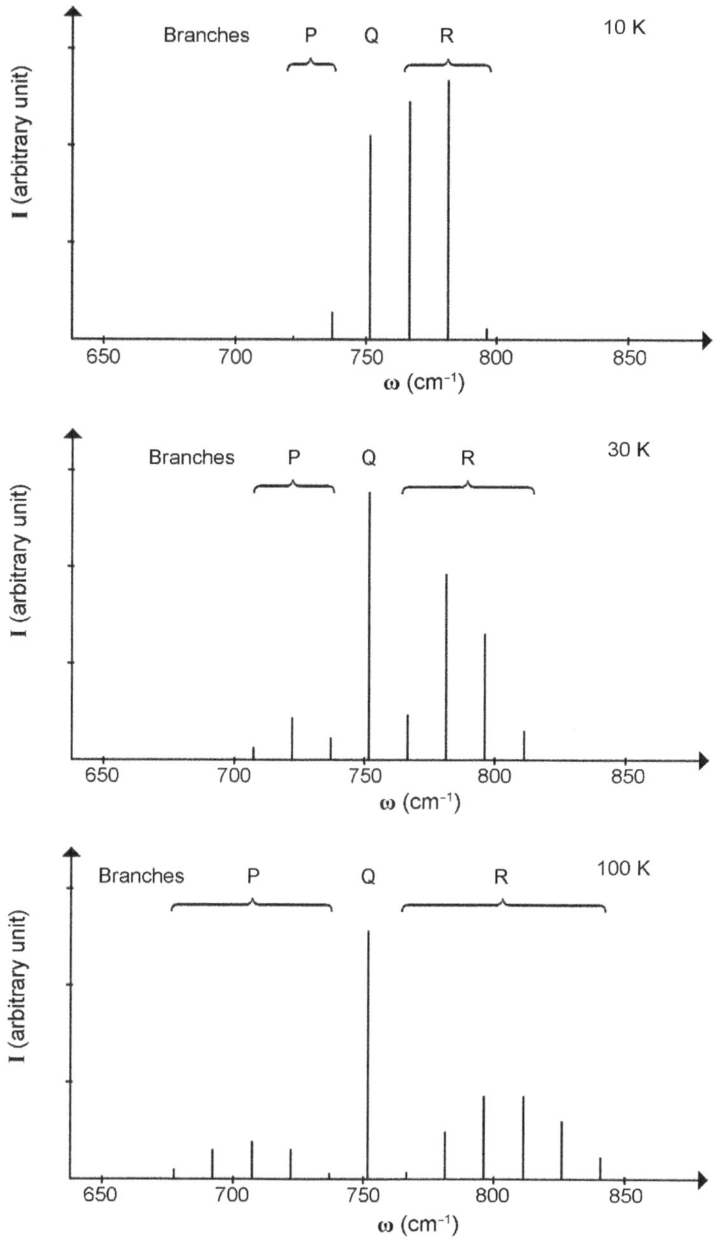

Figure 5.8. *Rovibrational bar spectra associated with vibration–inversion mode ν_2 at various temperatures*

The spectrum consists of a pure vibrational branch Q at 751.6 cm^{-1} frequency, flanked by P lines on the side of low frequencies, and R lines on the side of high frequencies, separated by 14.8 cm^{-1} or $2B^{eff}$.

At T = 10 K, only a few low rotational levels are populated. The bar spectrum shows i) a branch Q that is in fact a superposition of three lines Q(1_1), Q(2_1) and Q(2_2), ii) a branch R consisting of a line R(0_0) at the frequency 766.4 cm^{-1} and a superposition of two lines R(1_0) and R(1_1) at 781.2 cm^{-1}, which are more intense than the branch Q, and iii) a branch P consisting of only one line P(1_0) at 736.8 cm^{-1}.

As the temperature increases, more and more rotational levels are populated and consequently, more and more lines appear in the spectrum with a redistribution of line intensity.

At T = 30 K, the branch Q is the most intense and it is a superposition of several lines Q(J_K) with J = 1,…,4 and $1 \leq K \leq J$. Moreover, the lines R(0_0) and R(1_K) decrease in intensity, allowing the increase of the line R(2_K) at frequency 796.1 cm^{-1} and the appearance of new line R(3_K) at the frequency 810.5 cm^{-1}. Finally, as far as branch P is concerned, two lines P(2_K) and P(3_K) appear at 722 and 707.2 cm^{-1}, respectively.

At T = 100 K, branch Q, the most intense, is a superposition of six lines Q(J_K) with J up to 6, and new lines are observed on the side of branches P and R, with a decrease in intensity for the latter.

Finally, the modification of infrared spectra with temperature variation shows that the spectral signature of the trapped molecule can be used to probe the temperature of the medium in which the fullerene molecules are observed.

5.6. Appendices

5.6.1. *Appendix 1: FORTRAN program*

This program is used to calculate the interaction potential energy of the NH_3 molecule trapped in a fullerene C_{60} nanocage.

```
C*********************************************************************
C* PROGRAM <==> nh3c60.f ----> See below for C60 molecule
C* THIS PROGRAM CALCULATES THE POTENTIAL ENERGY
C* OF AN AMMONIA NH3 MOLECULE THAT IS RIGID OR
C* IN INVERSION MOTION TRAPPED IN A FULLERENE C60 NANOCAGE
C*********************************************************************
      PARAMETER (NC=60,MINI=0,IMOL=1)
C*********************************************************************
C     MINI=0 ----> NO MINIMISATION/A R0 and/or PHI0 and/or QUI0
C         =1 ----> MINIMISATION       /A R0 and/or PHI0 and/or QUI0
C     IMOL=1 ----> RIGID MOLECULE
C         =2 ----> MOLECULE IN INVERSION MOTION
C*********************************************************************
      IMPLICIT DOUBLE PRECISION (A-H,O-Z)
      DIMENSION RLM(3),ECG(3),ECMD(3),ECMQ(3),ECM(3)
      DIMENSION ALC(NC,3,3),QUAC(NC,3,3)
      DIMENSION XCA(NC),YCA(NC),ZCA(NC)
      DIMENSION XC(NC),YC(NC),ZC(NC),PHIC(NC),THEC(NC)
      DIMENSION HX(5),HY(5),HZ(5),DIPM(3),ALM(3,3),QUAM(3,3)
      DIMENSION EPS(5,1),SIG(5,1),C6(5,1),C12(5,1)
      DIMENSION VINTNU2(181),VINV(181)
C*********************************************************************
C      OPEN(1,FILE='VFITPHTHX1',FORM='FORMATTED')
      WRITE(*,555)
 555  FORMAT(34X,'NH3 IN THE FULLERENE  C60 MOLECULE'///)
C*********************************************************************
C*DEFINITION OF SEVERAL VALUES
C*********************************************************************
      PI= DACOS(-1.D0)
      S3= DSQRT(3.D0)
      ABOHR= 0.52917725D0
      UAMEV= 27211.4D0
C*********************************************************************
C*THE POSITIONS OF CARBON ATOMS IN C60 ARE CALCULATED
C*FROM DATA IN THE LITERATURE ----> Exp.
C*Hedberg, Hedberg, Bethune,... Science 254,410(1991)
C*  R1(C==C) = 1.398 Ang     R2(C--C) = 1.455 Ang
C******************************************************
C*RADIUS OF C60    in Ang  and in a.u.
C******************************************************
      RCA= 3.556D0
      RC= 3.556D0/ABOHR
C*********************************************************************
```

```
C*POSITIONS OF CARBON ATOMS OF FULLERENE C60
C*IN THE ABSOLUTE REFERENCE FRAME (O, X, Y, Z) D ORIGIN THE CENTER OF
C*C60 MOLECULE
C***********************************************************************
C*CARTESIAN COORDINATES IN  Ang.
C************************************************
      XCA(1)=  0.D0
      YCA(1)=  0.69911D0
      ZCA(1)=  3.48660D0
      XCA(2)=  0.D0
      YCA(2)= -0.69911D0
      ZCA(2)=  3.48660D0
      XCA(3)=  1.17712D0
      YCA(3)=  1.42650D0
      ZCA(3)=  3.03720D0
      XCA(4)=  1.17712D0
      YCA(4)= -1.42650D0
      ZCA(4)=  3.03721D0
      XCA(5)= -1.17712D0
      YCA(5)=  1.42650D0
      ZCA(5)=  3.03721D0
      XCA(6)= -1.17712D0
      YCA(6)= -1.42650D0
      ZCA(6)=  3.03721D0
      XCA(7)=  2.30814D0
      YCA(7)=  0.72750D0
      ZCA(7)=  2.60545D0
      XCA(8)=  2.30814D0
      YCA(8)= -0.72750D0
      ZCA(8)=  2.60545D0
      XCA(9)= -2.30814D0
      YCA(9)=  0.72750D0
      ZCA(9)=  2.60545D0
      XCA(10)= -2.30814D0
      YCA(10)= -0.72750D0
      ZCA(10)=  2.60545D0
      XCA(11)=  0.72750D0
      YCA(11)=  2.60545D0
      ZCA(11)=  2.30814D0
      XCA(12)=  0.72750D0
      YCA(12)= -2.60545D0
      ZCA(12)=  2.30814D0
      XCA(13)= -0.72750D0
```

```
YCA(13)= 2.60545D0
ZCA(13)= 2.30814D0
XCA(14)= -0.72750D0
YCA(14)= -2.60545D0
ZCA(14)= 2.30814D0
XCA(15)= 3.03721D0
YCA(15)= 1.17712D0
ZCA(15)= 1.42650D0
XCA(16)= -3.03721D0
YCA(16)= 1.17712D0
ZCA(16)= 1.42650D0
XCA(17)= 3.03721D0
YCA(17)= -1.17712D0
ZCA(17)= 1.42650D0
XCA(18)= -3.03721D0
YCA(18)= -1.17712D0
ZCA(18)= 1.42650D0
XCA(19)= 1.42650D0
YCA(19)= 3.03721D0
ZCA(19)= 1.17712D0
XCA(20)= 1.42650D0
YCA(20)= -3.03721D0
ZCA(20)= 1.17712D0
XCA(21)= -1.42650D0
YCA(21)= 3.03721D0
ZCA(21)= 1.17712D0
XCA(22)= -1.42650D0
YCA(22)= -3.03721D0
ZCA(22)= 1.17712D0
XCA(23)= 2.60545D0
YCA(23)= 2.30814D0
ZCA(23)= 0.72750D0
XCA(24)= -2.60545D0
YCA(24)= 2.30814D0
ZCA(24)= 0.72750D0
XCA(25)= 2.60545D0
YCA(25)= -2.30814D0
ZCA(25)= 0.72750D0
XCA(26)= -2.60545D0
YCA(26)= -2.30814D0
ZCA(26)= 0.72750D0
XCA(27)= 3.48660D0
YCA(27)= 0.D0
```

```
ZCA(27)= 0.69911D0
XCA(28)= -3.48660D0
YCA(28)= 0.D0
ZCA(28)= 0.69911D0
XCA(29)= 0.69911D0
YCA(29)= 3.48660D0
ZCA(29)= 0.D0
XCA(30)= 0.69911D0
YCA(30)= -3.48660D0
ZCA(30)= 0.D0
XCA(31)= -0.69911D0
YCA(31)= 3.48660D0
ZCA(31)= 0.D0
XCA(32)= -0.69911D0
YCA(32)= -3.48660D0
ZCA(32)= 0.D0
XCA(33)= 2.60545D0
YCA(33)= 2.30814D0
ZCA(33)= -0.72750D0
XCA(34)= -2.60545D0
YCA(34)= 2.30814D0
ZCA(34)= -0.72750D0
XCA(35)= 2.60545D0
YCA(35)= -2.30814D0
ZCA(35)= -0.72750D0
XCA(36)= -2.60545D0
YCA(36)= -2.30814D0
ZCA(36)= -0.72750D0
XCA(37)= 3.48660D0
YCA(37)= 0.D0
ZCA(37)= -0.69911D0
XCA(38)= -3.48660D0
YCA(38)= 0.D0
ZCA(38)= -0.69911D0
XCA(39)= 1.42650D0
YCA(39)= 3.03721D0
ZCA(39)= -1.17712D0
XCA(40)= 1.42650D0
YCA(40)= -3.03721D0
ZCA(40)= -1.17712D0
XCA(41)= -1.42650D0
YCA(41)= 3.03721D0
ZCA(41)= -1.17712D0
```

```
XCA(42)= -1.42650D0
YCA(42)= -3.03721D0
ZCA(42)= -1.17712D0
XCA(43)=  3.03721D0
YCA(43)=  1.17712D0
ZCA(43)= -1.42650D0
XCA(44)= -3.03721D0
YCA(44)=  1.17712D0
ZCA(44)= -1.42650D0
XCA(45)=  3.03721D0
YCA(45)= -1.17712D0
ZCA(45)= -1.42650D0
XCA(46)= -3.03721D0
YCA(46)= -1.17712D0
ZCA(46)= -1.42650D0
XCA(47)=  0.72750D0
YCA(47)=  2.60545D0
ZCA(47)= -2.30814D0
XCA(48)=  0.72750D0
YCA(48)= -2.60545D0
ZCA(48)= -2.30814D0
XCA(49)= -0.72750D0
YCA(49)=  2.60545D0
ZCA(49)= -2.30814D0
XCA(50)= -0.72750D0
YCA(50)= -2.60545D0
ZCA(50)= -2.30814D0
XCA(51)=  2.30814D0
YCA(51)=  0.72750D0
ZCA(51)= -2.60545D0
XCA(52)=  2.30814D0
YCA(52)= -0.72750D0
ZCA(52)= -2.60545D0
XCA(53)= -2.30814D0
YCA(53)=  0.72750D0
ZCA(53)= -2.60545D0
XCA(54)= -2.30814D0
YCA(54)= -0.72750D0
ZCA(54)= -2.60545D0
XCA(55)=  1.17712D0
YCA(55)=  1.42650D0
ZCA(55)= -3.03721D0
XCA(56)=  1.17712D0
```

```
      YCA(56)= -1.42650D0
      ZCA(56)= -3.03721D0
      XCA(57)= -1.17712D0
      YCA(57)=  1.42650D0
      ZCA(57)= -3.03721D0
      XCA(58)= -1.17712D0
      YCA(58)= -1.42650D0
      ZCA(58)= -3.03721D0
      XCA(59)=  0.D0
      YCA(59)=  0.69911D0
      ZCA(59)= -3.48660D0
      XCA(60)=  0.D0
      YCA(60)= -0.69911D0
      ZCA(60)= -3.48660D0
C****************************************
C*CARTESIAN COORDINATES in  a.u.
C****************************************
      DO 100 IC=1,NC
      XC(IC)= XCA(IC)/ABOHR
      YC(IC)= YCA(IC)/ABOHR
      ZC(IC)= ZCA(IC)/ABOHR
 100  CONTINUE
C****************************************
C* SPHERICAL COORDINATES in  rad
C****************************************
      PHIC(1)=  0.157080D+01
      THEC(1)=  0.197889D+00
      PHIC(2)=  0.471239D+01
      THEC(2)=  0.197889D+00
      PHIC(3)=  0.880889D+00
      THEC(3)=  0.546967D+00
      PHIC(4)=  0.540230D+01
      THEC(4)=  0.546962D+00
      PHIC(5)=  0.226070D+01
      THEC(5)=  0.546962D+00
      PHIC(6)=  0.402248D+01
      THEC(6)=  0.546962D+00
      PHIC(7)=  0.305333D+00
      THEC(7)=  0.748528D+00
      PHIC(8)=  0.597785D+01
      THEC(8)=  0.748528D+00
      PHIC(9)=  0.283626D+01
      THEC(9)=  0.748528D+00
```

PHIC(10)= 0.344693D+01
THEC(10)= 0.748528D+00
PHIC(11)= 0.129851D+01
THEC(11)= 0.864418D+00
PHIC(12)= 0.498468D+01
THEC(12)= 0.864418D+00
PHIC(13)= 0.184308D+01
THEC(13)= 0.864418D+00
PHIC(14)= 0.444010D+01
THEC(14)= 0.864418D+00
PHIC(15)= 0.369742D+00
THEC(15)= 0.115802D+01
PHIC(16)= 0.277185D+01
THEC(16)= 0.115802D+01
PHIC(17)= 0.591344D+01
THEC(17)= 0.115802D+01
PHIC(18)= 0.351133D+01
THEC(18)= 0.115802D+01
PHIC(19)= 0.113170D+01
THEC(19)= 0.123341D+01
PHIC(20)= 0.515148D+01
THEC(20)= 0.123341D+01
PHIC(21)= 0.200989D+01
THEC(21)= 0.123341D+01
PHIC(22)= 0.427329D+01
THEC(22)= 0.123341D+01
PHIC(23)= 0.724964D+00
THEC(23)= 0.136476D+01
PHIC(24)= 0.241663D+01
THEC(24)= 0.136476D+01
PHIC(25)= 0.555822D+01
THEC(25)= 0.136476D+01
PHIC(26)= 0.386656D+01
THEC(26)= 0.136476D+01
PHIC(27)= 0.000000D+00
THEC(27)= 0.137291D+01
PHIC(28)= 0.314159D+01
THEC(28)= 0.137291D+01
PHIC(29)= 0.137291D+01
THEC(29)= 0.157080D+01
PHIC(30)= 0.491028D+01
THEC(30)= 0.157080D+01
PHIC(31)= 0.176869D+01

THEC(31)= 0.157080D+01
PHIC(32)= 0.451450D+01
THEC(32)= 0.157080D+01
PHIC(33)= 0.724964D+00
THEC(33)= 0.177683D+01
PHIC(34)= 0.241663D+01
THEC(34)= 0.177683D+01
PHIC(35)= 0.555822D+01
THEC(35)= 0.177683D+01
PHIC(36)= 0.386656D+01
THEC(36)= 0.177683D+01
PHIC(37)= 0.000000D+00
THEC(37)= 0.176869D+01
PHIC(38)= 0.314159D+01
THEC(38)= 0.176869D+01
PHIC(39)= 0.113170D+01
THEC(39)= 0.190818D+01
PHIC(40)= 0.515148D+01
THEC(40)= 0.190818D+01
PHIC(41)= 0.200989D+01
THEC(41)= 0.190818D+01
PHIC(42)= 0.427329D+01
THEC(42)= 0.190818D+01
PHIC(43)= 0.369742D+00
THEC(43)= 0.198357D+01
PHIC(44)= 0.277185D+01
THEC(44)= 0.198357D+01
PHIC(45)= 0.591344D+01
THEC(45)= 0.198357D+01
PHIC(46)= 0.351133D+01
THEC(46)= 0.198357D+01
PHIC(47)= 0.129851D+01
THEC(47)= 0.227718D+01
PHIC(48)= 0.498468D+01
THEC(48)= 0.227718D+01
PHIC(49)= 0.184308D+01
THEC(49)= 0.227718D+01
PHIC(50)= 0.444010D+01
THEC(50)= 0.227718D+01
PHIC(51)= 0.305333D+00
THEC(51)= 0.239306D+01
PHIC(52)= 0.597785D+01
THEC(52)= 0.239306D+01

```
      PHIC(53)=  0.283626D+01
      THEC(53)=  0.239306D+01
      PHIC(54)=  0.344693D+01
      THEC(54)=  0.239306D+01
      PHIC(55)=  0.880889D+00
      THEC(55)=  0.259463D+01
      PHIC(56)=  0.540230D+01
      THEC(56)=  0.259463D+01
      PHIC(57)=  0.226070D+01
      THEC(57)=  0.259463D+01
      PHIC(58)=  0.402248D+01
      THEC(58)=  0.259463D+01
      PHIC(59)=  0.157080D+01
      THEC(59)=  0.294370D+01
      PHIC(60)=  0.471239D+01
      THEC(60)=  0.294370D+01
C*******************************************************************
C*ELECTRIC PARAMETERS, IN a.u., OF CARBON ATOMS IN
C*THEIR RESPECTIVE REFERENCE FRAMES (Cj, xj, yj, zj) j=1,60  (see Figure)
C*******************************************************************
      ALCxx= 1.68D0/ABOHR/ABOHR/ABOHR
      ALCyy= 1.68D0/ABOHR/ABOHR/ABOHR
      ALCzz= 0.48D0/ABOHR/ABOHR/ABOHR
      QUACxx= -0.7435D0/2.D0
      QUACyy= -0.7435D0/2.D0
      QUACzz=  0.7435D0
C*******************************************************************
C*COMPONENTS IN THE ABSOLUTE REFERENCE FRAME OF TENSORS: POLARIZABILITY
C*AND QUADRUPOLE MOMENT OF CARBON ATOMS OF FULLERENE
C*ALC(I,J,K) AND QUAC(I,J,K)    I = 1,NC     J = 1,3     K = 1,3
C*******************************************************************
      DO 110 I=1,NC
C****************************************
      COSPHIC= DCOS(PHIC(I))
      SINPHIC= DSIN(PHIC(I))
      COSTHEC= DCOS(THEC(I))
      SINTHEC= DSIN(THEC(I))
C*******************************************************************
C* TRANSFORMATION MATRIX(M.E. ROSE) REFERENCE FRAME ATTACHED
C*TO EACH CARBON ATOM ---------> ABSOLUTE REFERENCE FRAME
C*******************************************************************
      RCXx=  COSPHIC*COSTHEC
      RCXy= -SINPHIC
```

```
      RCXz= COSPHIC*SINTHEC
      RCYx= SINPHIC*COSTHEC
      RCYy= COSPHIC
      RCYz= SINPHIC*SINTHEC
      RCZx= -SINTHEC
      RCZy= 0.D0
      RCZz= COSTHEC
C****************************************
      ALC(I,1,1)= RCXx*RCXx*ALCxx+RCXy*RCXy*ALCyy+RCXz*RCXz*ALCzz
      ALC(I,1,2)= RCXx*RCYx*ALCxx+RCXy*RCYy*ALCyy+RCXz*RCYz*ALCzz
      ALC(I,1,3)= RCXx*RCZx*ALCxx+RCXy*RCZy*ALCyy+RCXz*RCZz*ALCzz
      ALC(I,2,1)= ALC(I,1,2)
      ALC(I,2,2)= RCYx*RCYx*ALCxx+RCYy*RCYy*ALCyy+RCYz*RCYz*ALCzz
      ALC(I,2,3)= RCYx*RCZx*ALCxx+RCYy*RCZy*ALCyy+RCYz*RCZz*ALCzz
      ALC(I,3,1)= ALC(I,1,3)
      ALC(I,3,2)= ALC(I,2,3)
      ALC(I,3,3)= RCZx*RCZx*ALCxx+RCZy*RCZy*ALCyy+RCZz*RCZz*ALCzz
C*****************************************************************
      QUAC(I,1,1)= RCXx*RCXx*QUACxx+RCXy*RCXy*QUACyy+RCXz*RCXz*QUACzz
      QUAC(I,1,2)= RCXx*RCYx*QUACxx+RCXy*RCYy*QUACyy+RCXz*RCYz*QUACzz
      QUAC(I,1,3)= RCXx*RCZx*QUACxx+RCXy*RCZy*QUACyy+RCXz*RCZz*QUACzz
      QUAC(I,2,1)= QUAC(I,1,2)
      QUAC(I,2,2)= RCYx*RCYx*QUACxx+RCYy*RCYy*QUACyy+RCYz*RCYz*QUACzz
      QUAC(I,2,3)= RCYx*RCZx*QUACxx+RCYy*RCZy*QUACyy+RCYz*RCZz*QUACzz
      QUAC(I,3,1)= QUAC(I,1,3)
      QUAC(I,3,2)= QUAC(I,2,3)
      QUAC(I,3,3)= RCZx*RCZx*QUACxx+RCZy*RCZy*QUACyy+RCZz*RCZz*QUACzz
C****************************************
  110 CONTINUE
C*****************************************************************
C* LENNARD-JONES POTENTIAL PARAMETERS IN cm-1 AND Ang
C*  Epsilon AND Sigma OF: Hydrogen , Nitrogen , Carbon
C*****************************************************************
      EPSN= 26.5D0
      SIGN= 3.38D0
      EPSH= 17.2D0
      SIGH= 2.53D0
      EPSC= 19.4D0
      SIGC= 3.40D0
C*****************************************************************
C*MIXED PARAMETERS Epsilon AND Sigma IN cm-1 AND Ang
C*****************************************************************
      EPS(1,1)= DSQRT(EPSN*EPSC)
```

```
      SIG(1,1)= (SIGN + SIGC)/2.D0
      EPS(2,1)= 0.D0
      SIG(2,1)= 0.D0
      EPS(3,1)= DSQRT(EPSH*EPSC)
      SIG(3,1)= (SIGH + SIGC)/2.D0
      EPS(4,1)= EPS(3,1)
      SIG(4,1)= SIG(3,1)
      EPS(5,1)= EPS(3,1)
      SIG(5,1)= SIG(3,1)
C*******************************************************************
C* LENNARD-JONES POTENTIAL PARAMETERS IN  a.u.
C*    C6 ET C12 (ATOM OF THE MOLECULE – ATOM OF CARBON FULLERENE)
C*******************************************************************
      DO LJ=1,5
      C6(LJ,1) = 4.D0*EPS(LJ,1)*SIG(LJ,1)**6/4820.167394D0
      C12(LJ,1)= 4.D0*EPS(LJ,1)*SIG(LJ,1)**12/105.832398D0
      END DO
C*******************************************************************
C*GEOMETRIC AND ELECTRIC PARAMETERS OF THE MOLECULE
C*IN ITS REFERENCE FRAME Gxyz
C*******************************************************************
C*ATOMIC MASSES  IN  gram
C*******************************************************************
      AMH=  1.00794D0
      AMN= 14.00674D0
      RO1= 3.D0*AMH/(AMN+3.D0*AMH)
C*****************************************************************
C* N—H bond length, Dipole, Quadrupole and Polarizability
C*of the rigid molecule ---> in a.u.
C*****************************************************************
      QNH=  1.91920571D0
      DIP=  0.58074D0
      QUA= -2.17853D0
      APA= 13.4827746D0
      APE= 16.1354606D0
C*****************************************************************
C* RIGID MOLECULE IMOL=1 or INVERSION MOTION IMOL=2
C*****************************************************************
      GOTO(50,51)IMOL
 50   INV= 1
      GO TO 53
 51   INV= 91
 53   DO 52 IB=1,INV
```

```
   IBB= 181+1 - IB
   GOTO(80,81)IMOL
C***************************************************
C*Beta Angle between N—H bond and NH3 axis
C***************************************************
 80 BETADEG= 67.97D0    ! Rigid molecule
   BETA= BETADEG*PI/180.D0
   GOTO 54
C***************************************************
C*Variation interval of Beta angle is 0.5 degrees
C***************************************************
 81 BETADEG= 90.D0 + (IB - 91)*0.5D0
   BETA= BETADEG*PI/180.D0
   VARNU2= PI/2.D0 - BETA
C***************************************************
C* VIB-INV POTENTIAL OF NH3 IN GAS PHASE
C*Ref: Swalen and Ibers  JCP vol 36, 1914 (1962)
C*    Modified by myself ---> In agreement with Exp. Values
C******************************************************************
   VINTNU2(IB)= 3728.D0 * VARNU2**2 +
   &          1546.D0 * (DEXP(-4.974D0*VARNU2**2) - 1.D0)
C******************************************************************
C*If Beta is < a Betaequilibrium, then the molecule-C60 interaction
C*energy is calculated with Betaequilibrium
C******************************************************************
   IF(IB.LE.47)GOTO 54  ! The value 47 <---> NH3 equilibrium
C                     as Beta interval is 0.5 degrees
   QNH= (1.8583112D0 + 0.15859D0 * DABS((IB-91)*0.5D0*PI/180.D0))
C
   DIP= 3.D0*QNH*DCOS(BETA)*(1.D0-5.16830D0/(QNH**3))
C
   QUA= 1.745694D0*QNH**2*(1.144877D0*DCOS(BETA)**2 - 0.5D0)
C
   APA= (1.9115494D0*QNH**3-0.2338837D0*(1.D0-3.D0*DCOS(BETA)**2))
   & *QNH**3/(QNH**6-0.0832D0*(1.D0+3.D0*DCOS(BETA)**2))/ABOHR**3
C
   APE= (1.9115494D0*QNH**3+0.1169419D0*(1.D0-3.D0*DCOS(BETA)**2))
   & *QNH**3/(QNH**6-0.0416D0*(5.D0-3.D0*DCOS(BETA)**2))/ABOHR**3
C******************************************************************
 54 QUAxx= -QUA/2.D0
   QUAyy= -QUA/2.D0
   QUAzz= QUA
   ALMxx= APE
```

```
      ALMyy= APE
      ALMzz= APA
C****************************************************
C*CONTROL OUTPUT
C****************************************************
C     WRITE(*,444)BETADEG,QNH,APA,APE,DIP,QUA,VINTNU2(91)
  444 FORMAT(5X,F10.6,6(5X,F10.6)//)
C*****************************************************************
C*POSITIONS OF ATOMS IN THE REFERENCE FRAME OF THE MOLECULE
C*ATOM NUMBER 2 IS FICTITIOUS!
C*****************************************************************
      QCB= QNH*DCOS(BETA)
      QSB= QNH*DSIN(BETA)
C*****************************************************************
      HX(1)=  0.D0
      HY(1)=  0.D0
      HZ(1)= -RO1*QCB
      HX(2)=  0.D0
      HY(2)=  0.D0
      HZ(2)=  0.D0
      HX(3)=  QSB
      HY(3)=  0.D0
      HZ(3)=  (1.D0-RO1)*QCB
      HX(4)= -QSB/2.D0
      HY(4)=  S3*QSB/2.D0
      HZ(4)=  HZ(3)
      HX(5)=  HX(4)
      HY(5)= -HY(4)
      HZ(5)=  HZ(3)
C*****************************************************************
C     WRITE(*,60)
   60 FORMAT(5X,'VEDQ (meV)',9X,'VEQQ (meV)',
     &10X,'VIM (meV)',11X,'VIG (meV)',12X,'VQUAN (meV)',
     &7X,'VTOT (meV)'//)
C*****************************************************************
      DO 1 IP= 1,1
      PHI0= (IP-1)*180.D0/72.D0
      PHI=  PHI0*PI/180.D0
      SA= DSIN(PHI)
      CA= DCOS(PHI)
C*****************************************************************
      DO 2 IT= 1,1
      TE0= (IT-1)*180.D0/72.D0 + 1.D0*78.6618 + 0.D0*20. + 0.D0*57.5
```

```
C                       along C27   along Hexa  along Penta
      TE= TE0*PI/180.D0
      SB= DSIN(TE)
      CB= DCOS(TE)
C     WRITE(*,70)PHI0,TE0
   70 FORMAT(/10X,'PHI = ',F7.1,5X,'THETA = ',F7.1/)
      TEVAR=TE-PI/2.D0
C*******************************************************************
C*POSITION OF THE CENTER OF MASS OF NH3  (Vector R0) ALONG THE
C*LINE DEFINED BY THE ORIENTATION  (Phi , Theta) OF THE MOLECULE
C*******************************************************************
      POTMIN= 0.D0
      DO 3 IR= -97,-87
      R0= IR * 0.002D0
C*******************************************
      X0= R0*CA*SB
      Y0= R0*SA*SB
      Z0= R0*CB
C*******************************************
C*Vector R0 is along the direction of C27
C*******************************************
C     X0= R0 *1.D0* DSIN(78.6618D0*PI/180.)
C     Y0= 0.D0
C     Z0= R0 *0.D0* DCOS(78.6618D0*PI/180.)
C*******************************************
      RM= R0/ABOHR
      XM= X0/ABOHR
      YM= Y0/ABOHR
      ZM= Z0/ABOHR
C*******************************************
C     WRITE(*,71)X0,Y0,Z0,R0
   71 FORMAT(/4X,'X0=',F8.4,4X,'Y0=',F8.4,4X,'Z0=',F8.4,4X,'R0=',F7.3)
C*******************************************************************
C     POTMIN= 100.D0
      DO 4 IK= 1,7
      QUI0= (IK-1)*180.D0/6.D0
      QUI= QUI0*PI/180.D0
      SC= DSIN(QUI)
      CC= DCOS(QUI)
C*******************************************************************
C* TRANSFORMATION MATRIX (M.E. ROSE) REFERENCE FRAME ATTACHED
C*TO THE MOLECULE -------> ABSOLUTE REFERENCE FRAME
C*******************************************************************
```

```
      RMXx= CA*CB*CC - SA*SC
      RMXy= -CA*CB*SC - SA*CC
      RMXz= CA*SB
      RMYx= SA*CB*CC + CA*SC
      RMYy= -SA*CB*SC + CA*CC
      RMYz= SA*SB
      RMZx= -SB*CC
      RMZy= SB*SC
      RMZz= CB
C***********************************************************************
C*COMPONENTS IN THE ABSOLUTE REFERENCE FRAME OF TENSORS: POLARIZABILITY,
C*DIPOLE MOMENT AND QUADRUPOLE MOMENT OF THE MOLECULE
C***********************************************************************
      ALM(1,1)= RMXx*RMXx*ALMxx+RMXy*RMXy*ALMyy+RMXz*RMXz*ALMzz
      ALM(1,2)= RMXx*RMYx*ALMxx+RMXy*RMYy*ALMyy+RMXz*RMYz*ALMzz
      ALM(1,3)= RMXx*RMZx*ALMxx+RMXy*RMZy*ALMyy+RMXz*RMZz*ALMzz
      ALM(2,1)= ALM(1,2)
      ALM(2,2)= RMYx*RMYx*ALMxx+RMYy*RMYy*ALMyy+RMYz*RMYz*ALMzz
      ALM(2,3)= RMYx*RMZx*ALMxx+RMYy*RMZy*ALMyy+RMYz*RMZz*ALMzz
      ALM(3,1)= ALM(1,3)
      ALM(3,2)= ALM(2,3)
      ALM(3,3)= RMZx*RMZx*ALMxx+RMZy*RMZy*ALMyy+RMZz*RMZz*ALMzz
C***********************************************************************
      DIPM(1) = DIP*RMXz
      DIPM(2) = DIP*RMYz
      DIPM(3) = DIP*RMZz
C***********************************************************************
      QUAM(1,1)= RMXx*RMXx*QUAxx+RMXy*RMXy*QUAyy+RMXz*RMXz*QUAzz
      QUAM(1,2)= RMXx*RMYx*QUAxx+RMXy*RMYy*QUAyy+RMXz*RMYz*QUAzz
      QUAM(1,3)= RMXx*RMZx*QUAxx+RMXy*RMZy*QUAyy+RMXz*RMZz*QUAzz
      QUAM(2,1)= QUAM(1,2)
      QUAM(2,2)= RMYx*RMYx*QUAxx+RMYy*RMYy*QUAyy+RMYz*RMYz*QUAzz
      QUAM(2,3)= RMYx*RMZx*QUAxx+RMYy*RMZy*QUAyy+RMYz*RMZz*QUAzz
      QUAM(3,1)= QUAM(1,3)
      QUAM(3,2)= QUAM(2,3)
      QUAM(3,3)= RMZx*RMZx*QUAxx+RMZy*RMZy*QUAyy+RMZz*RMZz*QUAzz
C***********************************************************************
C*CALCULATION OF VARIOUS CONTRIBUTIONS
C*   VEDQ ==> ELECT DIP MOL - QUAD OF ATOMS C OF C60
C*   VEQQ ==> ELECT QUACD MOL - QUAD OF ATOMS C OF C60
C*   VIG  ==> INDUC OF ATOMS C OF  C60 ---> MOL
C*   VIM  ==> INDUC MOL ---> ATOMS C OF C60
C*   VQUAN ==> QUANTUM
```

```
C*******************************************************************
C* ZEROING
C*******************************************************************
   VEDQ= 0.D0
   VEQQ= 0.D0
   VIG= 0.D0
   VIM= 0.D0
   VQUAN= 0.D0
C*******************************************************************
C*ZEROING OF ELECTRIC FIELD INDUCED BY THE C ATOMS OF C60
C*ON THE POSITION OF THE MOLECULE
C*******************************************************************
   DO 19 I=1,3
 19 ECG(I)= 0.D0
C*******************************************************************
C*     START LOOP ON THE CARBON ATOMS OF C60
   DO 120 JC=1,NC
C*******************************************************************
C*COMPONENTS OF THE MOLECULE – ATOM JC DISTANCE VECTOR
C*******************************************************************
   RLM(1)= XM-XC(JC)
   RLM(2)= YM-YC(JC)
   RLM(3)= ZM-ZC(JC)
C     WRITE(*,*)RLM(1),RLM(2),RLM(3)
C*******************************************************************
   RLM2=RLM(1)**2+RLM(2)**2+RLM(3)**2
   RLM1=DSQRT(RLM2)
   RLM4=RLM2*RLM2
   RLM5=RLM4*RLM1
   RLM7=RLM5*RLM2
   RLM9=RLM7*RLM2
C*******************************************************************
C*CALCULATION OF ELECTROSTATIC AND INDUCTION CONTRIBUTIONS
C*******************************************************************
C*ZEROING OF ELECTRIC FIELD INDUCED BY THE MOLECULE
C*ON THE ATOM POSITION CONSIDERED  ECM = ECMD + ECMQ
C*******************************************************************
   DO 30 I=1,3
   ECMD(I)= 0.D0
   ECMQ(I)= 0.D0
 30 CONTINUE
C*******************************************************************
   DO 31 I1=1,3
```

```
      DO 32 I2=1,3
      ECMD(I1)= ECMD(I1)-TAC2(I1,I2,RLM,RLM2,RLM5)*DIPM(I2)
      DO 33 I3=1,3
      TAC= TAC3(I1,I2,I3,RLM,RLM2,RLM7)
      ECMQ(I1)= ECMQ(I1)-TAC*QUAM(I2,I3)/3.D0
      ECG(I1)= ECG(I1) - TAC*QUAC(JC,I2,I3)/3.D0
      VEDQ= VEDQ + DIPM(I1)*TAC*QUAC(JC,I2,I3)/3.D0
      DO 34 I4=1,3
      VEQQ= VEQQ + QUAM(I1,I2)
     &    *TAC4(I1,I2,I3,I4,RLM,RLM2,RLM4,RLM9)*QUAC(JC,I3,I4)/9.D0
 34   CONTINUE
 33   CONTINUE
 32   CONTINUE
 31   CONTINUE
      DO 35 I=1,3
 35   ECM(I)= ECMD(I) + ECMQ(I)
      DO 36 I1=1,3
      DO 37 I2=1,3
      VIM= VIM - 0.5D0*ECM(I1)*ALC(JC,I1,I2)*ECM(I2)
 37   CONTINUE
 36   CONTINUE
C*********************************************************************
C*CALCULATION OF LENNARD-JONES (ATOM-ATOM) QUANTUM CONTRIBUTION
C*********************************************************************
      DO 40 I=1,5
      XI= RMXx*HX(I) + RMXy*HY(I) + RMXz*HZ(I)
      YI= RMYx*HX(I) + RMYy*HY(I) + RMYz*HZ(I)
      ZI= RMZx*HX(I) + RMZy*HY(I) + RMZz*HZ(I)
C*********************************************************************
      XIL= -RLM(1) - XI
      YIL= -RLM(2) - YI
      ZIL= -RLM(3) - ZI
      RIL2= XIL**2 + YIL**2 + ZIL**2
      RIL4= RIL2*RIL2
      RIL6= RIL4*RIL2
      RIL12= RIL6*RIL6
      VQUAN= VQUAN + (C12(I,1)/RIL12) - (C6(I,1)/RIL6)
 40   CONTINUE
C*********************************************************************
 120  CONTINUE
C************************* END LOOP ON THE CARBON ATOMS
C*********************************************************************
      DO 18 I1=1,3
```

```
   DO 17 I2=1,3
 17 VIG= VIG - 0.5D0*ECG(I1)*ALM(I1,I2)*ECG(I2)
 18 CONTINUE
C*******************************************************************
C* TOTAL POTENTIAL ENERGY
C*******************************************************************
   POT=VEDQ+VEQQ+VIM+VIG+VQUAN
C*******************************************************************
C-CHANGE OF UNIT OF VARIOUS CONTRIBUTIONS from a.u. to meV
C*******************************************************************
   VEDQ=VEDQ*UAMEV
   VEQQ=VEQQ*UAMEV
   VIG=VIG*UAMEV
   VIM=VIM*UAMEV
   VQUAN=VQUAN*UAMEV
   POT=POT*UAMEV
C*******************************************************************
C    WRITE(*,61)QUI0,VIM,VQUAN,POT
   WRITE(*,62)R0,QUI0,VIM,VQUAN,POT
 61 FORMAT(2X,F7.1,3(4X,D15.6))
 62 FORMAT(2X,F7.3,2X,F7.1,3(4X,D15.6))
C*******************************************************************
C*MINIMIZATION OF NH3-C60 POTENTIAL WITH RESPECT TO
C*CHOSEN VARIABLES ( Z0  AND/OR  PHI0  AND/OR  QUI0 )
C*   MINI = 0 ========> NO MINIMIZATION
C*   MINI = 1 ========> MINIMIZATION
C*******************************************************************
   IF(MINI.EQ.0)GO TO 66
   IF(POT.GT.POTMIN)GO TO 66
   POTMIN= POT
   R0MIN= R0
   QUIMIN= QUI0
   VEDQM= VEDQ
   VEQQM= VEQQ
   VIMM= VIM
   VIGM= VIG
   VQUANM= VQUAN
C*******************************************************************
 66 CONTINUE
  4 CONTINUE
C    WRITE(*,75)R0,QUI0,VIM,VQUAN,POT
C    WRITE(1,76)POT
C    WRITE(1,77)R0,POT
```

```
 75 FORMAT(5X,F9.4,5X,F9.2,5X,3(4X,D15.6))
 76 FORMAT(5X,D15.7)
 77 FORMAT(5X,F9.4,5X,F11.3)
  3 CONTINUE
C*********************************************************************
    IF(IMOL.EQ.1)GO TO 221
    IF(IB.LE.47)GO TO 220
    VINV(IB)= VINTNU2(IB) + POTMIN  + 127.839D0
    VINV(IBB)=VINV(IB)
    GO TO 221
220 VINV(IB)= VINTNU2(IB) - 180.169D0 + 127.839D0
    VINV(IBB)=VINV(IB)
221 CONTINUE
C*********************************************************************
C    WRITE(*,78)PHI0,TE0,QUIMIN,POTMIN
 78 FORMAT(3(5X,F9.2),5X,D15.6)
C    WRITE(*,72)PHI0,TE0,VIM,VQUAN,POT
C    WRITE(1,76)POTMIN
C    WRITE(1,92)PHI,TEVAR,POTMIN
 92 FORMAT(5X,F9.5,5X,F9.5,5X,F9.2)
C    WRITE(*,72)PHI0,TE0,R0MIN,QUIMIN,POTMIN
 72 FORMAT(4(4X,F9.3),8X,D15.6)
  2 CONTINUE
  1 CONTINUE
C    WRITE(*,73)IB,BETADEG,R0MIN,QUIMIN,POTMIN,VINTNU2(IB),VINV(IB)
 73 FORMAT(3X,I2,4X,F6.2,5X,F7.3,5X,F5.1,3(5X,D13.6))
 52 CONTINUE
C*********************************************************************
C*RECORDING OF  Vinversion
C*********************************************************************
    IF(IMOL.EQ.1)GO TO 223
    DO 55 INB=1,181
 55 WRITE(1,74)VINV(INB)
 74 FORMAT(F14.4)
223 CONTINUE
C*********************************************************************
C   CLOSE(1)
C*********************************************************************
    STOP
    END
C*********************************************************************
C
C
```

```
C*********************************************************************
C*CALCULATION OF COMPONENTS OF ACTION TENSORS OF ORDERS 2, 3, 4
C*********************************************************************
      FUNCTION TAC2(I,J,R,R2,R5)
      IMPLICIT DOUBLE PRECISION (A-H,O-Z)
      DIMENSION R(3)
      TAC2=(3.D0*R(I)*R(J)-R2*KR(I,J))/R5
      RETURN
      END
C*********************************************************************
      FUNCTION TAC3(I,J,K,R,R2,R7)
      IMPLICIT DOUBLE PRECISION (A-H,O-Z)
      DIMENSION R(3)
      TAC3=-(15.D0*R(I)*R(J)*R(K)-3.D0*R2*(R(I)*KR(J,K)
     &                  +R(J)*KR(I,K)+R(K)*KR(I,J)))/R7
      RETURN
      END
C*********************************************************************
      FUNCTION TAC4(I,J,K,L,R,R2,R4,R9)
      IMPLICIT DOUBLE PRECISION (A-H,O-Z)
      DIMENSION R(3)
      TAC4=(105.D0*R(I)*R(J)*R(K)*R(L)-15.D0*R2*(R(I)*R(J)*KR(K,L)
     &   +R(I)*R(K)*KR(J,L)+R(I)*R(L)*KR(J,K)+R(J)*R(K)*KR(I,L)
     &   +R(J)*R(L)*KR(I,K)+R(K)*R(L)*KR(I,J))
     &+3.D0*R4*(KR(I,J)*KR(K,L)+KR(I,K)*KR(J,L)+KR(I,L)*KR(J,K)))/R9
      RETURN
      END
C*********************************************************************
C* KRONECKER SYMBOL
C*********************************************************************
      FUNCTION KR(I,J)
      IF(I.EQ.J)GO TO 1
      KR=0
      GO TO 2
    1 KR=1
    2 RETURN
      END
C*********************************************************************
```

5.6.2. *Appendix 2: Expressions of the components of the dipole moment vector and its derivatives with respect to the normal vibrational coordinates*

Expressions [4.45] and [4.46] of section 4.3.7 and of Appendix A (section 4.7.1) of Chapter 4 of Volume 3 can be used to write the components of the dipole moment and its derivatives for a symmetric top molecule of type XY_3 with respect to the absolute reference frame as follows:

$$\text{Permanent moment: } \vec{\mu}_A^e = \mu^e \begin{pmatrix} \cos\varphi\sin\theta \\ \sin\varphi\sin\theta \\ \cos\theta \end{pmatrix}$$

$$\text{Mode } \nu_1 : \left(\frac{\partial\vec{\mu}_A}{\partial q_1}\right) = \left(\frac{\partial\mu_z}{\partial q_1}\right) \begin{pmatrix} \cos\varphi\sin\theta \\ \sin\varphi\sin\theta \\ \cos\theta \end{pmatrix}$$

$$\text{Mode } \nu_2 : \left(\frac{\partial\vec{\mu}_A}{\partial q_2}\right) = \left(\frac{\partial\mu_z}{\partial q_2}\right) \begin{pmatrix} \cos\varphi\sin\theta \\ \sin\varphi\sin\theta \\ \cos\theta \end{pmatrix}$$

$$\text{Mode } \nu_{3_a} : \left(\frac{\partial\vec{\mu}_A}{\partial q_{3_a}}\right) = \left(\frac{\partial\mu_x}{\partial q_{3_a}}\right) \begin{pmatrix} \cos\varphi\cos\theta\cos\chi - \sin\varphi\sin\chi \\ \sin\varphi\cos\theta\cos\chi + \cos\varphi\sin\chi \\ -\sin\theta\cos\chi \end{pmatrix}$$

$$\text{Mode } \nu_{3_b} : \left(\frac{\partial\vec{\mu}_A}{\partial q_{3_b}}\right) = \left(\frac{\partial\mu_y}{\partial q_{3_b}}\right) \begin{pmatrix} -\cos\varphi\cos\theta\sin\chi - \sin\varphi\cos\chi \\ -\sin\varphi\cos\theta\sin\chi + \cos\varphi\cos\chi \\ \sin\theta\sin\chi \end{pmatrix}$$

$$\text{Mode } \nu_{4_a} : \left(\frac{\partial\vec{\mu}_A}{\partial q_{4_a}}\right) = \left(\frac{\partial\mu_x}{\partial q_{4_a}}\right) \begin{pmatrix} \cos\varphi\cos\theta\cos\chi - \sin\varphi\sin\chi \\ \sin\varphi\cos\theta\cos\chi + \cos\varphi\sin\chi \\ -\sin\theta\cos\chi \end{pmatrix}$$

$$\text{Mode } \nu_{4_b} : \left(\frac{\partial\vec{\mu}_A}{\partial q_{4_b}}\right) = \left(\frac{\partial\mu_y}{\partial q_{4_b}}\right) \begin{pmatrix} -\cos\varphi\cos\theta\sin\chi - \sin\varphi\cos\chi \\ -\sin\varphi\cos\theta\sin\chi + \cos\varphi\cos\chi \\ \sin\theta\sin\chi \end{pmatrix}$$

For the NH_3 molecule:

$$\mu^e = 1.476 \text{ D}$$

$$\left(\frac{\partial \mu_z}{\partial q_1}\right)_0 = 7.83 \times 10^{-2} \text{ D}$$

$$\left(\frac{\partial \mu_z}{\partial q_2}\right)_0 = -41.17 \times 10^{-2} \text{ D}$$

$$\left(\frac{\partial \mu_x}{\partial q_{3a}}\right)_0 = -15.04 \times 10^{-2} \text{ D}$$

$$\left(\frac{\partial \mu_x}{\partial q_{4a}}\right)_0 = 16.06 \times 10^{-2} \text{ D}$$

$$\left(\frac{\partial \mu_y}{\partial q_{3b}}\right)_0 = 9.24 \times 10^{-2} \text{ D}$$

$$\left(\frac{\partial \mu_y}{\partial q_{4b}}\right)_0 = -20.71 \times 10^{-2} \text{ D}$$

6

Adsorption on a Graphite Substrate

The theoretical models presented in Volumes 1 and 2 of this set are applied to the construction of far infrared (orientational spectrum) and near-infrared spectra in the frequency ranges of vibrational modes of the symmetric top NH_3 molecule adsorbed on a graphite substrate. These vibrational modes can be subjected to frequency shifts and/or degeneracy lifting when the molecule is in this anisotropic environment. This can be used to determine the characteristics of this environment by comparing the spectrum of the nano-perturbed molecule to that of the isolated molecule.

6.1. Introduction

Observations in various cold regions of the interstellar medium, where the temperature nears -250°C (about 20 K), revealed the presence of several hundred molecules, spanning from the simplest ones, composed of several atoms, to complex molecules, among which are many organic molecules. These regions are also characterized by a high density of atoms and molecules ($n \geq 10^3$ cm^{-3}). Moreover, the signature at 2,175 Å in the observed extinction spectra of the interstellar medium is attributed to the presence of carbon grains that are small, rather spherical solids (graphite or diamond) or amorphous of dimension ~ 200 Å or of polycyclic aromatic hydrocarbon (PAH) molecules [DRA 88, MAT 89].

Chapter 5 of Volume 1 and Chapter 6 of Volume 2 [DAH 17, DAH 19a] presented the adsorption of diatomic and triatomic molecules on a graphite substrate (1000). This substrate is often used to model the surface of interstellar dust grains [FAR 00, REE 02, MOR 03, MOR 04a].

Moreover, several observations were conducted in the infrared spectral range, in search of the presence of small molecules in the interstellar dust clouds. In particular, the presence of ammonia was established in various regions with abundances of about ~5–30% compared to water, the most abundant element [WHI 96, LAC 98, CHI 00, DAR 01, GIB 01].

In this chapter, the theoretical models developed in Volumes 1 and 2 are applied to the case of the symmetric top NH_3 molecule adsorbed on the graphite substrate (1000), in order to determine the near and far infrared absorption spectra at very low temperature.

6.2. "NH_3 molecule–substrate" system interaction energy

6.2.1. *Description of the system*

The phenomenon of molecule adsorption on a graphite substrate was presented in Chapter 5 of Volume 1 and Chapter 6 of Volume 2. The objective was to determine the adsorption energies and the diffusion constants of the molecules on the substrate surface.

This chapter focuses on the calculation of near and far infrared absorption spectra of the molecule adsorbed on the substrate. For this purpose, it is important to first briefly recall the characteristics of the "NH_3–graphite" system and the expressions of the interaction energies governing them.

The graphite (1000) is a crystal composed of carbon atoms arranged in stacked planes of hexagonal structure, separated by a distance $L = 3.357$ Å, and alternatively shifted ones with respect to the others with a periodicity of 2.

To each plane p is associated a primitive cell, of parameter $\mathrm{a} = 2.460$ Å, containing two carbon atoms ($s = 1,2$) and primitive translational vectors $\vec{a}_1$ and $\vec{a}_2$ defined in the fixed reference frame $(O,\vec{X},\vec{Y},\vec{Z})$ associated with the substrate as:

$$\begin{aligned} \vec{a}_1 &= a \times \vec{X} \\ \vec{a}_2 &= \frac{a}{2} \times \vec{X} + \frac{a\sqrt{3}}{2} \times \vec{Y} \end{aligned} \qquad [6.1]$$

Figure 6.1 presents the geometric characteristics of the adsorption of the NH_3 molecule on the graphite substrate (1000), as well as the definition of the unit cell and its vectors.

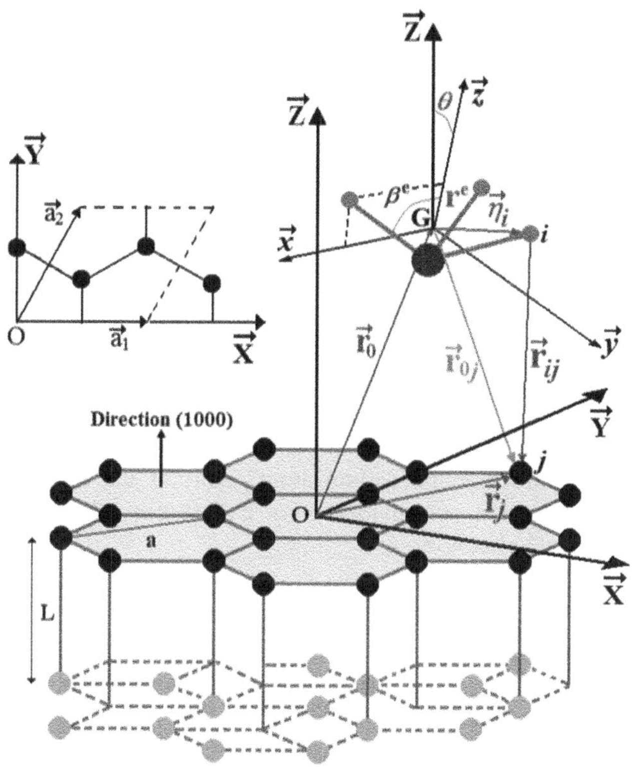

Figure 6.1. *Geometric characteristics of the adsorption of the NH_3 molecule on the graphite substrate (1000). The primitive cell and vectors are defined on the upper left. For a color version of this figure, see www.iste.co.uk/dahoo/infrared2.zip*

In this system, each carbon atom j is defined by its position vector $\vec{r}_j$:

$$\vec{r}_j = \frac{a}{2}\left(2l_1 + l_2 + s - \mathrm{mod}(\tfrac{p-1}{2})\right)\times\vec{X} + \frac{a}{2\sqrt{3}}\left(3l_2 + s - \mathrm{mod}(\tfrac{p-1}{2})\right)\times\vec{Y} - (p-1)L\times\vec{Z} \qquad [6.2]$$

where l_1 and l_2 are the relative integer numbers and the term $\mathrm{mod}(\frac{p-1}{2})$ means "the rest of the division of (p - 1) by 2".

Moreover, the covalent bonds between the carbon atoms of the same plane give rise to the distributions of aspheric atomic charges that generate an electric field outside the substrate. This phenomenon was modeled by placing a quadrupole moment $\vec{\Theta}^{c}$ on each atom in the direction (1000), whose non-zero matrix elements in the fixed reference frame are $\Theta_{XX}^{c} = \Theta_{YY}^{c} = -\frac{1}{2}\Theta^{c}$ and $\Theta_{ZZ}^{c} = \Theta^{c}$ and whose value was estimated as $\Theta^{c} = 1$ DÅ [VER 92].

On the contrary, the conduction property by plane of the graphite induces anisotropy of the dipole polarizability of carbon atoms whose non-zero matrix elements in the fixed reference frame are: $\alpha_{XX}^{c} = \alpha_{YY}^{c} = 1.444$ Å^3 and $\alpha_{ZZ}^{c} =$ 0.413 Å^3.

6.2.2. "NH$_3$ molecule–graphite substrate" interaction energy

The interaction potential energy V_{MS} between the molecule and the substrate is the sum, by "molecule–graphite carbon atom" pair, of the dispersion–repulsion, electrostatic and induction contributions. As a first step, the molecule and the substrate are assumed to be rigid.

Dispersion–repulsion contribution

In order to take into account the anisotropy of the polarizability of graphite (1000) planes, Carlos and Cole [CAR 80] proposed a modified form of Lennard-Jones atom–atom potential energy:

$$V_{\text{LJ}} = \sum_j \sum_i 4\varepsilon_{ij} \left\{ \left(\frac{\sigma_{ij}}{|\vec{r}_{ij}|}\right)^{12} \left[1+\gamma_R\left(1-\frac{6}{5}\cos^2\theta_{ij}\right)\right] - \left(\frac{\sigma_{ij}}{|\vec{r}_{ij}|}\right)^{6} \left[1+\gamma_A\left(1-\frac{3}{2}\cos^2\theta_{ij}\right)\right] \right\} \quad [6.3]$$

where γ_R and γ_A are coefficients associated with the anisotropy and θ_{ij} is the angle between the $\vec{Z}$ axis and the vector $\vec{r}_{ij}$ linking the atom i of the molecule and the atom j of carbon, which can be written as: $\vec{r}_{ij} = \vec{r}_{0j} - \vec{\eta}_i = \vec{r}_j - \vec{r}_0 - \vec{\eta}_i$. This highlights the dependency of the potential energy on various degrees of freedom of the system: internal vibrations of the substrate through $\vec{r}_j$, translational motion of the molecule through vector $\vec{r}_0$ and its internal vibrations and its orientational motion through vectors $\{\vec{\eta}_i\}$.

The values of the atom–atom parameters of expression [6.3] are presented in Table 6.1.

Atom–C	ε (meV)	σ (Å)	γ_A	γ_R
N–C	2.81	3.390	0.4	-1.05
H–C	2.27	2.965	0.4	-0.54

Table 6.1. *Lennard-Jones atom–atom parameters and parameters associated with graphite carbon atom anisotropy. Energy conversion factor 1 meV = 8.06554 cm^{-1}*

Electrostatic and induction contributions

These contributions describe the interaction between permanent multipole moments of molecule–carbon atom partners, and the interaction between these permanent moments and the electric field that they generate at the position of each partner (see Chapter 2):

$$V_{\mathrm{E}} = \sum_j \left\{ \begin{array}{l} -\dfrac{1}{3} L_p \times \vec{\mu}_0 . \vec{\nabla}\vec{\nabla}\vec{\nabla}\left(\dfrac{1}{\left|\vec{\mathbf{r}}_{0j}\right|}\right) : \vec{\vec{\Theta}}^{\mathrm{c}} \\ \qquad + \dfrac{1}{9} L_p \times \vec{\vec{\Theta}}_0 : \vec{\nabla}\vec{\nabla}\vec{\nabla}\vec{\nabla}\left(\dfrac{1}{\left|\vec{\mathbf{r}}_{0j}\right|}\right) : \vec{\vec{\Theta}}^{\mathrm{c}} \end{array} \right\} \qquad [6.4]$$

$$V_{\mathrm{I}} = -\frac{1}{2}\sum_j \vec{E}_{0\to j} . \vec{\vec{\alpha}}^{\mathrm{c}} . \vec{E}_{0\to j} - \frac{1}{2}\sum_{j,j'} \vec{E}_{j\to 0} . \vec{\vec{\alpha}}_0 . \vec{E}_{j'\to 0} \qquad [6.5]$$

where

$$\vec{E}_{0\to j} = L_p \times \left\{ \vec{\nabla}\vec{\nabla}\left(\frac{1}{\left|\vec{\mathbf{r}}_{0j}\right|}\right) . \vec{\mu}_0 - \frac{1}{3}\vec{\nabla}\vec{\nabla}\vec{\nabla}\left(\frac{1}{\left|\vec{\mathbf{r}}_{0j}\right|}\right) : \vec{\vec{\Theta}}_0 \right\} \qquad [6.6a]$$

$$\vec{E}_{j\to 0} = L_p \times \left\{ \frac{1}{3}\vec{\nabla}\vec{\nabla}\vec{\nabla}\left(\frac{1}{\left|\vec{\mathbf{r}}_{0j}\right|}\right) : \vec{\vec{\Theta}}^{\mathrm{c}} \right\} \qquad [6.6b]$$

are the electric fields generated, on the one hand, by the molecule at the site of atom j of the substrate, and, on the other hand, by the atom j at the site of the molecule.

In the above expressions, L_p is a screen factor associated with plane p: $L_1 = 1$ for the surface plane (p = 1) and $L_p = 2/(\varepsilon+1)$ for the internal planes, where $\varepsilon \approx 2.8$ is the graphite dielectric constant.

The non-zero components of the dipole moment vector and the non-zero elements of polarizability and quadrupole moment tensor matrices with respect to the reference frame $(G, \vec{x}, \vec{y}, \vec{z})$ attached to the equilibrium configuration of the molecule were presented in section 5.2.2 of Chapter 5.

On the other hand, the dependency of V_{MS} on internal coordinates $\{q_v\}$ of the molecule is contained in the vectors and tensors $\vec{\eta}_i$, $\vec{\mu}_0$, $\vec{\vec{\Theta}}_0$ and $\vec{\vec{\alpha}}_0$ (see Appendix 1 (section 2.5.1) of Chapter 2 and expressions [3.7] of Chapter 3). The dependency on orientational coordinates $\vec{\Omega}(\varphi,\theta,\chi)$ appears at the time of the change of the reference frame $(G,\vec{x},\vec{y},\vec{z}) \to (G,\vec{X},\vec{Y},\vec{Z})$.

6.3. Equilibrium configuration and potential energy surfaces

In practice, to provide a good convergence of the calculations of potential energy and prevent the edge effects that the molecule may be subjected to, a preliminary stage is the calculation of the potential energy depending on substrate dimension. This study allowed a balance between good convergence and calculation time. Hence, the optimized substrate is represented by a crystal formed of 10 planes of radius ~ 50 Å (~ 20a) each, therefore around 30,000 atoms.

Assuming that the NH_3 molecule and the substrate are rigid, the equilibrium configuration (position and orientation) results from the minimization of the potential energy V_{MS} with respect to Cartesian coordinates $\vec{r}_0(X,Y,Z)$ of the center of mass G of the molecule and its angular coordinates $\vec{\Omega}(\varphi,\theta,\chi)$.

The resulting minimum potential energy is $V_{MS}^{e} = -147$ meV, where the $\vec{z}$ axis of the molecule (C_3 symmetry axis) is collinear with the $\vec{Z}$ axis of the fixed reference frame, and therefore above the center of the hexagon ($X^e = 0$, $Y^e = 0$) at an approach distance $Z^e = 3.14$ Å (distance between the center of mass G of the molecule and the substrate surface) and in the top configuration ($\varphi^e = 0$, $\theta^e = 0$, $\chi^e = 0$). In this configuration, the rotational motion of the molecule about its axis (motions in φ and in χ) is nearly free, the energy barrier being ≤ 0.5 meV.

Two other specific positions of the molecule are worth highlighting:

– above the middle of a C–C bond, the local minimum energy is $V_{MS}^{C\text{-}C} = -145$ meV, for the same orientation ($\varphi = 0$, $\theta = 0$, $\chi = 0$) and an approach distance $Z = 3.19$ Å, the spinning motion being still quasi-free;

– above a carbon atom C, the minimum value is then $V_{MS}^{C} = -144$ meV, for the same orientation and the same approach distance, the spinning motion still quasi-free.

The potential energy surface $V_{MS}(X,Y)$ of the lateral motion of the molecule is then determined by minimizing $V_{MS}(\vec{r}_0, \vec{\Omega})$ with respect to the approach distance Z and the angular coordinates (φ, θ, χ). This potential energy surface is nearly flat, the energy variation being $\Delta V_{MS}(X,Y) \leq 3$ meV. The lateral translational motion of the molecule is therefore quasi-free.

Moreover, when the molecule is above the center of the hexagon in the "flying saucer" configuration ($\varphi = 0$, $\theta = \pi$, $\chi = 0$), the resulting value of the energy is $V_{MS} = -145$ meV for an approach distance Z = 3.25 Å, and a quasi-free spinning motion. The energy gap between the two configurations $\theta = 0$ and $\theta = \pi$ is ≤ 2 meV.

Moreover, in order to determine the energy barrier of the rotational motion in θ of the molecule, the energy value for ($\varphi = 0$, $\theta = \pi/2$, $\chi = \pi/3$) orientation was calculated and is $V_{MS} = -134$ meV for an approach distance Z = 3.31 Å. The rotational barrier is then 13 meV.

Assuming a motion of small-amplitude oscillations in θ in the vicinity of $\theta = 0$, a numerical resolution of the eigenvalue equation associated with this motion yields a frequency $\omega_\theta \approx 45$ cm^{-1}.

Finally, the variations of potential energy $V_{MS}(Z)$ depending on the approach distance for $\theta = 0$ and $\theta = \pi$, the other variables being at their equilibrium value, are quasi-identical. Figure 6.2 presents, for $\theta = 0$ (top configuration), the curve associated with $V_{MS}(Z)$ that characterizes the motion of an oscillator. The numerical resolution of the eigenvalue equation associated with such a motion of the molecule yields a frequency $\omega_Z = 155$ cm^{-1}.

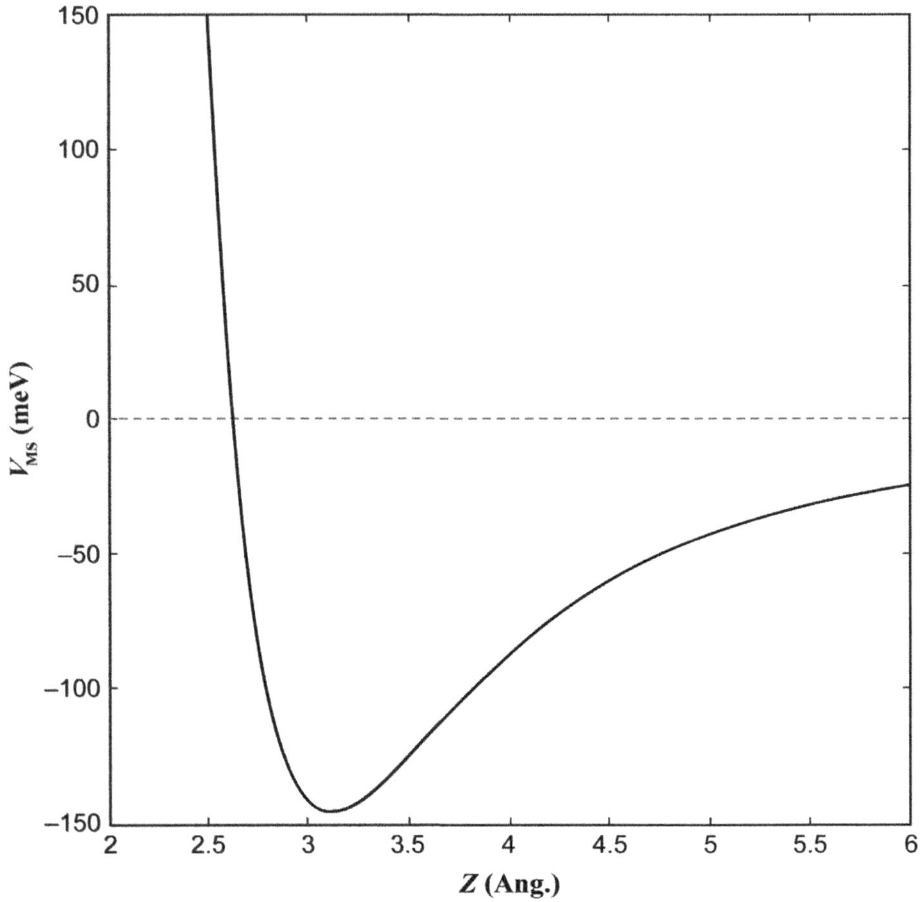

Figure 6.2. *Variations of potential energy depending on the approach distance, V_{MS} (Z), of the NH_3 molecule adsorbed on the graphite substrate (1000)*

6.3.1. *Adsorption energy*

The adsorption energy E_a of a molecule adsorbed on a surface, with a thermal energy $(\frac{n}{2}+1)k_BT$, is defined as the energy required to maintain this molecule in its equilibrium configuration on this surface. It is approximately given by:

$$E_a \approx \left(\frac{n}{2}+1\right)k_BT - V_{MS}^e - \sum_{s=1}^{n}\frac{\hbar\omega_s}{2}\coth\left(\frac{\hbar\omega_s}{2k_BT}\right) \qquad [6.7]$$

where n is the number of external degrees of freedom of the molecule (6 for a nonlinear polyatomic molecule) and ω_s (s = 1 to n) are the frequencies associated with the translational and orientational motions of the molecule in the vicinity of its equilibrium configuration.

For the NH_3 molecule, the values of frequencies ω_s and the adsorption energy E_a depending on the temperature are presented in Table 6.2.

ω_X	ω_Y	ω_Z	ω_θ	ω_ϕ	ω_χ	$E_a(10\,K)$	$E_a(30\,K)$	$E_a(195\,K)$
~0	~0	~155	~45	~0	~0	137	143	178

Table 6.2. *Frequencies (in cm^{-1}) of the translational and orientational motions and adsorption energy (in meV) of NH_3 adsorbed on the graphite surface*

At temperature T = 195 K, the value of the resulting adsorption energy $E_a = 178$ meV is in good agreement with the experimental value 172 meV, measured by Avgul and Kiselev [AVG 70].

6.4. Hamiltonian of the system

The Hamiltonian of the system composed of the adsorbed molecule, which is assumed to be in its fundamental electronic state, and the graphite substrate is:

$$H = H_{vib} + T_{rot} + \Delta T_{rot}^{vib} + T_{trans} + H_S + V_{MS} \qquad [6.8]$$

In this expression, H_{vib} is the Hamiltonian associated with the internal vibrational motions of the molecule in the gas phase, T_{rot} and T_{trans} are, respectively, the kinetic rotational and translational operators of the rigid molecule in the gas phase (expressions [4.7] and [4.8] of Chapter 4).

The much smaller term $\Delta T_{\text{rot}}^{\text{vib}}$ in expression [6.8] characterizes the vibration–rotation interaction due to the deformation of the molecule during its rotational motion in the gas phase. Its introduction is generally taken into account by the use of rotational constants and centrifugal distortion constants of the molecule, depending on its vibrational state.

Finally, H_S is the Hamiltonian associated with the vibrational motions of the atoms of the graphite substrate (1000), also termed lattice phonon modes.

This Hamiltonian was presented in Chapter 3 of this volume or, in greater detail, in Chapter 4 of Volume 1 [DAH 17], for face-centered cubic crystal structures of rare gases.

In this section, Figure 6.3 presents the normalized phonon density of states $\omega^2 \nu(\omega^2)$ of the graphite substrate (1000).

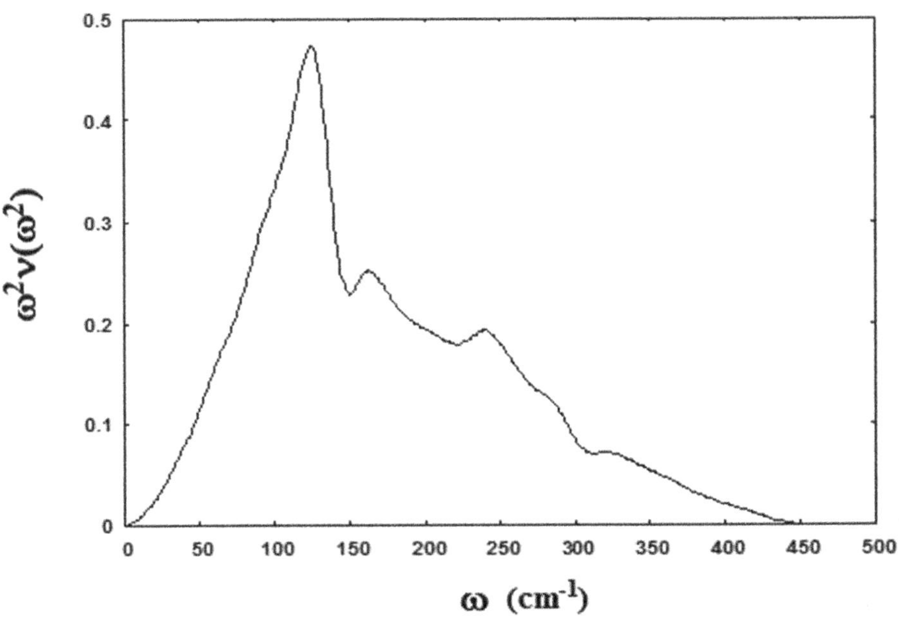

Figure 6.3. *Normalized phonon density of states $\omega^2 \nu(\omega^2)$ of the graphite substrate (1000)*

6.4.1. *Separation of movements*

The Born–Oppenheimer approximation (adiabatic approximation) can be used to separate the internal vibration modes $\{q_\nu\}$ of the molecule, high frequencies from the low-frequency modes, which are: vibrational motions of the center of mass G of instantaneous displacement vector $\vec{u}_0 = \vec{r}_0 - \vec{r}_0^{\,e}$ and of angular variable orientation (φ, θ, χ) of the molecule with respect to the fixed reference frame $(O, \vec{X}, \vec{Y}, \vec{Z})$, and the vibrational motion of the substrate (lattice) atoms of instantaneous displacement vectors $\vec{u}_j = \vec{r}_j - \vec{r}_j^{\,e}$.

After the detailed study of the molecule–substrate interaction potential energy surfaces, V_{MS} can be explicitly written as:

$$V_{\mathrm{MS}} = V_{\mathrm{MS}}^{\mathrm{e}} + V_{\mathrm{MS}}\left(\vec{u}_0, \{\vec{u}_j\}, \vec{\Omega}\right) + W_{\mathrm{M}}\left(\{q_\nu\}\right) + \Delta V_{\mathrm{MS}}\left(\vec{u}_0, \{\vec{u}_j\}, \vec{\Omega}, \{q_\nu\}\right) \qquad [6.9]$$

where $V_{\mathrm{MS}}^{\mathrm{e}}$ is the value of the energy associated with the equilibrium configuration of the molecule $V_{\mathrm{MS}}^{\mathrm{e}} = -147$ meV.

In expression [6.9], the term $W_{\mathrm{M}}\left(\{q_\nu\}\right)$ characterizes the vibrational dependency of the adsorbed molecule, in its equilibrium configuration, on the substrate assumed to be rigid. It is a perturbation term that can be written as a Taylor series expansion, similar to expression [4.10] of Chapter 4, and can induce frequency shifts and/or degeneracy lifting of the vibrational modes of the molecule.

As for the term $V_{\mathrm{MS}}\left(\vec{u}_0, \{\vec{u}_j\}, \vec{\Omega}\right)$, it characterizes the dependency of low-frequency modes: external modes (orientation and vibrations of the center of mass) of the molecule and vibrational modes of the atoms of the substrate. It can be written as:

$$V_{\mathrm{MS}}\left(\vec{u}_0, \{\vec{u}_j\}, \vec{\Omega}\right) = V_{\mathrm{MS}}^{\mathrm{e}}\left(\vec{\Omega}\right) + V_{\mathrm{MS}}^{\mathrm{e}}\left(\vec{u}_0, \{\vec{u}_j\}\right) + \Delta V_{\mathrm{MS}}\left(\vec{u}_0, \{\vec{u}_j\}, \vec{\Omega}\right) \qquad [6.10]$$

where $V_{\mathrm{MS}}^{\mathrm{e}}\left(\vec{\Omega}\right)$ characterizes the orientational dependency of the molecule in equilibrium position $\vec{r}_0^{\,e}$ on the rigid substrate, $V_{\mathrm{MS}}^{\mathrm{e}}\left(\vec{u}_0, \{\vec{u}_j\}\right)$ characterizes the dependency of vibrational modes of the center of mass of the molecule and of the substrate atoms, the molecule being at its equilibrium orientation, and

$\Delta V_{\text{MS}}\left(\vec{u}_0,\left\{\vec{u}_j\right\},\vec{\Omega}\right)$ is a term of dynamic coupling between the motion of the center of mass of the molecule and the substrate modes, constituting the "heat bath" and the orientational motion of the molecule. It is a perturbation term that allows the transfer of orientational energy to the "heat bath". Its introduction induces displacement and broadening of the orientational levels of the molecule.

Finally, in expression [6.9], the term $\Delta V_{\text{MS}}(\vec{u}_0,\left\{\vec{u}_j\right\},\vec{\Omega},\{q_\nu\})$ characterizes the dynamic coupling between the external motions of the molecule and the lattice modes, on the one hand, and the internal modes (vibrations) of the molecule, on the other hand. This last term may allow the transfer of vibrational energy. However, since the focus is here on the calculation of infrared spectra, this term is ignored in what follows.

6.4.2. *Renormalized Hamiltonians*

The separation of movements in the previous section allows us to write the total Hamiltonian in the form:

$$
\begin{aligned}
H &= \left[H_{\text{vib}} + W_{\text{M}}(\{q_\nu\})\right] + \left[T_{\text{rot}} + V_{\text{MS}}^{\text{e}}\left(\vec{\Omega}\right)\right] \\
&\quad + \left[T_{\text{trans}} + H_{\text{S}} + V_{\text{MS}}^{\text{e}}\left(\vec{u}_0,\left\{\vec{u}_j\right\}\right)\right] + \Delta V_{\text{MS}}\left(\vec{u}_0,\left\{\vec{u}_j\right\},\vec{\Omega}\right) \\
&= \tilde{H}_{\text{vib}} + \tilde{H}_{\text{orient}} + \tilde{H}_{\text{bath}} + \Delta V_{\text{MS}}\left(\vec{u}_0,\left\{\vec{u}_j\right\},\vec{\Omega}\right)
\end{aligned}
\qquad [6.11]
$$

where $\tilde{H}_{\text{vib}}$ and $\tilde{H}_{\text{orient}}$ are, respectively, the renormalized Hamiltonians of the internal vibration and orientational motions of the molecule, their sum $\tilde{H}_{\text{a}} = \tilde{H}_{\text{vib}} + \tilde{H}_{\text{orient}}$ characterizes the optically active system; and $\tilde{H}_{\text{bath}}$ is the renormalized Hamiltonian of the heat bath, including the vibrational motions of the center of mass of the molecule and the substrate atoms.

6.4.3. *Translational motions*

The analysis of the potential energy surface $V_{\text{MS}}^{\text{e}}\left(\vec{u}_0,\left\{\vec{u}_j\right\}\right)$ allows us to solve the equations of motion associated with the renormalized Hamiltonian $\tilde{H}_{\text{bath}}$ and determine the associated phonon density of states. Such a resolution was previously presented in section 4.5.6 of Chapter 4 of Volume 1 [DAH 17].

In the harmonic approximation, the coupling between the vibrational motion of the center of mass (translation) of the molecule and the substrate modes is then negligible. When the lateral translational motion of the molecule is quasi-free, the associated frequencies are $\omega_X \approx 0$ and $\omega_Y \approx 0$ (Table 6.2), and that associated with the approach translational motion (Figure 6.2) is $\omega_Z = 155 \text{ cm}^{-1}$. As for the substrate, the vibrational modes of the atoms are given by the phonon density of states $\omega^2 \nu(\omega^2)$ presented in Figure 6.3. It can be noted that the mode associated with the molecule is resonant, with the frequency being within the range of frequencies of the phonon density of the substrate.

6.4.4. *Vibrational motions*

6.4.4.1. *Vibrational modes ν_1, ν_3 and ν_4 of the NH_3 molecule*

As already mentioned, the potential energy term $W_\mathrm{M}(\{q_\nu\})$ reflecting the internal vibrational dependency of the molecule is considered to be a perturbation. It is developed as a Taylor series according to expression [4.10] of Chapter 4 and is introduced by the stationary perturbations method at the first order for the calculation of the frequency shift of a vibrational mode ν of the molecule:

$$\Delta\omega_\nu = \hbar^{-1}\left[\langle 0...1_\nu...0|W_\mathrm{M}(\{q_\nu\})|0...1_\nu...0\rangle - \langle 0...0...0|W_\mathrm{M}(\{q_\nu\})|0...0...0\rangle\right] \qquad [6.12]$$

$|0...0...0\rangle$ and $|0...1_\nu...0\rangle$ being the eigenvectors of the vibrational Hamiltonian H_vib of the molecule in the gas phase associated, respectively, with the fundamental state and with the first excited state of mode ν (see Appendix 4 (section 4.7.4) of Volume 3).

It should also be noted that when the molecule is adsorbed above the center of the hexagon, the top configuration $(\theta = 0)$ and the flying saucer configuration $(\theta = \pi)$ are quasi-equienergetic. The frequency shifts should therefore be calculated for these two configurations. The values obtained are presented in Table 6.3.

It can be noted that the frequency shifts are different for the two adsorption configurations. Moreover, the degeneracy of modes ν_3 and ν_4 is lifted only in the flying saucer configuration ($\theta = \pi$).

Vibrational modes	ω_v^{gas} (exp.) (cm^{-1})	$\Delta\omega_v$ (cm^{-1}) $\theta^e = 0$	$\Delta\omega_v$ (cm^{-1}) $\theta^e = \pi$
v_1	3,337	5.5	4.4
v_{3a}	3,448	5.9	3.3
v_{3b}		5.9	2.4
v_{4a}	1,627	- 2.7	2.2
v_{4b}		- 2.7	0.1

Table 6.3. *Vibrational frequency shifts of NH_3 adsorbed on a graphite substrate (1000) above the center of the hexagon, for $\theta = 0$ and $\theta = \pi$*

6.4.4.2. *Vibration–inversion mode v_2 of the NH_3 molecule*

In the gas phase, this mode was presented in detail in Chapter 4 of Volume 3, and its characteristics were recalled in the previous chapters of this volume. Figure 6.4 presents the shape of the double well potential and the energy levels and eigenvector solutions of the Schrödinger equation of the Hamiltonian associated with this mode.

An eigenvector $|v_2\alpha\rangle$ ($\alpha = s, a$; s for symmetric and a for antisymmetric) associated with level v_2 (for $v_2 = 0, 1$) can be written as a linear combination of eigenvectors $|v_2\lambda\rangle$ ($\lambda = L, R$; L for the left well and R for the right well):

$$\begin{aligned} |v_2 s\rangle &= \frac{1}{\sqrt{2}}|v_2 L\rangle + \frac{1}{\sqrt{2}}|v_2 R\rangle \\ |v_2 a\rangle &= \frac{1}{\sqrt{2}}|v_2 L\rangle - \frac{1}{\sqrt{2}}|v_2 R\rangle \end{aligned} \quad [6.13]$$

The selection rules are determined from the calculation of the transition elements of the normal coordinate q_2 of this vibration–inversion mode, hence $|\langle v_2\alpha|q_2|v_2'\alpha'\rangle|$. The resulting values and the associated experimental frequencies (see Table 4.9 of Volume 3) are presented in Table 6.4. Hence, in the near-infrared range, the only allowed transitions are: $|0s\rangle \to |1a\rangle$ and $|0a\rangle \to |1s\rangle$, which are shown in Figure 6.4.

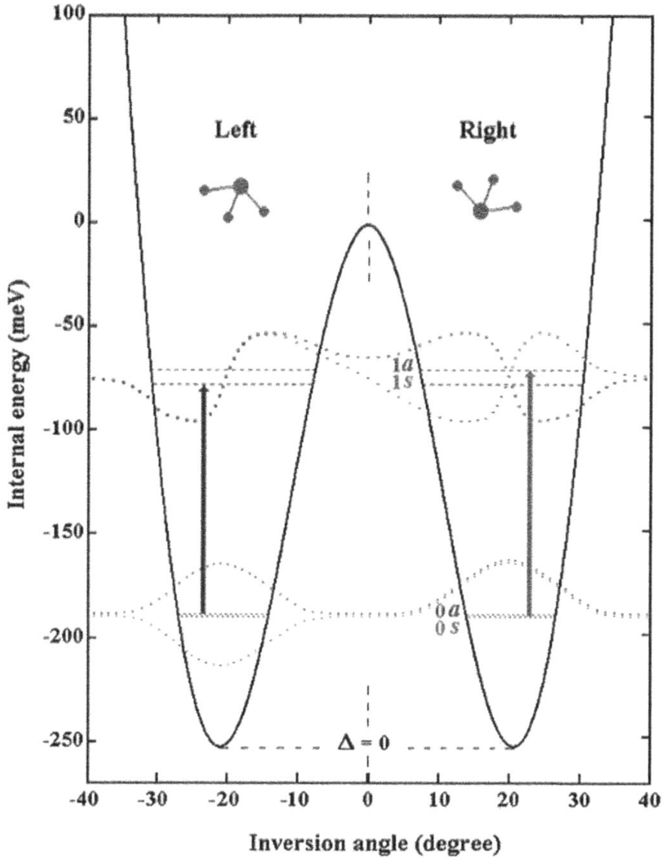

Figure 6.4. *Potential energy of the vibration–inversion mode of the NH_3 molecule in the gas phase, depending on the inversion angle, the angle between the N–H bond and the plane formed by the H atoms. For a color version of this figure, see www.iste.co.uk/dahoo/infrared2.zip*

As already seen, when the ammonia molecule is adsorbed on the graphite surface, above the center of the hexagon, the energy gap between the $\theta = 0$ and $\theta = \pi$ configurations is ≤ 2 meV. Therefore, this gap corresponds to the gap Δ between the two potential wells associated with the inversion motion of the molecule. The symmetry of the inversion potential function is then broken, as presented in Figure 6.5. Moreover, there is an increase in the height of the inversion barrier of 4 meV.

$\vert v_2\alpha\rangle$	$\vert v_2'\alpha'\rangle$	$\omega_2^{gas}(v_2\alpha \to v_2'\alpha')$	$\vert\langle v_2\alpha \vert q_2 \vert v_2'\alpha'\rangle\vert$
$\vert 0s\rangle$	$\vert 0a\rangle$	0.8	3.978
	$\vert 1s\rangle$	933.3	0
	$\vert 1a\rangle$	968.3	0.866
$\vert 0a\rangle$	$\vert 1s\rangle$	932.5	0.888
	$\vert 1a\rangle$	967.5	0
$\vert 1s\rangle$	$\vert 1a\rangle$	35.8	3.454

Table 6.4. *Experimental frequencies (cm^{-1}) and transition elements of the dimensionless normal coordinate q_2 between the vibration–inversion levels of the NH_3 molecule in the gas phase*

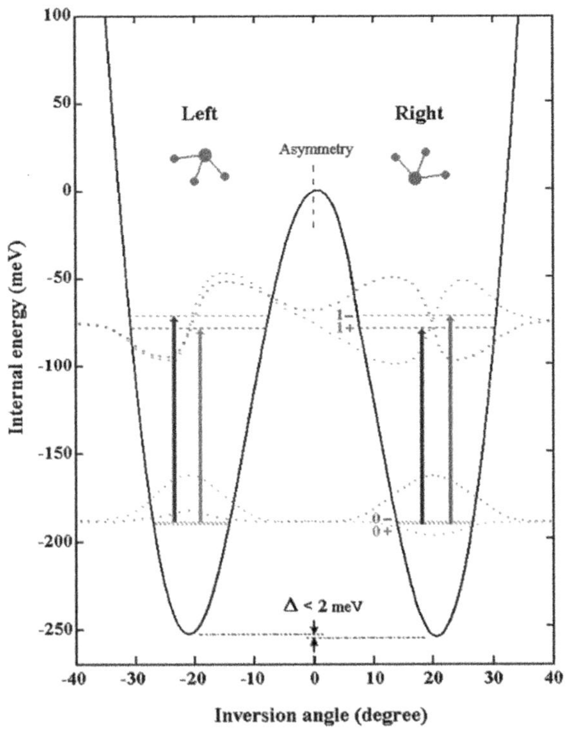

Figure 6.5. *Potential energy of the vibration–inversion mode of the NH_3 molecule adsorbed on the graphite surface (1000). Δ represents the energy gap between the potential wells. For a color version of this figure, see www.iste.co.uk/dahoo/infrared2.zip*

The Schrödinger eigenvalue equation associated with the vibration–inversion Hamiltonian is solved numerically. The resulting eigenvectors are no longer symmetric or antisymmetric. An eigenvector $|v_2\alpha\rangle$ ($v_2 = 0,1$ and $\alpha = +,-$) can be written as a linear combination of eigenvectors $|v_2\lambda\rangle$ ($\lambda = L$, R; L for the left well and R for the right well) as:

$$\begin{aligned} |0+\rangle &= 0,095|0L\rangle + 0,995|0R\rangle \\ |0-\rangle &= 0,995|0L\rangle - 0,095|0R\rangle \\ |1+\rangle &= 0,683|1L\rangle + 0,730|1R\rangle \\ |1-\rangle &= 0,730|1L\rangle - 0,683|1R\rangle \end{aligned} \qquad [6.14]$$

It is interesting to note that the wave functions of levels $v_2 = 0$ are strongly localized, and the probability of tunneling from one well to another is then low. In the case of levels $v_2 = 1$, the functions are weakly localized, where the associated coefficients are close to $1/\sqrt{2}$.

The frequency values and the transition elements of the normal coordinate q_2 of this mode are presented in Table 6.5. Hence, in the near-infrared range, unlike the molecule in the gas phase, the four transitions are allowed (with or without inversion): $|0+\rangle \to |1+\rangle$, $|0+\rangle \to |1-\rangle$, $|0-\rangle \to |1+\rangle$ and $|0-\rangle \to |1-\rangle$, which are shown in Figure 6.5.

$\|v_2\alpha\rangle$	$\|v_2'\alpha'\rangle$	$\omega_2(v_2\alpha \to v_2'\alpha')$	$\Delta\omega_2(v_2\alpha \to v_2'\alpha')$	$\|\langle v_2\alpha\|q_2\|v_2'\alpha'\rangle\|$
$\|0+\rangle$	$\|0-\rangle$	3.1	–	1.080
	$\|1+\rangle$	942.9	- 0.9	0.567
	$\|1-\rangle$	977.4	18.7	0.674
$\|0-\rangle$	$\|1+\rangle$	939.8	- 5.4	0.684
	$\|1-\rangle$	974.3	14.2	0.556
$\|1+\rangle$	$\|1-\rangle$	34.5	–	3.454

Table 6.5. *Frequencies, frequency shifts (cm^{-1}) and transition elements of the dimensionless normal coordinate q_2 between the vibration–inversion levels of the NH_3 molecule adsorbed on a graphite substrate (1000)*

Finally, similar to the modes ν_1, ν_3 and ν_4, the term $W_{\mathrm{M}}(\{q_\nu\})$ containing the internal vibrational dependency of the molecule can be introduced to determine the frequency shifts of the four previous transitions. The resulting values are also presented in Table 6.5.

6.4.5. *Orientational motion*

The study of the orientational potential energy hypersurface $V_{\mathrm{MS}}^{\mathrm{e}}(\varphi,\theta,\chi)$ of the molecule adsorbed on the graphite surface shows that the rotational motion of the latter is rather perturbed.

In order to solve the Schrödinger eigenvalue equation associated with the renormalized orientational Hamiltonian $\tilde{H}_{\text{orient}}$

$$\tilde{H}_{\text{orient}}\left|jmk\right\rangle=\left[T_{\text{rot}}+V_{\mathrm{MC}}^{\mathrm{e}}(\varphi,\theta,\chi)\right]\left|jmk\right\rangle=E_{jmk}\left|jmk\right\rangle \tag{6.15}$$

– the approach is similar to that adopted in section 3.4.2 of Chapter 3 of this volume, namely:

– the numerical adjustment of the potential energy hypersurface $V_{\mathrm{MS}}^{\mathrm{e}}(\varphi,\theta,\chi)$ based on the elements of the rotation matrix $D_{p,q}^{l}(\varphi,\theta,\chi)$ (expression [3.20] of Chapter 3)

$$V_{\mathrm{MS}}(\varphi,\theta,\chi)=\sum_{l=0}^{l_{\max}}\sum_{\substack{p=-l\\q=-l}}^{+l}A_{p,q}^{l}\times D_{p,q}^{l}(\varphi,\theta,\chi) \tag{6.16}$$

– and the use of a strong perturbation method by writing the $\left|jmk\right\rangle$ vectors to be determined as linear combinations of the eigenvectors $\left|JMK\right\rangle$ of free rotation of the molecule in the gas phase (expression [3.22] of Chapter 3)

$$\left|jmk\right\rangle=\sum_{J,M,K}\Lambda_{JMK}^{jmk}\left|JMK\right\rangle \tag{6.17}$$

The fitting coefficients $A_{p,q}^{l}$ (see Appendix 2 (section 3.6.2) of Chapter 3) are calculated using the Gauss–Legendre quadrature method up to the order $l_{\max}=8$ (the influence of the higher orders is negligible) with $-l\le p\le +l$ and $-l\le q\le +l$. It is important to note that these coefficients can be real or complex, and that the values of number q are multiples of 3 (including 0), with the symmetry axis of the rigid molecule being of order 3.

The non-negligible values of coefficients $A^{l}_{p,q}$ obtained by the numerical adjustment are presented in Table 6.6.

p \ l	q	0	1	2	3	4	5	6
0	**0**	- 130.3	- 5.6	- 13.8	+ 3.7	- 1.3	+ 2.5	- 0.6
0	**+3**	--	--	--	+ 10.8	- 0.2	+ 1.6	- 1.3
0	**-3**	--	--	--	- 10.8	+ 0.2	- 1.6	+ 1.3
0	**+6**	--	--	--	--	--	--	- 2.5
0	**-6**	--	--	--	--	--	--	- 2.5

Table 6.6. *Non-negligible values of fitting coefficients A (meV) [6.16] of the orientational potential energy hypersurface of NH_3 adsorbed on the graphite surface*

Finally, the coefficients Λ^{jmk}_{JMK} of expression [6.17] are determined by solving the secular equation attached to the Hamiltonian $\tilde{H}_{\text{orient}}$ that is rewritten in the form of a matrix to be diagonalized, as presented in Chapter 3, equations [3.28] and [3.29].

According to the explanations presented in Chapter 3, the dimensions of this matrix depend on the maximum value taken by the number l in expression [6.16] and on the orientational energy levels involved in the spectral lines.

However, unlike the case of the NH_3 molecule trapped in the argon matrix (Chapter 3), in the present case, this matrix is sparse and cannot be arranged as independent submatrices. The matrix to be diagonalized has minimum dimensions 1771 x 1771.

Figure 6.6 presents the diagram of orientational energy levels obtained for the NH_3 molecule adsorbed on the graphite surface (1000). This figure also presents the possible transitions, non-negligible at temperature T = 10 K, where only the low energy levels are populated.

The results show that the obtained eigenvectors are strong mixtures of eigenvectors of free rotation of the molecule. Hence, the quantum numbers of each state are redefined according to the weights of the involved free rotational states. Moreover, the energy levels are subjected to splitting due to the degeneracy lifting of the magnetic quantum number m and spinning quantum number k.

Finally, the eigenvectors obtained will be used to determine the elements of the orientational transitions of the dipole moment, in order to build the infrared spectra.

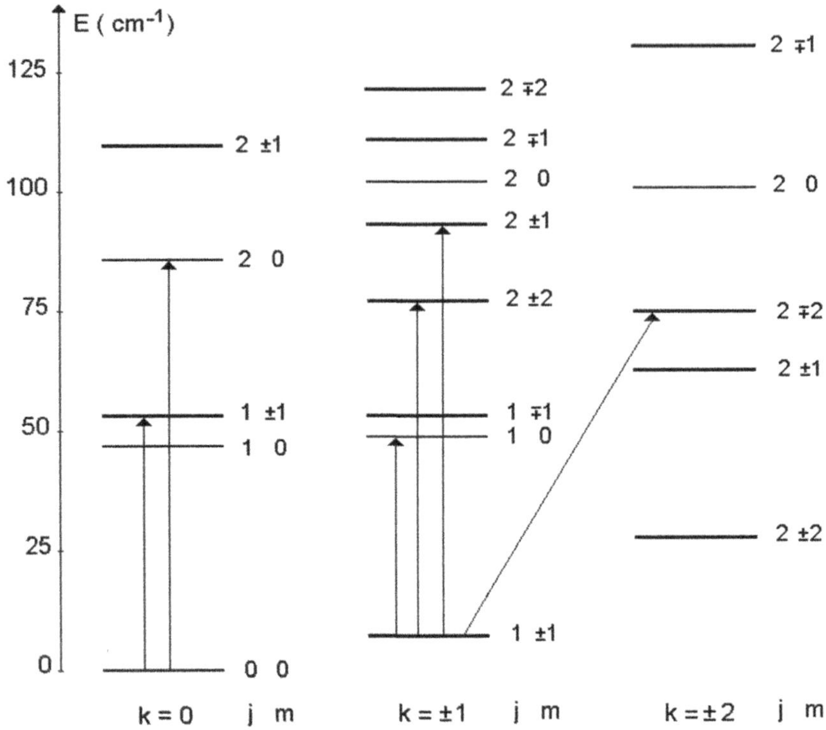

Figure 6.6. *Orientational energy level scheme of the NH_3 adsorbed on the graphite surface. It also presents the non-negligible allowed orientational transitions*

As already mentioned in Chapter 3, during the orientational motion, the molecule is not rigorously rigid. Then, using the adiabatic approximation, its rotational constants, on the one hand, and the expansion coefficients of the expression [6.16], on the other hand, depend on the vibrational state of the molecule. The Hamiltonian $\tilde{H}_{\text{orient}}$ can be written as $\tilde{H}^{\text{v}}_{\text{orient}}$, and the previous matrices must be built and diagonalized in each vibrational state. However, the modifications to the levels scheme remain negligible.

Finally, in the adiabatic approximation, the eigenenergies and eigenvectors of the Hamiltonian of the optical system can be written as:

$$\begin{aligned} &|\text{v}_\nu;jmk\rangle = |\text{v}_\nu\rangle \otimes |jmk\rangle \\ &E_{\text{v}_\nu;jmk} = E_{\text{v}_\nu} + E_{jmk} \end{aligned} \qquad [6.18]$$

Figures 6.7 and 6.8 present the vibration–orientation energy level schemes attached, respectively, to the symmetric stretching ν_1 and vibration–inversion ν_2 modes, on the one hand, and to the antisymmetric angular deformation ν_3 and ν_4 modes, on the other hand. These figures also present the possible (those with non-zero elements of transition of the dipole moment operator of the active molecule) rovibrational (or oriento-vibrational) transitions whose intensities are non-negligible. In fact, at temperature T = 10 K, only a few low energy levels are populated.

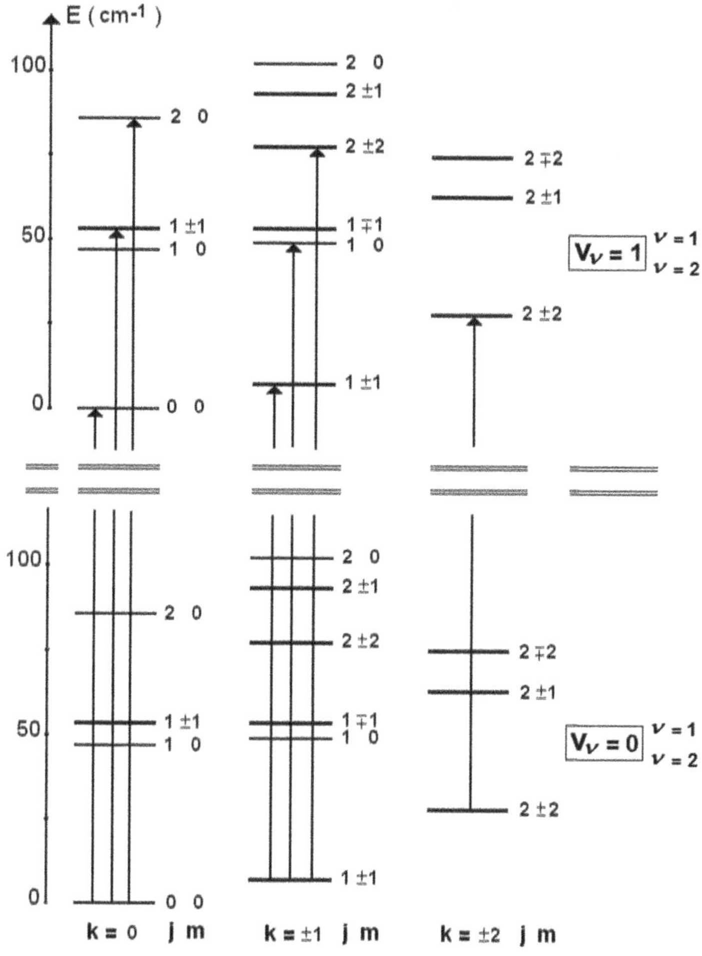

Figure 6.7. *Vibration–orientation energy level schemes of NH_3 adsorbed on the graphite surface, attached to modes ν_1 and ν_2. The figure also presents the non-negligible allowed transitions of the dipole moment operator*

These figures show that the near-infrared spectra associated with modes ν_1 and ν_2 are composed of pure vibration and vibration–orientation lines taking into account the selection rule $\Delta k = 0$, while those associated with modes ν_3 and ν_4 are only composed of vibration–orientation lines taking into account the selection rule $\Delta k = \pm 1$.

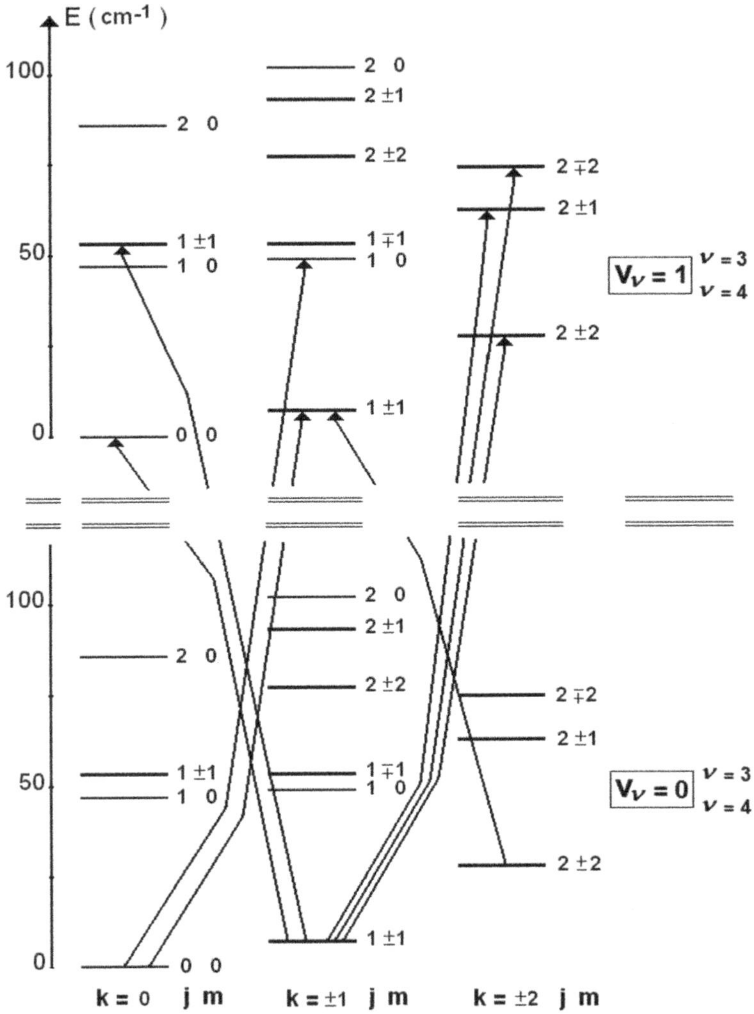

Figure 6.8. *Vibration–orientation energy level schemes of NH_3 adsorbed on the graphite surface, attached to modes ν_3 and ν_4. The figure also presents non-negligible allowed transitions of the dipole moment operator*

6.4.6. *Orientational motion – heat bath dynamic coupling*

In the harmonic description of the substrate and the adsorbed molecule, the perturbation Hamiltonian $\Delta V_{\text{MS}}\left(\vec{u}_0,\{\vec{u}_j\},\vec{\Omega}\right)$ can be written as a Taylor series expansion in the vicinity of the equilibrium positions of the center of mass of the molecule and of the atoms of the substrate, as a sum of two terms:

$$\begin{aligned}\Delta V_{\text{MS}}\left(\vec{u}_0,\{\vec{u}_j\},\vec{\Omega}\right) &= \sum_j \left[\vec{\nabla} V_{\text{MS}}(\vec{\Omega})\right](\vec{u}_j-\vec{u}_0) \\ &\quad + \frac{1}{2}\sum_{j,j'}\left[\vec{\nabla}\vec{\nabla} V_{\text{MS}}(\vec{\Omega})\right](\vec{u}_j-\vec{u}_0)(\vec{u}_{j'}-\vec{u}_0) \\ &= \Delta\tilde{H}_{1\text{p}}\left(\vec{\Omega},\vec{u}\right)+\Delta\tilde{H}_{2\text{p}}\left(\vec{\Omega},\vec{u}\right)\end{aligned} \qquad [6.19]$$

coupling the orientational motion of the molecule to the translational vibrations of its center of mass and the atoms in the substrate, the "heat bath". This may allow the transfer of orientational energy to the heat bath through a one-phonon and/or two-phonon process.

In expression [6.19], the angular terms characterize the first and second derivatives with respect to the coordinate distances between the molecule and the carbon atoms of the substrate, their decrease being rapid.

Hence, in order to determine the expressions of $\Delta\tilde{H}_{1\text{p}}\left(\vec{\Omega},\vec{u}\right)$ and $\Delta\tilde{H}_{2\text{p}}\left(\vec{\Omega},\vec{u}\right)$, the adopted hypothesis is that of the coupling between the orientation of the molecule and the translational vibration modes of its center of mass and of the nearest-neighbor carbon atoms (atoms belonging to the hexagon under the molecule). The energy transferred to the latter naturally dissipates in the substrate through the phonons of density of states $\omega^2\text{v}(\omega^2)$ (Figure 6.3).

The relatively long analytical calculation is used to write these coupling terms as products of functions depending on dynamic coordinates u_0 and u_j and of rotation matrix elements $D^l_{p,q}(\varphi,\theta,\chi)$ truncated at the order $l_1 = 2$ for the first term and $l_2 = 2$ for the second term. The resulting expressions are presented in Appendix 2 (section 6.7.2).

6.5. Infrared spectra of the NH_3 molecule adsorbed on the graphite substrate

The far- and near-infrared absorption coefficients, attached to a number $\mathcal{N}$ of optically active molecules per unit volume (here, per unit surface), without interactions between them, at a temperature T as well as the conditions (hypotheses and approximations) of their calculation, were presented in detail in Chapter 4 of Volume 1 [DAH 17].

Nevertheless, in order to present the infrared absorption spectra in the case of the adsorption of NH_3 on the graphite substrate, it is useful to recall below expressions [4.106]–[4.110] of Volume 1 adapted to the present case.

There are two steps in building the spectrum:

– the first step is to calculate the bar spectrum or integrated absorption coefficients of the active molecule at temperature T, following the renormalization of the system Hamiltonian and not take into consideration the dynamic coupling, considered a perturbation, between the various degrees of freedom of the total system. It offers information on the quality of the chosen basis of the theoretical model;

– the second step is to introduce, within the framework of the adiabatic approximation, the dynamic coupling between the orientational motion of the optically active molecule and the heat bath, $\Delta\tilde{H}_{1p}\left(\vec{\Omega},\vec{u}\right)$ at one phonon and $\Delta\tilde{H}_{2p}\left(\vec{\Omega},\vec{u}\right)$ at two phonons. Then, by applying the model developed in sections 4.6.1–4.6.3 of Volume 1, the displacements and broadenings of the orientational energy levels are calculated on the basis of temporal correlation functions and means of $\Delta\tilde{H}_{1p}\left(\vec{\Omega},\vec{u}\right)$ and $\Delta\tilde{H}_{2p}\left(\vec{\Omega},\vec{u}\right)$, which can then be used to determine the frequency shifts $\Delta^{(1)},\Delta^{(2)}$ and widths at mid-height $\Gamma^{(2)}$ of the lines associated with $i \to f$ transitions between vibration–orientation levels.

The term $\Delta^{(1)}$ is the frequency shift of the line due to the first-order term in the cumulant series expansion of the time evolution operator. At this order, the broadening is still zero.

On the contrary, the terms $\Delta^{(2)}$ and $\Gamma^{(2)}$ are the frequency shift and width at mid-height of this same line and they are due to the second-order term of the expansion of the time evolution operator.

Table 6.7 presents the frequencies, the frequency shifts $\Delta^{(1)},\Delta^{(2)}$ and the widths at mid-height $\Gamma^{(2)}$ associated with possible transitions between orientation energy levels of NH_3 adsorbed on the graphite surface, for a temperature $T = 10$ K. At this temperature, only the first three levels are significantly populated.

These results can be used to determine the bar spectra and the near- and far-infrared profile spectra.

$j_i m_i k_i \to j_f m_f k_f$	$\omega_{j_i m_i k_i \to j_f m_f k_f}$ (cm^{-1})	$\Delta^{(1)}$ (cm^{-1})	$\Delta^{(2)}$ (cm^{-1})	$\Gamma^{(2)}$ (cm^{-1})
$0\ 0\ 0 \to 0\ 0\ 0$	0	0	0	0.1
$0\ 0\ 0 \to 1\ \pm1\ \pm1$	7.4	-0.5	0.4	0.1
$0\ 0\ 0 \to 1\ 0\ \pm1$	49.2	-4.2	2.1	1.9
$0\ 0\ 0 \to 1\ \pm1\ 0$	54.5	-3.7	2.3	1.0
$0\ 0\ 0 \to 2\ 0\ 0$	86.4	-5.2	2.6	0.4
$1\ \pm1\ \pm1 \to 0\ 0\ 0$	-7.4	0.5	-0.4	0.1
$1\ \pm1\ \pm1 \to 1\ \pm1\ \pm1$	0	0	0	0.1
$1\ \pm1\ \pm1 \to 2\ \pm2\ \pm2$	21.8	-1.0	1.5	0.1
$1\ \pm1\ \pm1 \to 1\ 0\ \pm1$	41.8	-4.7	1.7	1.9
$1\ \pm1\ \pm1 \to 1\ \pm1\ 0$	47.1	-3.2	1.9	1.0
$1\ \pm1\ \pm1 \to 2\ \pm1\ \pm2$	54.3	-5.7	2.2	0.3
$1\ \pm1\ \pm1 \to 2\ \mp2\ \pm2$	67.5	-8.7	2.2	0.3
$1\ \pm1\ \pm1 \to 2\ \pm2\ \pm1$	70.4	-4.2	2.2	1.9
$2\ \pm2\ \pm2 \to 1\ \pm1\ \pm1$	-21.8	1.0	-1.5	0.1
$2\pm2\pm2 \to 2\pm2\pm2$	0	0	0	0.1

Table 6.7. *Frequencies, frequency shifts and widths at mid-height associated with the orientational transitions of NH_3 adsorbed on the graphite surface at temperature $T = 10$ K*

6.5.1. *Far-infrared spectrum*

When the spectral transitions involve only orientational states, $j_i m_i k_i \to j_f m_f k_f$ (or $i \to f$) of the fundamental vibrational state, the integrated absorption coefficient is given by:

$$I^0(\omega) = \frac{8\pi^2 \mathcal{N} \omega}{3hc} \sum_{\substack{j_i m_i k_i \\ j_f m_f k_f}} \frac{e^{-E_{j_i m_i k_i}/k_B T} - e^{-E_{j_f m_f k_f}/k_B T}}{Z} \times \left| \left\langle j_i m_i k_i \left| \vec{\mu}_A^e \right| j_f m_f k_f \right\rangle \right|^2 \times \delta(\omega - \omega_{j_i m_i k_i \to j_f m_f k_f}) \quad [6.20]$$

where Z is the canonical partition function of the orientational states, $\omega_{j_i m_i k_i \to j_f m_f k_f} = \hbar^{-1}(E_{j_f m_f k_f} - E_{j_i m_i k_i})$ is the frequency associated with the transition and $\left\langle j_i m_i k_i \left| \vec{\mu}_A^e \right| j_f m_f k_f \right\rangle$ is the dipole transition moment. Figure 6.6 presents the energy levels E_{jmk} and the allowed transitions of a non-negligible transition moment.

After introduction of the dynamic coupling between orientation and heat bath, the far-infrared profile spectrum, in the hypothesis of the Lorentzian model, can be obtained from the expression:

$$I(\omega) = \frac{8\pi^2 \mathcal{N} \omega}{3hc} \sum_{\substack{j_i m_i k_i \\ j_f m_f k_f}} \frac{e^{-E_{j_i m_i k_i}/k_B T} - e^{-E_{j_f m_f k_f}/k_B T}}{Z} \times \left| \left\langle j_i m_i k_i \left| \vec{\mu}_A^e \right| j_f m_f k_f \right\rangle \right|^2 \times \frac{\Gamma^{(2)}}{(\omega - \omega_{j_i m_i k_i \to j_f m_f k_f} - \Delta^{(1)} - \Delta^{(2)})^2 + (\Gamma^{(2)})^2} \quad [6.21]$$

Table 6.8 presents, for $T = 10$ K, the far-infrared spectral lines of non-negligible intensity I^0 (in %), their frequency as well as their frequency shifts $\Delta^{(1)}, \Delta^{(2)}$ and width at mid-height $\Gamma^{(2)}$ obtained after the introduction of the dynamic coupling between the orientational motion of the active molecule and the "heat bath" phonon modes.

$j_i m_i k_i \to j_f m_f k_f$	$\omega_{j_i m_i k_i \to j_f m_f k_f}$ (cm^{-1})	I^0 (%)	$\Delta^{(1)}$ (cm^{-1})	$\Delta^{(2)}$ (cm^{-1})	$\Gamma^{(2)}$ (cm^{-1})
0 0 0 → 1 ±1 0	54.5	100	-3.7	2.3	1.0
0 0 0 → 2 0 0	86.4	15.3	-5.2	2.6	0.4
1 ±1 ±1 → 1 0 ±1	41.8	25.9	-4.7	1.7	1.9
1 ±1 ±1 → 2 ∓2 ±2	67.5	9.4	-8.7	2.2	0.3
1 ±1 ±1 → 2 ±2 ±1	70.4	36.5	-4.2	2.2	1.9

Table 6.8. *Frequencies, intensities, frequency shifts and widths at mid-height of lines associated with orientational transitions of ammonia adsorbed on the graphite surface at temperature T = 10 K*

Finally, Figure 6.9 presents the bar spectrum and the far-infrared profile spectrum of NH_3 adsorbed on the graphite substrate at a temperature $T = 10$ K.

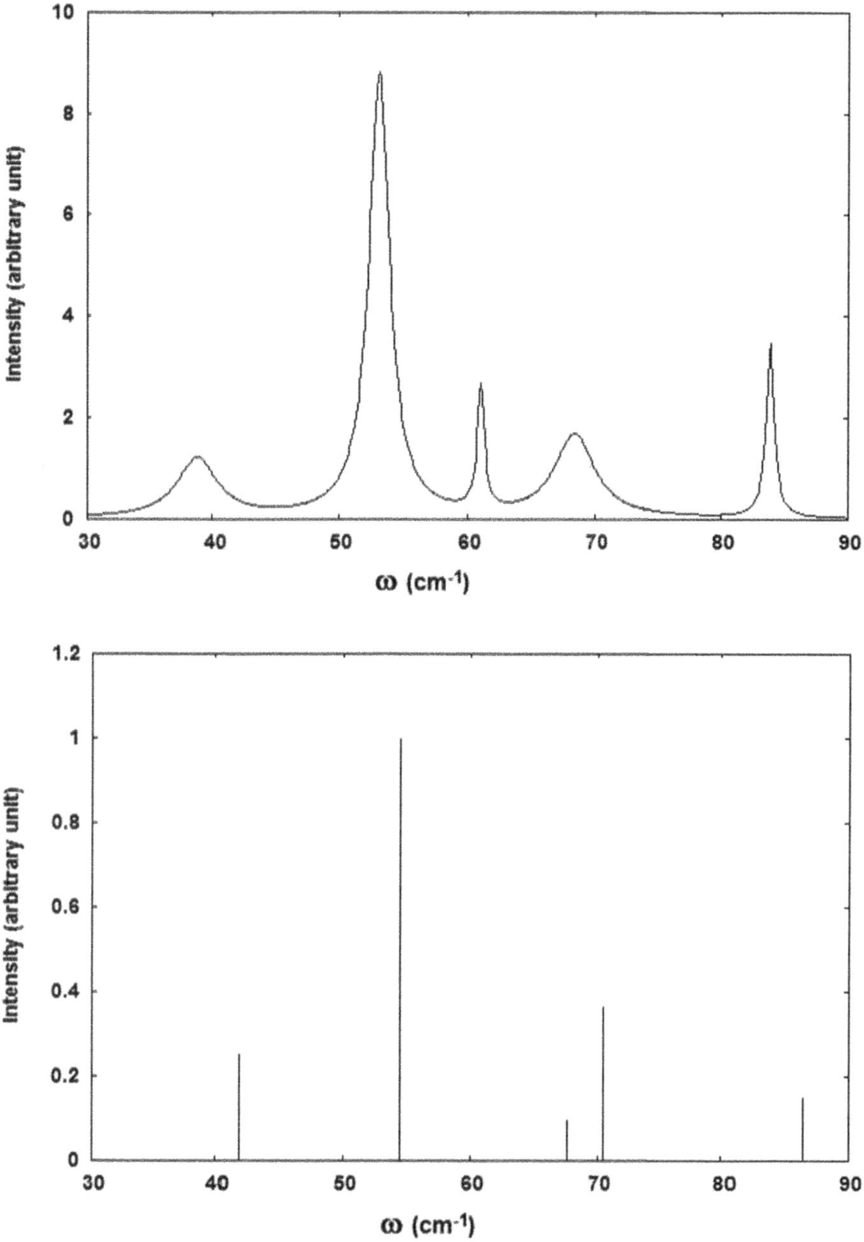

Figure 6.9. *Far-infrared bar and profile spectra of the ammonia adsorbed on the graphite surface at temperature T = 10 K*

6.5.2. *Near-infrared spectrum*

As previously noted, the molecule can have two orientational equilibrium configurations ($\theta^e = 0$ and $\theta^e = \pi$) that are quasi-equienergetic. The number $\mathcal{N}$ of optically active molecules, per unit surface and without interactions between them, can be written as: $\mathcal{N} = \mathcal{N}^{(0)} + \mathcal{N}^{(\pi)}$, with: $\mathcal{N}^{(0)} \approx \mathcal{N}^{(\pi)}$.

In the near-infrared range associated with the transition $0_\nu \rightarrow 1_\nu$ of the vibration mode ν of the molecule, the bar spectrum is given by the expression:

$$I_\nu^0(\omega) = \frac{8\pi^2\omega}{3hc} \sum_{\theta^e=0,\pi} \mathcal{N}^{(\theta^e)} \left|\left\langle 0_\nu \left| q_\nu^{(\theta^e)} \right| 1_\nu \right\rangle\right|^2 \sum_{\substack{j_i m_i k_i \\ j_f m_f k_f}} \frac{e^{-E_{0_\nu j_i m_i k_i}/k_B T} - e^{-E_{1_\nu j_f m_f k_f}/k_B T}}{Z^{(\theta^e)}} \times$$
$$\left|\left\langle j_i m_i k_i \left| \frac{d\vec{\mu}_A}{dq_\nu} \right| j_f m_f k_f \right\rangle\right|^2 \times \delta(\omega - \omega_{0_\nu j_i m_i k_i \rightarrow 1_\nu j_f m_f k_f}^{(\theta^e)}) \quad [6.22]$$

where $Z^{(\theta^e)}$, $\omega_{0_\nu j_i m_i k_i \rightarrow 1_\nu j_f m_f k_f}^{(\theta^e)} = \omega_\nu + \Delta\omega_\nu^{(\theta^e)} + \hbar^{-1}(E_{j_f m_f k_f} - E_{j_i m_i k_i})$ and $\left\langle 0_\nu \left| q_\nu^{(\theta^e)} \right| 1_\nu \right\rangle$ are, respectively, the canonic partition function of vibration–orientation states, the frequency associated with the considered vibration–orientation transition and the element of vibrational transition of the dimensionless normal coordinate associated with the vibrational mode ν, and they depend on the angular equilibrium position θ^e. Finally, $\left\langle j_i m_i k_i \left| \frac{d\vec{\mu}_A}{dq_\nu} \right| j_f m_f k_f \right\rangle$ is the moment of the considered orientational transition.

In the case of vibrational modes ν_1, ν_3 and ν_4, ω_ν are the vibration frequencies of the molecule in the gas phase ω_ν^{gas}, and $\Delta\omega_\nu^{(\theta^e)}$ are the frequency shifts due to the adsorption of the molecule. Their values are presented in Table 6.3.

Moreover, in the harmonic approximation of vibrational modes, the values of the transition elements $\left\langle 0_\nu \left| q_\nu^{(\theta^e)} \right| 1_\nu \right\rangle$ are equal to $1/\sqrt{2}$.

In the case of the vibration–inversion mode ν_2, the values of ω_ν, $\Delta\omega_\nu^{(\theta^e)}$ and $\left\langle 0_\nu \left| q_\nu^{(\theta^e)} \right| 1_\nu \right\rangle$ are presented in Table 6.5.

Figures 6.7 and 6.8 present, respectively, the vibration–orientation energy $E_{v_\nu jmk}$ level schemes and the allowed transitions of non-negligible orientational transition moments, associated with modes ν_1, ν_2, and ν_3, ν_4.

Finally, as previously, when the "orientation–heat bath" dynamic coupling is introduced the near-infrared profile spectrum, in the hypothesis of the Lorentzian model, can be obtained with the following expression:

$$I_\nu(\omega) = \frac{8\pi^2\omega}{3hc} \sum_{\theta^e=0,\pi} \mathcal{N}^{(\theta^e)} \left|\left\langle 0_\nu \left| q_\nu^{(\theta^e)} \right| 1_\nu \right\rangle\right|^2 \sum_{\substack{j_i m_i k_i \\ j_f m_f k_f}} \frac{e^{-E_{0_\nu j_i m_i k_i}/k_B T} - e^{-E_{1_\nu j_f m_f k_f}/k_B T}}{Z^{(\theta^e)}} \times$$
$$\left|\left\langle j_i m_i k_i \left| \frac{d\bar{\mu}_A}{dq_\nu} \right| j_f m_f k_f \right\rangle\right|^2 \times \qquad [6.23]$$
$$\frac{\Gamma^{(2)}}{(\omega - \omega^{(\theta^e)}_{0_\nu j_i m_i k_i \to 1_\nu j_f m_f k_f} - \Delta^{(1)} - \Delta^{(2)})^2 + (\Gamma^{(2)})^2}$$

6.5.2.1. *Symmetric stretching mode* ν_1

As previously noted, the equilibrium orientation θ^e of the optically active molecule adsorbed on the surface of the graphite substrate, the frequency of the symmetric stretching mode ν_1, which is non-degenerate in the gas phase, undergoes a different shifting. Table 6.3 indicates a frequency shifting gap between the orientation $\theta^e = 0$ and $\theta^e = \pi$ of 1.1 cm^{-1}. The near-infrared spectrum associated with this mode then presents two groups of distant vibration–orientation lines of 1.1 cm^{-1}, respectively.

Figure 6.10 presents the bar spectrum and the profile spectrum in the frequency range of mode ν_1 at temperature $T = 10$ K.

It is important to note that the spectral lines at ~ 3,342 cm^{-1} are due to pure vibrational transitions, while those at ~ 3,384, ~ 3,397 and ~ 3,412 cm^{-1} are due to vibration–orientation transitions and, as we can see, the intensity of the latter is low. Moreover, since the orientational transition elements of the dipole moment do not depend on the spinning angle χ of the molecule, the only allowed transitions are those with $\Delta k = 0$.

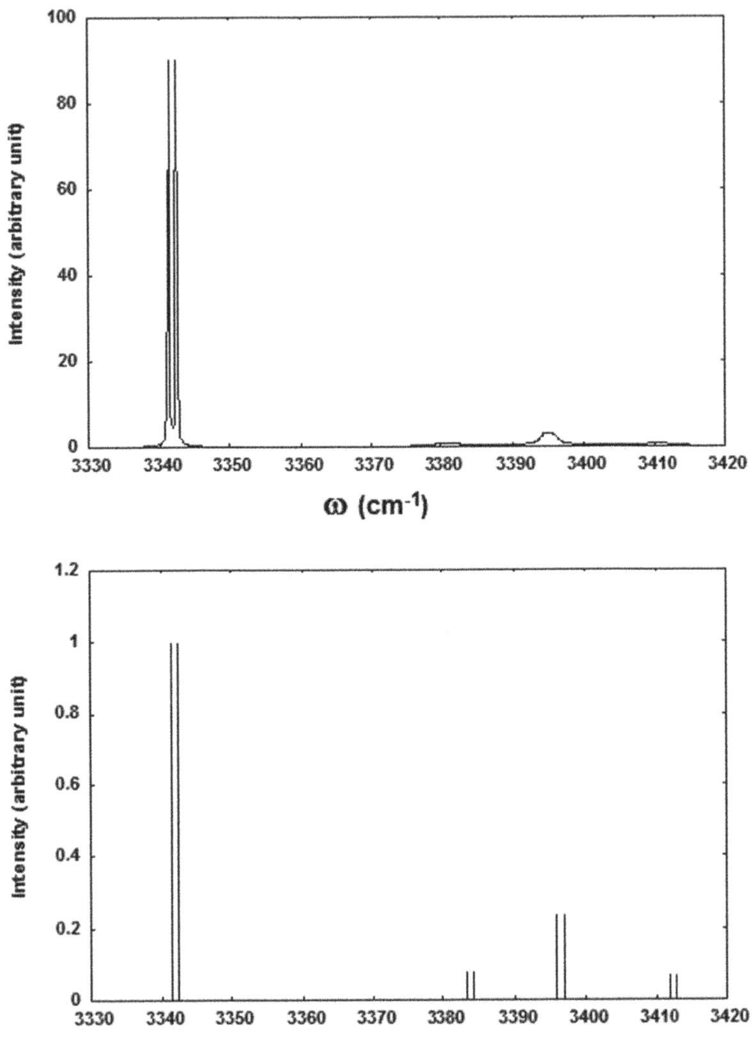

Figure 6.10. *Near-infrared bar and profile spectra in the range of mode ν_1 of NH_3 adsorbed on the graphite surface at temperature $T = 10$ K*

6.5.2.2. *Antisymmetric deformation modes ν_3 and ν_4*

In the gas phase, these two vibration modes are doubly degenerate. When the molecule is adsorbed on the substrate surface, as noted in Table 6.3, the two modes undergo frequency shifts without degeneracy lifting for the equilibrium orientation $\theta^e = 0$ and with degeneracy lifting for $\theta^e = \pi$. Each mode then has three vibration frequencies:

– 3,450.4 cm^{-1}, 3,451.3 cm^{-1} and 3,453.9 cm^{-1} in the frequency range of mode ν_3;

– 1,624.3 cm^{-1}, 1,627.1 cm^{-1} and 1,629.2 cm^{-1} in the frequency range of mode ν_4.

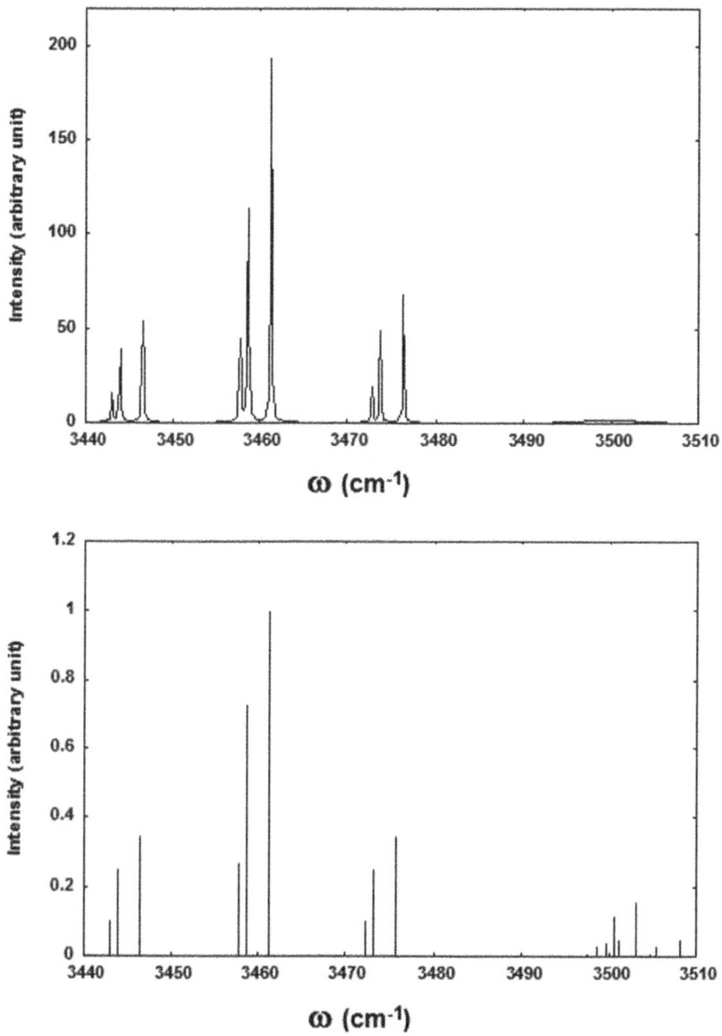

Figure 6.11. *Near-infrared bar and profile spectra in the range of mode ν_3 of NH_3 adsorbed on the graphite surface at temperature $T = 10$ K*

The near-infrared spectrum, in the frequency range of the two modes ν_3 and ν_4, is therefore composed of three groups of lines associated only with vibration–orientation transitions, as shown in Figures 6.11 and 6.12. Indeed, the orientational transition elements of the dipole moment of these modes depend on the spinning angle χ of the molecule, which forbids transitions with $\Delta k = 0$.

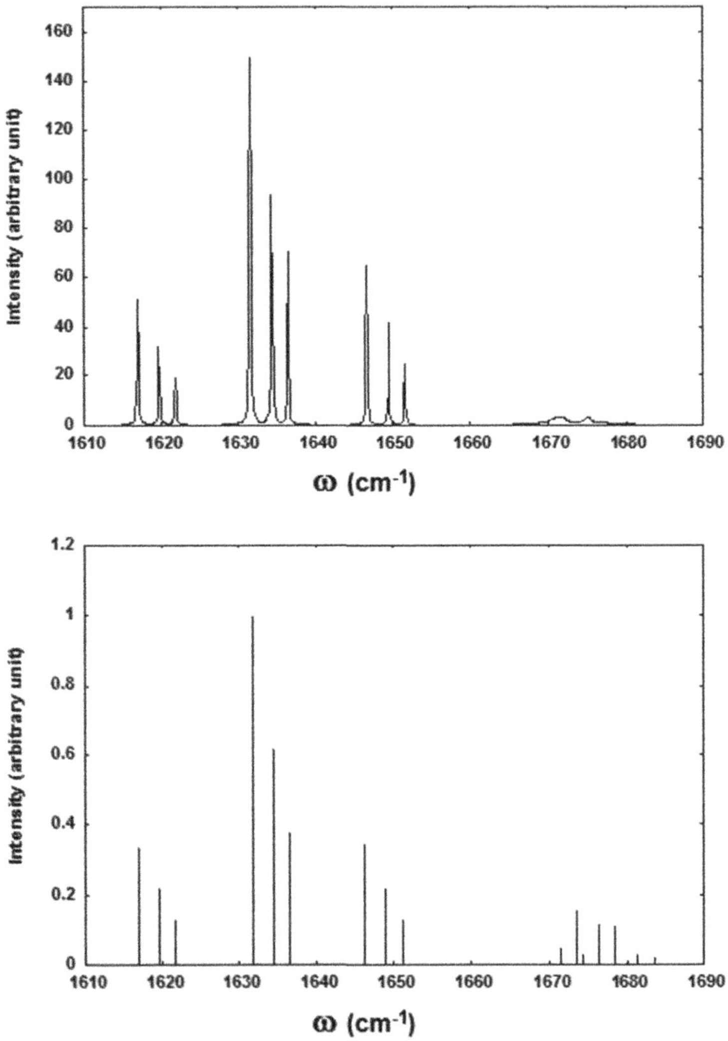

Figure 6.12. *Near-infrared bar and profile spectra in the range of mode ν_4 of NH_3 adsorbed on the graphite surface at temperature $T = 10$ K*

6.5.2.3. *Vibration–inversion mode* ν_2

As previously noted in the case of mode ν_1, the vibration mode ν_2 is governed by the selection rule $\Delta k = 0$. The near-infrared spectrum in the frequency range of this mode features lines associated with pure vibration and vibration–orientation transitions. Moreover, these transitions can occur with or without inversion of the molecule, because of the difference of ~ 2 meV and the appearance of asymmetry between the two vibration–inversion potential wells in Figure 6.5.

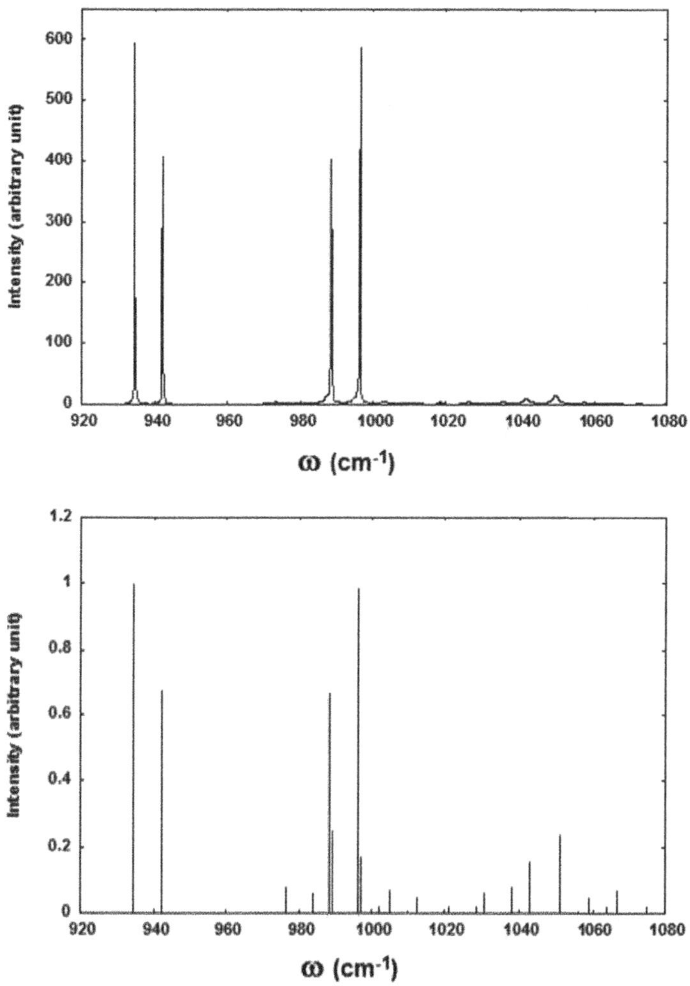

Figure 6.13. *Near-infrared bar and profile spectra in the range of mode* ν_2 *of* NH_3 *adsorbed on the graphite surface at temperature* $T = 10$ *K*

Figure 6.13 presents two lines at 942 and 988.5 cm^{-1} due to pure vibration transitions without inversion, while the two most intense lines at 934.4 and 996.1 cm^{-1} are due to vibration–inversion transitions. A similar group of four lines, less intense than the four previous ones, occurs, on the one hand, at 997 and 1,041 cm^{-1} due to vibration–orientation transitions and, on the other hand, at 989.5 and 1,049.5 cm^{-1} due to vibration–inversion–orientation transitions. Other low intensity lines are also present but, due to their broadening, they disappear in the profile spectrum.

6.6. Conclusion

A molecule adsorbed on a substrate surface is subjected to interaction forces with the atoms of the latter. Consequently, there is a perturbation of the vibrational and rotational energy levels of the molecule. The vibrational transition frequencies are shifted, while the rotational spectrum can be completely modified. In this chapter, the extended Lakhlifi–Dahoo theoretical model was used to calculate the modified spectrum with the approach described in Chapter 4 of Volume 1 and applied in Chapter 3 of this volume in the case of the ammonia molecule trapped in a nanocage of a rare gas matrix. Relying on the density matrix, an atom–atom potential, the Liouville space, the theorem of cumulants and the Born–Oppenheimer approximation applied to separate the slow modes, associated with the degrees of freedom attached to the phonons of the substrate, with the translational motion of the center of mass of the molecule and with the restricted rotation, from the vibration modes, line shifts and linewidths are calculated.

6.7. Appendices

6.7.1. *Appendix 1: FORTRAN program*

This program is used to calculate the interaction potential energy experienced by the NH_3 molecule adsorbed on the graphite substrate (1000).

```
C*******************************************************************
C*PROGRAM <==> nh3grap.f
C*THIS PROGRAM CALCULATES THE POTENTIAL ENERGY
C*OF AN AMMONIA NH3 MOLECULE
C*ADSORBED ON A GRAPHITE SURFACE (1000)
C*******************************************************************
      PARAMETER (MINI=1,IMOL=1,INVMAX=181)
C*******************************************************************
C     MINI=1 ----> NO MINIMIZATION/AT Z0 and/or PHI0 and/or QUI0
C     MINI=2 ----> MINIMIZATION     /AT Z0 and/or PHI0 and/or QUI0
```

```
C    IMOL=1 ----> RIGID MOLECULE
C    IMOL=2 ----> MOLECULE IN INVERSION MOTION
C*******************************************************************
      IMPLICIT DOUBLE PRECISION (A-H,O-Z)
      DIMENSION RLM(3),ECG(3),ECMD(3),ECMQ(3),ECM(3)
      DIMENSION CL(10),ALG(3,3),QUG(3,3)
      DIMENSION HX(5),HY(5),HZ(5),DIPM(3),ALM(3,3),QUM(3,3)
      DIMENSION C6(5,1),C12(5,1),GA(5,1),GR(5,1)
      DIMENSION XJOHN(INVMAX),VJOHN(INVMAX)
      DIMENSION VINTNU2(INVMAX),VPAIR(INVMAX),VIMPAIR(INVMAX)
      OPEN(1,FILE='INVNH3GR',FORM='FORMATTED')
C     OPEN(2,FILE='JOHNH3GR',FORM='FORMATTED')
      WRITE(*,100)
100   FORMAT(34X,'NH3/GRAPHITE'///)
C*******************************************************************
C*DEFINITION OF SEVERAL VALUES
C*******************************************************************
      PI=DACOS(-1.D0)
      S2=DSQRT(2.D0)
      S3=DSQRT(3.D0)
      DRPARA=0.1D0
      ABOHR=0.52917725D0
      UAMEV=27211.4D0
C*******************************************************************
C*GEOMETRIC AND ELECTRIC PARAMETERS OF GRAPHITE    IN a.u.
C*******************************************************************
      AA=2.46D0/ABOHR
      TOX=AA/2.D0
      TOY=AA/S3/2.D0
      DL=3.357D0/ABOHR
*******************************************************************
C
C*Take RADIUS = 15*AA , IL = 15 and NPLAN = 5 ===> 7497 atoms
C*Take RADIUS = 15*AA , IL = 15 and NPLAN = 3 ===> 4499 atoms
C*Take RADIUS = 15*AA , IL = 15 and NPLAN = 2 ===> 2998 atoms
C*Take RADIUS = 10*AA , IL = 10 and NPLAN = 3 ===> 2033 atoms
C*Take RADIUS = 10*AA , IL = 10 and NPLAN = 2 ===> 1354 atoms
C
C*  The difference of energy between the 1st case and the last case is of
C*  about  -3.2 meV  =====> Consequently it is possible
C*  to take the last case corresponding to
C*  1354 atoms  and add the value  Vabais = -3.2 meV to
```

```
C*  the calculated energy values====> a reduction of the
C*                              potential surface
C
C* Important ---> To obtain a graphene surface (sheet)
C*  it is sufficient to take  NPLAN = 1          =============
C
C*******************************************************************
C    IL=15
C    NPLAN=5
C    RADIUS=IL*AA+0.20D0
C    VABAIS=0.D0
     IL=10
     NPLAN=2
     RAYON=IL*AA+0.20D0
     VABAIS=-3.2D0
C*******************************************************************
     ALGPA= 2.7837D0
     ALGPE= 9.7432D0
     QUGR= 0.7435D0
     EPS=2.8D0
     CL(1)=1.D0
     CL(2)=2.D0/(EPS+1.D0)
     DO IC=3,10
     CL(IC)=CL(2)
     ENDDO
C*******************************************************************
C*COMPONENTS IN THE ABSOLUTE REFERENCE FRAME OF TENSORS:
C*POLARIZABILITY AND QUADRUPOLE MOMENT OF GRAPHITE
C*******************************************************************
     DO 13 I=1,3
     DO 13 J=1,3
     ALG(I,J)=0.D0
     QUG(I,J)=0.D0
13   CONTINUE
     ALG(1,1)=ALGPE
     ALG(2,2)=ALGPE
     ALG(3,3)=ALGPA
     QUG(1,1)=- QUGR/2.
     QUG(2,2)=- QUGR/2.
     QUG(3,3)= QUGR
C*******************************************************************
C* LENNARD-JONES POTENTIAL PARAMETERS          IN a.u.
```

```
C* C6 AND C12 (MOLECULE ATOM - GRAPHITE CARBON ATOM)
C* GA AND GR OF Carlos AND Cole
C*******************************************************************
   C6(1,1)=28.59D0
   C6(2,1)=0.D0
   C6(3,1)=10.21D0
   C6(4,1)=C6(3,1)
   C6(5,1)=C6(3,1)
   C12(1,1)=1976348.74D0
   C12(2,1)=0.D0
   C12(3,1)=312890.92D0
   C12(4,1)=C12(3,1)
   C12(5,1)=C12(3,1)
C*****************************
   GA(1,1)=0.4D0
   GA(2,1)=0.D0
   GA(3,1)=0.4D0
   GA(4,1)=0.4D0
   GA(5,1)=0.4D0
   GR(1,1)=-1.05D0
   GR(2,1)=0.D0
   GR(3,1)=-0.54D0
   GR(4,1)=-0.54D0
   GR(5,1)=-0.54D0
C*******************************************************************
C* GEOMETRIC AND ELECTRIC PARAMETERS OF THE MOLECULE (a.u.)
C*******************************************************************
C* ATOMIC MASSES (g/mole)
C*******************************************************************
   AMH= 1.00794D0
   AMN=14.00674D0
   RO1=3.D0*AMH/(AMN+3.D0*AMH)
C**********************************************
C*DIPOLE POLARIZABILITY ASSUMED CONSTANT
C*DURING THE INVERSION MOTION
C**********************************************
   APE=14.71D0
   APA=16.33D0
C*******************************************************************
   GOTO(24,25)IMOL
C   IMOL=1 --> RIGID MOLECULE
C   IMOL=2 --> MOLECULE IN INVERSION MOTION
```

```
C*******************************************************************
 24  INV=1
     GO TO 26
 25  INV=181
C*******************************************************************
 26  DO 81 IB=1,INV
C     WRITE(*,*)INV
     GOTO(27,28)IMOL
C**************************** RIGID MOLECULE ************************
 27  BETADEG=67.97D0
     BETA=67.97D0*PI/180.D0
     QNH=1.919206D0
     DIP=0.58074D0
     QUA=-2.17853D0
     GO TO 29
C********************MOLECULE IN INVERSION MOTION*****************
 28  BETADEG=90.D0+(IB-91)*0.5D0
     BETA=(90.D0+(IB-91)*0.5D0)*PI/180.D0
     VARNU2 =  PI/2.D0 - BETA
     XJOHN(IB)  =  BETA - PI/2.D0
C*******************************************************************
C*INTERNAL POTENTIAL ATTACHED TO THE VIBRATION–INVERSION MOTION OF NH3
C*IN GAS PHASE     in meV
C*Ref. J.D. Swalen and J.A. Ibers,J.Chem.Phys.vol.36,1914(1962)
C*******************************************************************
     VINTNU2(IB)=3728.D0*VARNU2**2+1546.D0*
    &                    (DEXP(-4.974D0*VARNU2**2)-1.D0)
C*******************************************************************
     QNH=1.8583112D0+0.15859D0*DABS((IB-91)*0.5D0*PI/180.D0)
     DIP=3.D0*QNH*DCOS(BETA)*(1.D0-5.16830D0/(QNH**3))
     QUA=1.745694D0*QNH**2*(1.144877D0*DCOS(BETA)*DCOS(BETA)-0.5D0)
C*******************************************************************
 29  CONTINUE
C********************************
C*CONTROL OUTPUT IN (a.u.)
C********************************
C     WRITE(*,444)BETADEG,QNH,DIP,QUA
444  FORMAT(5X,F10.6,10X,F10.6,10X,F10.6,10X,F10.6)
C*******************************************************************
C*POSITIONS OF ATOMS IN THE REFERENCE FRAME OF THE MOLECULE
C*******************************************************************
     QCB=QNH*DCOS(BETA)
     QSB=QNH*DSIN(BETA)
```

```
C***********************************************************************
     HX(1)=0.D0
     HY(1)=0.D0
     HZ(1)=-RO1*QCB
     HX(2)=0.D0
     HY(2)=0.D0
     HZ(2)=0.D0
     HX(3)=QSB
     HY(3)=0.D0
     HZ(3)=(1.D0-RO1)*QCB
     HX(4)=-QSB/2.D0
     HY(4)=S3*QSB/2.D0
     HZ(4)=HZ(3)
     HX(5)=HX(4)
     HY(5)=-HY(4)
     HZ(5)=HZ(3)
C***********************************************************************
C*POSITION OF THE CENTER OF MASS IN a.u. AND ORIENTATION OF THE MOLECULE
C***********************************************************************
     XPAS=TOX/20.D0
     YPAS=XPAS*S3
     DO 1 IX=1,1
     XM=(IX-1)*XPAS+TOX*1.D0
     X0=XM*ABOHR
     DO 2 IY=1,1
     YM=(IY-1)*YPAS+TOY*1.D0
     Y0=YM*ABOHR
C    WRITE(*,52)X0,Y0
 52  FORMAT(25X,'X0 = ',F6.3,'A',5X,'Y0 = ',F6.3,'A'//)
C***********************************************************************
C    WRITE(*,65)
 65  FORMAT(3X,'THETA',8X,'VEDQ (meV)',8X,'VEQQ (meV)',
    &8X,'VIM (meV)',9X,'VIG (meV)',8X,'VQUAN (meV)',
    &8X,'VTOT (meV)'//)
C***********************************************************************
     DO 3 IT=1,1
     TE0=(IT-1)*180.D0/36.D0
     TE=PI*(IT-1)/36.D0
     SB=DSIN(TE)
     CB=DCOS(TE)
     S2B=SB*SB
     C2B=CB*CB
C***********************************************************************
```

```
C     WRITE(*,59)
C     WRITE(*,60)
 59  FORMAT(2X,'Z0 (Ang)',5X,'VEDQ (meV)',8X,'VEQQ (meV)',
    &8X,'VIM (meV)',9X,'VIG (meV)',8X,'VQUAN (meV)',
    &8X,'VTOT (meV)'//)
 60  FORMAT(2X,'KHI (deg)',4X,'VEDQ (meV)',8X,'VEQQ (meV)',
    &8X,'VIM (meV)',9X,'VIG (meV)',8X,'VQUAN (meV)',
    &8X,'VTOT (meV)'//)
C*******************************************************************
C     POTMIN=0.D0
      DO 4 IP=1,1
      PHI0=(IP-1)*180.D0/18.D0
      PHI=PI*(IP-1)/18.D0
      SA=DSIN(PHI)
      CA=DCOS(PHI)
      S2A=SA*SA
      C2A=CA*CA
C*******************************************************************
C*COMPONENTS IN THE ABSOLUTE REFERENCE FRAME OF TENSORS: POLARIZABILITY,
C*DIPOLE MOMENT AND QUADRUPOLE MOMENT OF THE MOLECULE
C*******************************************************************
      ALM(1,1)=APE*(C2A*C2B+S2A)+APA*C2A*S2B
      ALM(1,2)=(APA-APE)*CA*SA*S2B
      ALM(1,3)=(APA-APE)*CA*SB*CB
      ALM(2,1)=ALM(1,2)
      ALM(2,2)=APE*(S2A*C2B+C2A)+APA*S2A*S2B
      ALM(2,3)=(APA-APE)*SA*SB*CB
      ALM(3,1)=ALM(1,3)
      ALM(3,2)=ALM(2,3)
      ALM(3,3)=APE*S2B+APA*C2B
C********************************************
      DIPM(1)=DIP*SB*CA
      DIPM(2)=DIP*SB*SA
      DIPM(3)=DIP*CB
C********************************************
      QUM(1,1)=QUA*(3.D0*S2B*C2A-1.D0)/2.D0
      QUM(1,2)=QUA*(3.D0*S2B*CA*SA)/2.D0
      QUM(1,3)=QUA*(3.D0*SB*CB*CA)/2.D0
      QUM(2,1)=QUM(1,2)
      QUM(2,2)=QUA*(3.D0*S2B*S2A-1.D0)/2.D0
      QUM(2,3)=QUA*(3.D0*SB*CB*SA)/2.D0
      QUM(3,1)=QUM(1,3)
      QUM(3,2)=QUM(2,3)
```

```
      QUM(3,3)=QUA*(3.D0*C2B-1.D0)/2.D0
C*******************************************************************
C     POTMIN=0.D0
      DO 5 IK=1,1
      QUI0=(IK-1)*180.D0/18.D0
      QUI=PI*(IK-1)/18.D0
      SC=DSIN(QUI)
      CC=DCOS(QUI)
      S2C=SC*SC
      C2C=CC*CC
C*******************************************************************
      POTMIN=0.D0
      DO 6 IZ=1,121
      Z0=3.1D0+(IZ-1)*0.003D0
      ZM=Z0/ABOHR
C*******************************************************************
C*CALCULATION OF VARIOUS CONTRIBUTIONS
C*   VEDQ  ==> ELECT DIP MOL - QUAD GRAP
C*   VEQQ  ==> ELECT QUAD MOL - QUAD GRAP
C*   VIG   ==> INDUC GRAP ---> MOL
C*   VIM   ==> INDUC MOL ---> GRAP
C*   VQUAN ==> QUANTUM
C*******************************************************************
C*ZEROING
C*******************************************************************
      VEDQ=0.D0
      VEQQ=0.D0
      VIG=0.D0
      VIM=0.D0
      VQUAN=0.D0
C*******************************************************************
C*ZEROING OF THE ELECTRIC FIELD INDUCED BY THE SUBSTRATE
C*ON THE POSITION OF THE MOLECULE
C*******************************************************************
      DO 19 I=1,3
19    ECG(I)=0.D0
C*******************************************************************
C*CONSTRUCTION OF SUBSTRATE PLANES
C*******************************************************************
      IPLS=0
      DO 21 KP=1,NPLAN
      ZL=-(KP-1)*DL
```

```
      IF(MOD(KP-1,2).EQ.0)THEN
      TX=0.D0
      TY=0.D0
      ELSE
      TX=+TOX
      TY=+TOY
      ENDIF
C*****************************************************************
C*POSITIONS OF CARBON ATOMS IN THE KP PLANE
C*   L1, L2 ----> THE CELL L(L1,L2)
C*   IS     ----> IS ATOM OF L CELL
C*****************************************************************
      DO 22 L1=-IL,IL
      DO 23 L2=-IL,IL
      DO 20 IS=1,2
C*****************************************************************
      XL=AA*((L1+L2/2.D0)+(IS-1))+TX
      YL=AA*((L2*S3/2.D0)+((IS-1)/S3))+TY
      RPARA=DSQRT(XL**2+YL**2)
      IF(RPARA.GE.RAYON)GO TO 20
C     WRITE(*,*)RPARA
      IPLS=IPLS+1
C*****************************************************************
C*COMPONENTS OF THE MOLECULE M - ATOM (L1,L2,KP,IS) DISTANCE VECTOR
C*****************************************************************
      RLM(1)=XM-XL
      RLM(2)=YM-YL
      RLM(3)=ZM-ZL
C     WRITE(*,*)RLM(1),RLM(2),RLM(3)
C*****************************************************************
      RLM2=RLM(1)**2+RLM(2)**2+RLM(3)**2
      RLM1=DSQRT(RLM2)
      RLM4=RLM2*RLM2
      RLM5=RLM4*RLM1
      RLM7=RLM5*RLM2
      RLM9=RLM7*RLM2
C*****************************************************************
C*CALCULATION OF ELECTROSTATIC AND INDUCTION CONTRIBUTIONS
C*****************************************************************
C*ZEROING OF THE ELECTRIC FIELD INDUCED BY THE MOLECULE
C*ON THE POSITION OF THE CONSIDERED ATOM  ECM = ECMD + ECMQ
C*****************************************************************
      DO 30 I=1,3
```

```
      ECMD(I)=0.D0
      ECMQ(I)=0.D0
30    CONTINUE
C*******************************************************************
      DO 31 I1=1,3
      DO 32 I2=1,3
      ECMD(I1)=ECMD(I1)-TAC2(I1,I2,RLM,RLM2,RLM5)*DIPM(I2)
      DO 33 I3=1,3
      TAC=TAC3(I1,I2,I3,RLM,RLM2,RLM7)
      ECMQ(I1)=ECMQ(I1)-TAC*QUM(I2,I3)/3.D0
      ECG(I1)=ECG(I1)-CL(KP)*TAC*QUG(I2,I3)/3.D0
      VEDQ=VEDQ+CL(KP)*DIPM(I1)*TAC*QUG(I2,I3)/3.D0
      DO 34 I4=1,3
      VEQQ=VEQQ+CL(KP)*QUM(I1,I2)
     &      *TAC4(I1,I2,I3,I4,RLM,RLM2,RLM4,RLM9)*QUG(I3,I4)/9.D0
34    CONTINUE
33    CONTINUE
32    CONTINUE
31    CONTINUE
      DO 35 I=1,3
35    ECM(I)=(ECMD(I)+ECMQ(I))*CL(KP)
      DO 36 I1=1,3
      DO 37 I2=1,3
      VIM=VIM-0.5D0*ECM(I1)*ALG(I1,I2)*ECM(I2)
37    CONTINUE
36    CONTINUE
C*******************************************************************
C*CALCULATION OF THE LENNARD-JONES(ATOM-ATOM) QUANTUM CONTRIBUTION
C*******************************************************************
      DO 40 I=1,5
      XI=(CA*CB*CC-SA*SC)*HX(I)-(CA*CB*SC+SA*CC)*HY(I)+CA*SB*HZ(I)
      YI=(SA*CB*CC+CA*SC)*HX(I)+(-SA*CB*SC+CA*CC)*HY(I)+SA*SB*HZ(I)
      ZI=-SB*CC*HX(I)+SB*SC*HY(I)+CB*HZ(I)
C*******************************************************************
      XIL=-RLM(1)-XI
      YIL=-RLM(2)-YI
      ZIL=-RLM(3)-ZI
      RIL2=XIL**2+YIL**2+ZIL**2
      VQUAN=VQUAN+(C12(I,1)/RIL2**6)
     &      *(1.D0+GR(I,1)*(1.D0 -6.D0*ZIL*ZIL/RIL2/5.D0))
     &    -(C6(I,1)/RIL2**3)
     &      *(1.D0+GA(I,1)*(1.D0 -3.D0*ZIL*ZIL/RIL2/2.D0))
40    CONTINUE
```

```
20 CONTINUE
23 CONTINUE
22 CONTINUE
21 CONTINUE
C     WRITE(*,*)IPLS
      DO 18 I1=1,3
      DO 17 I2=1,3
17 VIG=VIG-0.5D0*ECG(I1)*ALM(I1,I2)*ECG(I2)
18 CONTINUE
C*******************************************************************
C*TOTAL POTENTIAL ENERGY
C*******************************************************************
      POT=VEDQ+VEQQ+VIM+VIG+VQUAN
C*******************************************************************
C-CHANGE OF UNIT OF VARIOUS TERMS FROM a.u.  TO  meV
C*******************************************************************
      VEDQ=VEDQ*UAMEV
      VEQQ=VEQQ*UAMEV
      VIG=VIG*UAMEV
      VIM=VIM*UAMEV
      VQUAN=VQUAN*UAMEV
      POT=POT*UAMEV  +  VABAIS
C*******************************************************************
C     WRITE(1,63)Z0,POT
C     WRITE(*,61)Z0,VEDQ,VEQQ,VIM,VIG,VQUAN,POT
C     WRITE(*,61)QUI0,VEDQ,VEQQ,VIM,VIG,VQUAN,POT
61 FORMAT(1X,F7.3,6(5X,E13.6))
C*******************************************************************
C*MINIMIZATION OF NH3-SURFACE POTENTIAL WITH RESPECT TO
C*CHOSEN VARIABLES ( Z0  AND/OR  PHI0  AND/OR  QUI0 )
C*   MINI = 1 ========> NO MINIMIZATION
C*   MINI = 2 ========> MINIMIZATION
C*******************************************************************
      IF(MINI.EQ.1)GO TO 6
      IF(POT.GT.POTMIN)GO TO 6
      POTMIN=POT
      Z0MIN=Z0
      PHIMIN=PHI0
      QUIMIN=QUI0
 6 CONTINUE
 5 CONTINUE
 4 CONTINUE
C*******************************************************************
```

```
323 FORMAT(3(5X,F7.1),10X,F8.3,10X,G14.5)
154 FORMAT(G13.6)
C*******************************************************************
    IF(IMOL.EQ.1)GO TO 333
C*******************************************************************
C*CALCULATION OF THE TOTAL INVERSION POTENTIAL AFTER MINIMIZATION WITH
C*RESPECT TO THE APPROACH DISTANCE  Z0
C*******************************************************************
C     IF(IB.LT.24.OR.IB.GT.68)GO TO 222
    IF(IB.LT.47.OR.IB.GT.135)GO TO 222
    VTOTAL=VINTNU2(IB) + POTMIN + 142.5463222D0
    GO TO 333
222 VTOTAL=VINTNU2(IB) - 146.2074650D0 + 142.5463222D0
333 CONTINUE
    VJOHN(IB) = VTOTAL - VINTNU2(IB)
    WRITE(*,62)BETADEG,Z0MIN,VINTNU2(IB),POTMIN,VTOTAL
 62 FORMAT(5X,F7.1,10X,F8.4,3(10X,F12.4))
C     WRITE(1,63)Z0,POT
 63 FORMAT(F8.3,2X,F8.3,2X,F12.4)
    WRITE(1,64)VTOTAL
 64 FORMAT(D20.10)
C*******************************************************************
 3 CONTINUE
 2 CONTINUE
 1 CONTINUE
81 CONTINUE
    CLOSE(1)
C     DO 845 JOHN=1,INV
C     VPAIR(JOHN)   = (VJOHN(JOHN)+VJOHN(INV+1-JOHN))/2.D0
C     VIMPAIR(JOHN) = (VJOHN(JOHN)-VJOHN(INV+1-JOHN))/2.
C     WRITE(*,439)XJOHN(JOHN),XJOHN(INV+1-JOHN),VINTNU2(JOHN),
C    &                 VPAIR(JOHN),VIMPAIR(JOHN)
C     WRITE(2,440)VINTNU2(JOHN),VPAIR(JOHN),VIMPAIR(JOHN)
439 FORMAT(5(5X,D15.7))
440 FORMAT(3(3X,D15.7))
C 845 CONTINUE
C     CLOSE(2)
    STOP
    END
C*******************************************************************
C
C
C
```

```
C
C*******************************************************************
C*CALCULATION OF THE COMPONENTS OF ACTION TENSORS OF ORDERS 2, 3, 4
C*******************************************************************
      FUNCTION TAC2(I,J,R,R2,R5)
      IMPLICIT DOUBLE PRECISION (A-H,O-Z)
      DIMENSION R(3)
      TAC2=(3.D0*R(I)*R(J)-R2*KR(I,J))/R5
      RETURN
      END
C*******************************************************************
      FUNCTION TAC3(I,J,K,R,R2,R7)
      IMPLICIT DOUBLE PRECISION (A-H,O-Z)
      DIMENSION R(3)
      TAC3=-(15.D0*R(I)*R(J)*R(K)-3.D0*R2*(R(I)*KR(J,K)
     &                +R(J)*KR(I,K)+R(K)*KR(I,J)))/R7
      RETURN
      END
C*******************************************************************
      FUNCTION TAC4(I,J,K,L,R,R2,R4,R9)
      IMPLICIT DOUBLE PRECISION (A-H,O-Z)
      DIMENSION R(3)
      TAC4=(105.D0*R(I)*R(J)*R(K)*R(L)-15.D0*R2*(R(I)*R(J)*KR(K,L)
     &   +R(I)*R(K)*KR(J,L)+R(I)*R(L)*KR(J,K)+R(J)*R(K)*KR(I,L)
     &   +R(J)*R(L)*KR(I,K)+R(K)*R(L)*KR(I,J))
     & +3.D0*R4*(KR(I,J)*KR(K,L)+KR(I,K)*KR(J,L)+KR(I,L)*KR(J,K)))/R9
      RETURN
      END
C*******************************************************************
C* KRONECKER SYMBOL
C*******************************************************************
      FUNCTION KR(I,J)
      IF(I.EQ.J)GO TO 1
      KR=0
      GO TO 2
    1 KR=1
    2 RETURN
      END
C*******************************************************************
```

6.7.2. *Appendix 2: Expressions of the molecule orientation – heat bath phonons coupling terms*

$$
\begin{aligned}
\Delta\tilde{H}_{1p}\left(\vec{\Omega},\vec{u}\right) \approx & -62(u_j - u_0)D^1_{0,0} - 81(u_j - u_0)D^2_{0,0} \\
& -8u_j\left[(D^1_{-1,0} - D^1_{1,0}) - i(D^1_{-1,0} + D^1_{1,0})\right] \\
& -128u_j\left[(D^2_{-1,0} - D^2_{1,0}) - i(D^2_{-1,0} + D^2_{1,0})\right]
\end{aligned}
$$

$$
\begin{aligned}
\Delta\tilde{H}_{2p}\left(\vec{\Omega},\vec{u}\right) \approx & -350(u_j^2 + u_0^2)D^1_{0,0} - 475(u_j^2 + u_0^2)D^2_{0,0} \\
& -54u_j^2\left[(D^1_{-1,0} - D^1_{1,0}) - i(D^1_{-1,0} + D^1_{1,0})\right] \\
& -113u_j^2\left[(D^2_{-1,0} - D^2_{1,0}) - i(D^2_{-1,0} + D^2_{1,0})\right]
\end{aligned}
$$

In these expressions, the numerical coefficients are expressed, respectively, in (meV.Å^{-1}) and (meV.Å^{-2}).

6.7.3. *Appendix 3: Expressions of the components of the dipole moment vector and its derivatives with respect to the normal vibration coordinates*

Expressions [4.45] and [4.46] in section 4.3.7 and Appendix A of Chapter 4, Volume 3 (section 4.7.1.) can be used to write the components of the dipole moment and its derivatives for a symmetric top molecule of type XY_3, with respect to the absolute reference frame as follows:

Permanent moment: $\vec{\mu}_A^e = \mu^e \begin{pmatrix} \cos\varphi\sin\theta \\ \sin\varphi\sin\theta \\ \cos\theta \end{pmatrix}$

Mode ν_1 : $\left(\dfrac{\partial\vec{\mu}_A}{\partial q_1}\right) = \left(\dfrac{\partial\mu_z}{\partial q_1}\right)\begin{pmatrix} \cos\varphi\sin\theta \\ \sin\varphi\sin\theta \\ \cos\theta \end{pmatrix}$

Mode ν_2 : $\left(\dfrac{\partial\vec{\mu}_A}{\partial q_2}\right) = \left(\dfrac{\partial\mu_z}{\partial q_2}\right)\begin{pmatrix} \cos\varphi\sin\theta \\ \sin\varphi\sin\theta \\ \cos\theta \end{pmatrix}$

$$\text{Mode } \nu_{3a}: \left(\frac{\partial \vec{\mu}_A}{\partial q_{3a}}\right) = \left(\frac{\partial \mu_x}{\partial q_{3a}}\right) \begin{pmatrix} \cos\varphi\cos\theta\cos\chi - \sin\varphi\sin\chi \\ \sin\varphi\cos\theta\cos\chi + \cos\varphi\sin\chi \\ -\sin\theta\cos\chi \end{pmatrix}$$

$$\text{Mode } \nu_{3b}: \left(\frac{\partial \vec{\mu}_A}{\partial q_{3b}}\right) = \left(\frac{\partial \mu_y}{\partial q_{3b}}\right) \begin{pmatrix} -\cos\varphi\cos\theta\sin\chi - \sin\varphi\cos\chi \\ -\sin\varphi\cos\theta\sin\chi + \cos\varphi\cos\chi \\ \sin\theta\sin\chi \end{pmatrix}$$

$$\text{Mode } \nu_{4a}: \left(\frac{\partial \vec{\mu}_A}{\partial q_{4a}}\right) = \left(\frac{\partial \mu_x}{\partial q_{4a}}\right) \begin{pmatrix} \cos\varphi\cos\theta\cos\chi - \sin\varphi\sin\chi \\ \sin\varphi\cos\theta\cos\chi + \cos\varphi\sin\chi \\ -\sin\theta\cos\chi \end{pmatrix}$$

$$\text{Mode } \nu_{4b}: \left(\frac{\partial \vec{\mu}_A}{\partial q_{4b}}\right) = \left(\frac{\partial \mu_y}{\partial q_{4b}}\right) \begin{pmatrix} -\cos\varphi\cos\theta\sin\chi - \sin\varphi\cos\chi \\ -\sin\varphi\cos\theta\sin\chi + \cos\varphi\cos\chi \\ \sin\theta\sin\chi \end{pmatrix}$$

References

[ABD 93] ABDELAOUI O., Etude en matrice d'argon des réactions du trichlorofluorométhane (fréon 11) et du trichlorobromométhane avec l'oxygène atomique généré par irradiation UV de l'ozone, PhD thesis, Université Pierre et Marie Curie, Paris, 1993.

[ABE 50] ABELES F., "La théorie générale des couches minces", *Journal de Physique et de Radium*, vol. 11, pp. 307–309, 1950.

[ABO 73a] ABOUAF-MARGUIN L., Etude du mouvement et de la relaxation vibrationnelle de l'ammoniac isolé en matrice à basse température, PhD thesis, Université Pierre et Marie Curie, Paris, 1973.

[ABO 73b] ABOUAF-MARGUIN L., DUBOST H., LEGAY F., "Vibrational relaxation of matrix-isolated ammonia studied by infrared double resonance", *Chem. Phys. Lett.*, vol. 22, pp. 603–607, 1973.

[ABO 77] ABOUAF-MARGUIN L., JACOX M.E., MILLIGAN D.E., "The rotation and inversion of normal and deuterated ammonia in inert matrices", *J. Mol. Spectrosc.*, vol. 67, pp. 34–61, 1977.

[ACH 89] ACHER O., BIGAN E., DREVILLON B., "Improvements of phase-modulated ellipsometry", *Rev. Sci. Instr.*, vol. 60, p. 65, 1989.

[ACH 96] ACHIMSKY L., Etude cinétique de la thermo-oxydation du polypropylène, PhD thesis, ENSAM, Paris, 1996.

[ADL 09] ADLER F., COSSEL K.C., THORPE M.J. *et al.*, "Phase stabilized, 1.5W frequency comb at 2.8–4.8 m", *Opt. Lett.*, vol. 34, pp. 1330–1332, 2009.

[ALA 09] ALAVI S., SUSILO R., RIPMEESTER J.A., "Linking microscopic guest properties to macroscopic observables in clathrates hydrates: Guest-host hydrogen bonding", *J. Chem. Phys.*, vol. 130, p. 174501, 2009.

[ALA 11] ALAYLI N., Frittage de pâte de nano et micro grains d'argent pour l'interconnexion dans un module de mécatronique de puissance : élaboration, caractérisation et mise en œuvre, PhD thesis, UVSQ, Versailles, 2011.

[ALL 17] ALLAL S.E., HARLE C., SOHIER D. *et al.*, "Three stable states simulated for 1D spin-crossover nanoparticles using the Ising-like model", *Eur. J. Inorg. Chem.*, vol. 2017, no. 36, pp. 4196–4201, 2017.

[AMA 53] AMAT G., *Contribution à l'étude de l'intensité des Bandes d'Absorption Infrarouge*, Publications scientifiques et techniques du ministère de l'Air, Paris, 1953.

[AMA 62] AMAT G., ARSAC A., BROCHARD J. *et al.*, "Optics in France", *Appl. Opt.*, vol. 1, pp. 260–278, 1962.

[AND 49] ANDERSON P.W., "Pressure broadening in the microwave and infra-red regions", *Phys. Rev.*, vol. 76, pp. 647–661, 1949.

[AND 84] ANDERSON D.Z., FRISCH J.C., MASSER C.S., "Mirror reflectometer based on optical cavity decay time", *Appl. Opt.*, vol. 23, pp. 1238–1245, 1984.

[APK 84] APKARIAN V.A., "Stimulated emission as a major channel for the depopulation of laser-excited states in condensed media", *Chem. Phys. Lett.*, vol. 110, pp. 168–172, 1984.

[ARE 72] ARECCHI F.T., COURTENS E., GILMORE R. *et al.*, "Atomic coherent states in quantum optics", *Phys. Rev. A*, vol. 6, pp. 2221–2237, 1972.

[ASA 93] ASAKI M.T., HUANG C.P., GARVEY D. *et al.*, "Generation of 11-fs pulse from a self-mode-locked Ti-sapphire laser", *Opt. Lett.*, vol. 18, pp. 997–979, 1993.

[ASK 85] ASKEY R., WILSON J., "Some basic hypergeometric orthogonal polynomials that generalize Jacobi polynomials", *Mem. Am. Math. Soc.*, vol. 319, 1985.

[ASP 74] ASPNES D.E., "Optimizing precision of rotating-analyzer ellipsometers", *J.O.S.A.*, vol. 64, pp. 639–646, 1974.

[ASP 75] ASPNES D.E., STUDNA A.A., "High precision scanning ellipsometer", *Appl. Opt.*, vol. 14, pp. 220–228, 1975.

[AVG 70] AVGUL N.N., KISELEV A.V., *Chemistry and Physics of Carbon*, vol. 6, WALKER P.L. (ed.), Marcel Dekker, New York, 1970.

[AZZ 77] AZZAM R.M.A., BASHARA N.M., *Ellipsometry and Polarized Light*, North Holland, Amsterdam, 1977.

[BAE 99] BAEV V.M., LATZ T., TOSCHEK P.E., "Laser intracavity absorption spectroscopy", *Applied Physics B.*, vol. 69, pp. 171–202, 1999.

[BAH 97] BAHOU M., Photodissociation de l'ozone et de ses complexes avec des molécules d'halogènes X_2 ($X_2 = Cl_2$, Br_2, BrCl) en matrices, PhD thesis, UPMC, Paris, 1997.

[BAR 58] BARANGER M., "General impact theory of pressure broadening", *Phys. Rev.*, vol. 112, pp. 855–865, 1958.

[BAR 61] BARCHEWITZ P., *Spectroscopie infrarouge. 1. Vibrations moléculaires*, Gauthier-Villars, Paris, 1961.

[BAR 66] BARCHEWITZ P., *Spectroscopie infrarouge. 2. Fonction potentielle, Moment dipolaire*, Gauthier-Villars, Paris, 1966.

[BAR 73] BARNETT J.D., BLOCK S., PIERMARINI G.J., "An optical fluorescence system for quantitative pressure measurement in the diamond-anvil cell", *Rev. Sci. Instr.*, vol. 44, pp. 1–9, 1973.

[BAS 11] BASAVAPOORNIMA C., JAYASANKAR C.K., TRÖSTER T. *et al.*, "Pressure-dependent fluorescence studies of Sm^{3+}-doped fluorophosphate glass", *High Press. Res.*, vol. 31, pp. 121–125, 2011.

[BAU 66] BAUR M.E., SALZMAN W.R., "Motional states of substitutional species in lattices", *Phys. Rev.*, vol. 151, pp. 710–720, 1966.

[BEL 72a] BELL R.J., *Introductory Fourier Transform Spectroscopy*, Academic Press, New York, 1972.

[BEL 72b] BELIKOVA T.P., SVIRIDENKOV E.A., SUCHKOV A.F. *et al.*, "Detection of weak absorption lines with the aid of a neodymium-glass laser", *JETP Lett.*, vol. 35, pp. 1076–1079, 1972.

[BEN 75] BEN-REUVEN A., "Spectral line shapes in gases in the binary-collision approximation", *Adv. Chem. Phys.*, vol. 20, pp. 235–293, 1975.

[BER 77] BERKOWITZ M., GERBER R.B., "Vibrational relaxation of molecules in solids: The role of rotational and of translational local modes", *Chem. Phys. Lett.*, vol. 49, pp. 260–264, 1977.

[BER 78] BERKOWITZ M., GERBER R.B., "Rotational mode participation in long-distance vibrational energy transfer in solids", *Chem. Phys. Lett.*, vol. 56, pp. 105–108, 1978.

[BER 96] BERRODIER I., Recherche de la configuration d'équilibre des molécules de N_2O et CO_2 isolées en matrices d'argon à basse température et calcul des déplacements de fréquence et étude de l'émission stimulée de N_2O et $^{13}CO_2$ piégées en matrice d'argon, Master's internship at LPMA-UPMC, Université Marne-la-Vallée, Paris, 1996.

[BER 05] BERTOLOTTI M., *Masers and Lasers, an Historical Approach*, IOP Publishing, CRC Press, New York, 2005.

[BER 06] BERTAUX J.L., KOROBALEV O., PERRIER S. *et al.*, "SPICAM on mars express: Observing modes and overview of UV spectrometer data and scientific results", *J. Geophys. Res.*, vol. 111, E10S90, pp. 1–40, 2006.

[BER 07] BERTAUX J.L., NEVEJANS D., KOROBALEV O. *et al.*, "SPICAV on Venus express: Three spectrometers to study the global structure and composition of the Venus atmosphere", *Planet. Space. Science*, vol. 22, pp. 120–129, 2007.

[BER 08] BERTAUX J.L., VANDEAELE A.C., WILQUET V. *et al.*, "First observation of 628 CO2 isotopologue band at 3.3 μm in the atmosphere of Venus by solar occultation from Venus Express", *Icarus*, vol. 195, pp. 28–33, 2008.

[BET 40] BETHE H., TELLER E., Ballistic laboratory of the Aberdeen proving ground report X-117, 1940.

[BEY 74] BEYELER H.U., "Vibration-rotation absorption spectra of a rotator hindered by fields of cubic symmetry", *J. Chem. Phys.*, vol. 60, pp. 4123–4129, 1974.

[BLO 46] BLOCH F., "Nuclear induction", *Phys. Rev. 70*, pp. 460–474, 1946.

[BLU 78] BLUMEN A., LIN S.H., MANZ J., "Theory of vibrational energy transfer among diatomic molecules in inert matrices", *J. Chem. Phys.*, vol. 69, p. 881, 1978.

[BOC 93] BOCCARA A.C., PICKERING C., RIVORY J., "Spectroscopic ellipsometry", *Proceedings of the 1st ICSE*, Paris, 1993.

[BOI 85] BOISSEL P., Relaxation vibrationnelle en matrice: SF_6 et NH_3 en matrice de gaz rare et d'azote, PhD thesis, UPSUD, Orsay, 1985.

[BOR 99] BORN M., WOLF E., *Principles of Optics*, Harvard University Press, Cambridge, 1999.

[BOU 95] BOUDON V., Spectroscopie des hexafluorures à nombre impair d'électrons, PhD thesis, Université de Bourgogne, Dijon, 1995.

[BRA 65] BRACEWELL R.M., *Fourier Transform and its Applications*, McGraw-Hill Book Co., New York, 1965.

[BRA 68] BRADLEY D.J., DURRANT A.J.F., "Generation of ultrashort dye laser pulses by mode locking", *Phys. Lett.*, vol. 27A, pp. 73–74, 1968.

[BRA 95] BRAUN T., DIOT C., HOGLANDER A. *et al.*, An Experimental User Level of Implementation, Research Report, INRIA, 1995.

[BRE 73] BRENT R. (ED.), "Section 7.3, Powell's Algorithm", *Algorithms for Minimization Without Derivatives*, Prentice Hall, Englewood Cliffs, 1973.

[BRO 49] BROSSEL J., KASTLER A., "The detection of magnetic resonance of excited levels: Depolarization effect of optical resonance and fluorescence radiations", *C.R. Acad. Sci.*, vol. 229, p. 1213, 1949.

[BRO 69] BROYDEN C.G., "A new double-rank minimization algorithm", *Appl. Phys. A, Mat. Sci. Process*, vol. 16, p. 670, 1969.

[BRO 93a] BROSSET P., Position et mouvement de la molécule d'ozone isolée en matrice inerte à basse température : Spectre d'absorption infrarouge et modèle théorique, PhD thesis, Université Pierre et Marie Curie, Paris, 1993.

[BRO 93b] BROSSET P., DAHOO R., GAUTHIER ROY B. *et al.*, "Analysis of IR absorption spectrum of O_3 in inert matrices : spectroscopic evidence for two trapping sites", *Chem. Phys.*, vol. 172, pp. 315–324, 1993.

[BRU 65] BRUHAT G., *Cours de physique générale optique*, Masson & Cie, Paris, 1965.

[BUB 72] BUBE R.H., *Electronic Properties of Crystalline Solids*, Academic Press, New York, 1972.

[BUC 81] BUCK B., SUKUMAR C.V., "Exactly soluble model of atom–phonon coupling showing periodic decay and revival", *Phys. Lett. A*, vol. 81, pp. 132–135, 1981.

[BUZ 90] BUZEK V.J., "SU(1,1) squeezing of SU(1,1) generalized coherent states", *J. Mod. Opt.*, vol. 37, pp. 303–316, 1990.

[CAM 90] CAMPARGUE A., STOECKEL F., CHENEVIER M., "High sensitivity intracavity laser spectroscopy: Applications to the study of overtone transitions in the visible range", *Spectrochim. Acta Rev.*, vol. 13, pp. 69–88, 1990.

[CAM 91] CAMPARGUE A., CHENEVIER M., DELON A. *et al.*, "Intracavity laser absorption spectroscopy of NO_2 and CH_4 cooled in a slit jet supersonic expansion", *Journal de Physique*, vol. 4, no. C7, pp. 471–476, 1991.

[CAM 10] CAMI, J., BERNARD-SALAS, J., PEETERS, E., MALEK, S.E., "Detection of C60 and C70 in a young planetary nebula", *Science*, doi: 10.1126/science.1192035, 2010.

[CAR 12] CARRASCO N., MAHJOUB A., GAUTHIER T. *et al.*, "Amines and optical properties of Titan's aerosols", *EPSC 2012 (European Planetary Science Congress) Sep 2012*, Madrid, Spain, EPSC2012-260, 2012.

[CAS 81] CASPER L.W., "Stability criteria for non-Doppler lasers", *Phys. Rev. A*, vol. 23, pp. 248–260, 1981.

[CAS 13] CASSAM-CHENAÏ P., LIEVIN J., "An improved third order dipole moment surface for methane", *J. Mol. Spectrosc.*, vol. 291, pp. 77–84, 2013.

[CAT 20] CATANESE A., RUTLEDGE J., SILFIES M.C. *et al.*, "Mid-infrared frequency comb with 6.7 W average power based on difference frequency generation", *Opt.. Lett.*, vol. 45, pp. 1248–1251, 2020.

[CHA 50] CHANDRASEKHAR S., *Radiative Transfer*, Clarendon Press, Oxford, 1950.

[CHA 56] CHANDRASEKHAR S., "The illumination and polarization of the sunlight sky on Rayleigh scattering", *Trans. Am. Phil. Soc.*, vol. 44, p. 6, 1956.

[CHA 73] CHAMPENEY D.C., *Fourier Transform and their Physical Applications*, Academic Press, ST Louis, MO, 1973.

[CHA 78] CHAMPION J.P., Développement complet de l'hamiltonien de vibration–rotation des molécules toupies sphériques : application : étude des interactions vibrationnelles et analyse des bandes harmoniques et de combinaison, exemple du méthane CH_4, PhD thesis Université de Bourgogne, Dijon, France, 1978.

[CHA 98] CHABBI H., DAHOO P.R., GAUTHIER R.B. *et al.*, "Vibrational stimulated emission of $^{13}CO_2$ in solid argon", *Chem. Phys. Lett.*, vol. 285, pp. 252–258, 1998.

[CHA 00a] CHABBI H., Dynamique moléculaire en matrice de gaz rare : mouvements et relaxation vibrationnelle de $^{13}CO_2$ et d'autres molécules triatomiques, PhD thesis, UPMC, Paris, 2000.

[CHA 00b] CHABBI H., DAHOO P.R., DUBOST H. *et al.*, "Experimental analysis of $^{13}CO_2$ infrared stimulated emissions in solid argon", *Low. Temp. Phys.*, vol. 26, pp. 719–727, 2000.

[CHA 00c] CHABBI H., DAHOO P.R., GAUTHIER R.B. *et al.*, "Infrared stimulated emission of N_2O trapped in argon matrices: Comparison with O_3 and CO_2", *J. Phys. Chem. A*, vol. 104, pp. 1670–1673, 2000.

[CHA 01] CHAABOUNI H., Etude par spectroscopie infrarouge de l'ozone et du dioxyde de soufre adsorbés sur la surface de la glace ou piégés dans son volume entre 10 et 170 k. Effets d'irradiation à 266 nm sur le mélange $O_3 + SO_2$ piégé en matrice d'argon ou dans la glace, PhD thesis, UPMC, Paris, 2001.

[CHA 02] CHABBI H., GAUTHIER ROY B., VASSEROT A. *et al.*, "Vibrational emissions of $^{13}CO_2$ trapped in rare gas matrices: Kinetic model for the intramolecular relaxation", *J. Chem. Phys.*, vol. 117, p. 4436, 2002.

[CHA 05] CHATILLON MM., Méthodologie de conception robuste appliquée aux trains de véhicules de tourisme, PhD thesis, ECDL, Lyon, 2005.

[CHE 19] CHEN Z., HÄNSCH T.W., PICQUE N., "Mid-infrared feed-forward dual-comb spectroscopy", *Proc. Natl. Acad. Sci. USA*, vol. 116, pp. 3454–3459, 2019.

[CHI 00] CHIAR J.E., TIELENS A.G.G.M., WHITTET D.C.B. *et al.*, "The composition and distribution of dust along the line of sight toward the galactic center", *Astrophys. J.*, vol. 537, pp. 749–762, 2000.

[CHI 12] CHIRUTA D., LINARES J., DAHOO P.R. *et al.*, "Analysis of long-range interaction effects on phase transitions in two-step spin-crossover chains by using Ising-type systems and Monte Carlo entropic sampling technique", *J. Appl. Phys.*, vol. 112, pp. 074906-1–074906-7, 2012.

[CHI 14a] CHIRUTA D., LINARES J., GARCIA Y. *et al.*, "Analysis of multi-step transitions in spin-crossover nanochains", *Phys. B: Condens. Matter*, vol. 434, pp. 134–138, 2014.

[CHI 14b] CHIRUTA D., LINARES J., DAHOO P.R. *et al.*, "Influence of pressure and interactions strength on hysteretic behavior in two-dimensional polymeric spin crossover compounds", *Phys. B Condens. Matter*, vol. 435, pp. 76–79, 2014.

[CHI 15] CHIRUTA D., JURESCHI C.M., LINARES J. *et al.*, "On the origin of multi-step spin transition behaviour in 1D nanoparticles", *Eur. Phys. J. B*, vol. 88, pp. 233-1–233-5, 2015.

[COA 03] COANGA J.M., Caractéristiques vibrationnelles du chlorure de nitrile, du diméthyl-éther et de l'oxyde d'éthylène absorbés sur la surface de la glace ou piégés dans son volume entre 10 et 170 K : étude de leur photodissociation à 30 K en matrices, phase solide et dans la glace, PhD thesis, UPMC, Paris, 2003.

[COA 12] COANGA J.M., ALAYLI N., DAHOO P.R., "Metrological applications of ellipsometry with temperature change", *4th International Metrology Conference (CAFMET 2012)*, April 2012, Marrakech, Morocco, pp. 262–266, 2012.

[COD 84] CODY G.D., "The optical absorption edge of a-Si:H", in *Semiconductors and Semimetals*, PANKOVE J.I. (ed.), vol. 21, p. 11, 1984.

[COH 73] COHEN-TANNOUDJI C., DIU B., LALOË F., *Mécanique quantique*, Herman, Paris, 1973.

[COH 87] COHEN-TANNOUDJI C., DUPONT-ROC J., GRYNBERG G., *Photons et atomes. Introduction à l'électrodynamique quantique*, Inter-Editions/CNRS, Paris, 1987.

[COL 68] COLLIS R.T.H., FERNALD F.G., ALDER J., "Lidar observations of Sierra wave conditions", *J. Appl. Meteor.*, vol. 7, pp. 227–233, 1968.

[COL 90] COLLINS R.W., "Automatic rotating element ellipsometers: Calibration, operation, and real-time applications", *Rev. Sci. Instrum.*, vol. 61, p. 2029, 1990.

[CON 61a] CONNES J., "Recherches sur la spectroscopie par transformation de Fourier", *Rev. Opt.*, vol. 40, pp. 45–79, 1961.

[CON 61b] CONNES J., "Recherches sur la spectroscopie par transformation de Fourier. IV. Considérations générales sur le bruit dans la méthode par transformation de Fourier", *Rev. Opt.*, vol. 40, pp. 116–140, 1961.

[CON 61c] CONNES J., "Recherches sur la spectroscopie par transformation de Fourier. V. bruit dans les transformations de Fourier numériques. Diverses méthodes pour obtenir le rapport s/b Maximum", *Rev. Opt.*, vol. 40, pp. 171–190, 1961.

[CON 61d] CONNES J., "Recherches sur la spectroscopie par transformation de Fourier. VI. Expériences et résultats", *Rev. Opt.*, vol. 40, pp. 231–235, 1961.

[CON 66] CONNES J., CONNES P., "Near-infrared planetary spectra by fourier spectroscopy. I. Instruments and results", *J. Opt. Soc. Am.*, vol. 56, pp. 896–910, 1966.

[CON 75] CONNES P., MICHEL G., "Astronomical Fourier spectrometer", *Appl. Opt.*, vol. 14, pp. 2067–2084, 1975.

[CON 92] CONNES P., "Pierre Jacquinot and the beginnings of Fourier transform spectrometry", *J. Phys. II*, vol. 2, pp. 565–571, 1992.

[CON 95] CONNES P., "Fourier transform spectrometry at Laboratoire Aimé Cotton 1964–1974", *Spectrochim. Acta A Mol. Biomol. Spectrosc.*, vol. 51, pp. 1097–1104, 1995.

[COO 65] COOLEY J.W., TUCKEY J.W., "An algorithm for the machine calculation of complex Fourier series", *Maths. Comput.*, vol. 19, pp. 297–301, 1965.

[COQ 94] COQUARD M.P., Effet stark dans les molécules tétraédriques, PhD thesis, Université de Bourgogne, Dijon, 1994.

[COU 96] COUDERT L., Etude des molécules possédant des potentiels à plusieurs minima. Application aux dérivés deutérés de l'ammoniac et du méthanol, PhD thesis, Université Pierre et Marie Curie, Paris, 1996.

[CSE 04] CSELE M., *Fundamentals of Light, Sources and Lasers*, John Wiley & Sons, New York, 2004.

[CUG 72] CUGLEY J.A., PULLIN A.D.E., "Far infrared rotational transitions of ammonia in an argon matrix", *Chem. Phys. Lett.*, vol. 17, pp. 406–409, 1972.

[CUG 73] CUGLEY J.A., PULLIN A.D.E., "The infrared spectrum of ammonia isolated in argon and nitrogen matrices", *Spectrochim. Acta Part A*, vol. 29, pp. 1665–1672, 1973.

[CUI 67] CUISENIER L., PINARD J., "Spectromètre de Fourier à 'oeils de chat' et à balayage rapide", *J. de Physique Colloques.*, vol. 28, pp. 97–104, 1967.

[CUN 01] CUNDIFF S.T., YE J., HALL J.L. "Optical frequency synthesis based on mode locked lasers", *Rev. Sci. Instrum.*, vol. 72, pp. 3749–3771, 2001.

[DAH 88] DAHOO P.R., Sur l'intensité et l'élargissement par la pression des raies de vibration-rotation des bandes ν_3 , $\nu_1+\nu_3$ et $\nu_1+\nu_3-2\nu_2^0$ de $^{12}C^{16}O_2$ et ν_3 de $^{14}N_2^{16}O$: étude expérimentale et interprétation des paramètres d'élargissement au moyen de modèles théoriques semi-classiques, PhD thesis, Université Pirre et Marie Curie, Paris, 1988.

[DAH 97] DAHOO P.R., Dynamique moléculaire en phase condensée, à basse température : molécules d'intérêt atmosphérique piégées en matrice inerte – spectroscopie résolue en temps et en fréquence et étude de complexes faiblement liés, HDR, Université de Versailles St Quentin en Yvelines, Versailles, 1997.

[DAH 98] DAHOO P.R., JASMIN D., BROSSET P. *et al.*, "Vibrational relaxation study of O_3 in rare gas and nitrogen matrices by time resolved infrared–infrared double resonance spectroscopy", *J. Chem. Phys.*, vol. 108, p. 8541, 1998.

[DAH 99a] DAHOO P.R., BERRODIER I., RADUCU V. *et al.*, "Splitting of ν_2 vibrational mode of CO_2 isotopic species in the unstable trapping site in argon matrix", *Eur. Phys. J. D*, vol. 5, pp. 71–81, 1999.

[DAH 99b] DAHOO P.R., CHABBI H., LAKHLIFI A., "Spectroscopy of O_3 trapped in rare gas matrices I. Theoretical model for low lying vibrational levels", *J. Chem. Phys.*, vol. 111, pp. 10192–10201, 1999.

[DAH 03] DAHOO P.R., HAMON T., SCHNEIDER M. *et al.*, "Ellipsometry: Principles, signal processing and applications to metrology", *Proceedings CIMNA*, Lebanon, 2003.

[DAH 04a] DAHOO P.R., HAMON T., NEGULESCU B. *et al.*, "Evidence by spectroscopic ellipsometry of optical property change in pulsed laser deposited NiO films when heated in air at Neel temperature", *Appl. Phys. A, Mat. Sci. Process*, vol. 79, pp. 1439–1443, 2004.

[DAH 04b] DAHOO P.R., GIRARD A., TESSEIR M. *et al.*, "Characterizaton of pulsed laser deposited SmFeO3 morphology: Effect of fluence, substrate temperature and oxygen pressure", *Appl. Phys. A, Mat. Sci. Process*, vol. 79, pp. 1399–1403, 2004.

[DAH 06] DAHOO P.R., LAKHLIFI A., CHABBI H. *et al.*, "Matrix effect on triatomic CO_2 molecule: Comparison between krypton and xenon", *J. Mol. Struct.*, vol. 786, pp. 157–167, 2006.

[DAH 10a] DAHOO P.R., ALAYLI N., GIRARD A. *et al.*, "Reliabilty in mechatronic systems from TEM, SEM and SE Material Analysis", in UEDA O., FUKUDA M., PEARTON S. *et al.* (eds), *Reliability and Materials Issues of Semiconductor Optical and Electrical Devices and Materials, Mater. Res. Soc. Symp. Proc.*, Warrendale, 2010.

[DAH 10b] DAHOO P.R., GIRARD A., "Thin films of $SmFeO_3$ on quartz substrate modelled as anisotropic optical material", *Vth International Conference on Spectroscopic Ellipsometry, (ICSE-V)*, May 2010, Albany, United States, 2010.

[DAH 10c] DAHOO P.R., GIRARD A., "Real time characterization of pulsed laser deposition of SmFeO3 thin films by ellipsometry: Anomalous temperature effects and growth rates", *Vth International Conference on Spectroscopic Ellipsometry, (ICSE-V)*, May 2010, Albany, United States, 2010.

[DAH 16a] DAHOO P.R., POUGNET P., EL HAMI A., *Nanometer-scale Defect Detection Using Polarized Light*, ISTE Ltd, London, and John Wiley & Sons, New York, 2016.

[DAH 16b] DAHOO P.R., PUIG R., LAKHLIFI A. *et al.*, "Simulation of relaxation channels of CO_2 in clathrate nanocages", *Journal of Physics: Conference Series, IOP Publishing*, no. 738, p. 012072, 2016.

[DAH 17] DAHOO P.R., LAKHLIFI A., *Infrared Spectroscopy of Diatomics for Space Observation*, ISTE Ltd, London, and John Wiley & Sons, New York, 2017.

[DAH 19a] DAHOO P.R., LAKHLIFI A., *Infrared Spectroscopy of Triatomics for Space Observation*, ISTE Ltd, London, and John Wiley & Sons, New York, 2019.

[DAH 19b] DAHOO P.R., KHETTAB M., LINARES H. *et al.* "Non-destructive characterization by spectroscopic ellipsometry of interfaces in mechatronic devices", in EL HAMI, A., POUGNET, P. (eds). *Embedded Mechatronic Systems 1: Analysis of Failures, Predictive Reliability*, ISTE Press, London, and Elsevier, Oxford, 2019.

[DAH 21] DAHOO P.R., LAKHLIFI A., *Infrared Spectroscopy of Symmetric and Spherical Top Molecules for Space Observation 1*, ISTE Ltd, London, and John Wiley & Sons, 2021.

[DAN 90] DANTZIG G.B., *Origins of the Simplex Method, History of Scientific Computing*, ACM Press, Reading, 1990.

[DAR 40] DARLING B.T., DENISSON D.M., "The water vapor molecule", *Phys. Rev.*, vol. 57, pp. 128–139, 1940.

[DAR 01] DARTOIS E., D'HENDECOURT L., "Search for NH_3 ice in cold dust envelopes around YSOs", *Astronomy and Astrophysics*, vol. 365, pp. 144–156, 2001.

[DAR 02] DARMO J., BRATSCHITSCH R., MULLER T. *et al.*, "Few-cycle THz generation for imaging and tomography applications", *Phys. Med. Biol.*, vol. 47, pp. 3691–3697, 2002.

[DAR 08] DARTOIS E., DEBOFFLE D., "Methane clathrate hydrate FTIR spectrum. Implications for its cometary and planetary detection", *Astron. Astrophys.*, vol. 490, pp. L19–L22, 2008.

[DAR 10] DARTOIS E., DEBOFFLE D., BOUZIT M., "Methane clathrate hydrate infrared spectrum. II. Near-infrared overtones, combination modes and cages assignments", *Astron. Astrophys.*, vol. 514, A49, 2010.

[DAV 59] DAVIDON N.C., *Variable Metric Methods for Minimization*, AEC Research and Development, Argonne Lab, 59 National 90 (Rev), 1959.

[DAV 96] DAVIDOVICH L., "Sub-Poissonian processes", *Rev. Mod. Phys.*, vol. 68, pp. 127–173, 1996.

[DAV 02] DAVIES A.G., LINFIELD E.H., JOHNSTON M.B., "The development of terahertz sources and their applications", *Phys. Med. Biol.*, vol. 47, pp. 3679–3689, 2002.

[DEB 68] DE BROGLIE L., *Ondes électromagnétiques et photons*, Gauthier-Villars, Paris, 1968.

[DEC 68] DE CADDES D.E., OSTERNIK L.M., TARG R., "Laser pulse-shaping and modelocking with acoustic waves", *Appl. Phys. Lett.*, vol. 12, pp. 74–76, 1968.

[DEM 65] DE MARIA A.J., STETSER DA., "Laser pulse-shaping and modelocking with acoustic waves", *Appli. Phys. Lett.*, vol. 7, pp. 71–73, 1965.

[DEM 66] DE MARIA A.J., FERRAR C.M., DANIELSON G.E., "Mode locking of a Nd^{3+} doped glass laser", *Appli. Phys. Lett.*, vol. 8, pp. 22–24, 1966.

[DEM 94] DEMEURE I., FARHAT J., "Systèmes de processus légers : concepts et exemples", *Technique et science informatiques*, vol. 13, no. 6, pp. 765–795, 1994.

[DEM 96] DEMTRÖDER W., *Laser Spectroscopy: Basic Concepts and Instrumentation*, 2nd ed., Springer-Verlag, Berlin, 1996.

[DEL 07] DEL'HAYE P., SCHLIESSER A., ARCZET O. *et al.*, "Optical frequency comb generation from a monolithic microresonator", *Nature*, vol. 450, pp. 1214–1217, 2007.

[DES 94] DE SAXCE A., Etude en matrice de la photodissociation du nitrate de chlore, molécule réservoir stratosphérique. Identification des complexes entre $ClONO_2$, ClNO, NO et HCl ou HBr, PhD thesis, UPMC, Paris, 1994.

[DEU 65] DEUTSCH T., "Mode-locking effects in an internally modulated ruby laser", *Appl. Phys. Lett.*, vol. 7, pp. 80–82, 1965.

[DEV 36] DEVONSHIRE A.F., "The rotation of molecules in fields of octahedral symmetry", *Proc. R. Soc. A*, vol. 153, pp. 601–621, 1936.

[DEX 53] DEXTER D.L., "A theory of sensitized luminescence in solids", *J. Chem. Phys.*, vol. 21, p. 836, 1953.

[DID 64] DI DOMENICO M., "Small-signal analysis of internal (coupling type) modulation of lasers", *J. Appl. Phys.*, vol. 35, pp. 2870–2876, 1964.

[DID 66] DI DOMENICO M., MARCOS H.M., GEUSIC J.E. *et al.*, "Generation of ultrashort optical pulses by mode locking the YAG: Nd laser", *Applied Phys. Lett.*, vol. 8, pp. 180–181, 1966.

[DID 10] DIDDAMS S.A., "The evolving optical frequency comb", *J.O.S.A. B*, vol. 27, pp. B51–B62, 2010.

[DID 20] DIDDAMS S.A., VAHALA K., UDEM T., "Optical frequency combs: Coherently uniting the electromagnetic spectrum", *Science*, vol. 369, no. 6501, Science 17 July 2020.

[DOE 86] DOERNER M., NIX W., "A method for interpreting the data from depth-sensing indentation instruments", *J. Mater. Res.*, vol. 1, pp. 601–609, 1986.

[DOS 16] DOOSE L.R., KARKOSCHKA E., TOMASKO M.G. *et al.*, "Vertical structure and optical properties of Titan's aerosols from radiance measurements made inside and outside he atmosphere", *Icarus*, vol. 270, pp. 355–375, 2016.

[DOW 54] DOWS A., WHITTLE E., PIMENTEL G.C., "Matrix isolation method for the experimental study of unstable species", *J. Chem. Phys.*, vol. 22, p. 1943, 1954.

[DRA 88] DRAINE B.T., "The discrete-dipole approximation and its application to interstellar graphite grains", *Astrophys. J.*, vol. 333, pp. 848–872, 1988.

[DRE 82] DREVILLON B., PERRIN J., MAROT R. *et al.*, "Fast polarization modulated ellipsometer using a microprocessor system for digital Fourier analysis", *Rev. Sci. Instrum.* vol. 53, p. 969, 1982.

[DRE 03] DRESSEL M., GRUNER G., *Optical Properties of Electrons in Matter*, 2nd ed., Cambridge University Press, Cambridge, 2003.

[DRU 87] DRUDE P., "On the Laws of Reflection and Refraction of light at the Interface to Absorbing Crystals", *Ann. Phys.*, vol. 32, pp. 584–625, 1987.

[DUB 75] DUBOST H., Étude des mouvements moléculaires, de la relaxation et des transferts d'énergie vibrationnelle de l'oxyde de carbone isolé en matrice de gaz rare à basse température, PhD thesis, Université Paris SUD, Orsay, 1975.

[DUB 76] DUBOST H., CHARNEAU R., "Infrared absorption spectra of carbon monoxide in rare gas matrices", *Chem. Phys.*, vol. 12, p. 407, 1976.

[EIN 15] EINSTEIN A., "The field equations of gravitation", *Sitzungsberichte der Preussischen Akademie der Wissenschaften: Proceedings of the Prussian Academy of Sciences*, pp. 844–847, 25 November 1915.

[ELH 13a] EL HAMI A., RADI B., *Incertitude et optimisation et fiabilité des structures*, Hermes-Lavoisier, Paris, 2013.

[ELH 13b] EL HAMI A., RADI B., *Uncertainty and Optimization in Structural Mechanics*, ISTE, London and John Wiley & Sons, New York, 2013.

[ERK 03] ERKOÇ S., TÜRKER L., "Ammonia deposition in fullerene: $(NH_3)_n@C_{60}$", *J. Mol. Struct.*, vol. 640, pp. 57–61, 2003.

[EVT 19] EVENT HORIZON TELESCOPE COLLABORATION, AKIYAMA K. *et al.*, "The first M87 event horizon telescope results. I. The shadow of the supermassive black hole", *Astrophys. J.*, vol. 875, no. L1, p. 17, 2019.

[FAN 63] FANO U., "Pressure broadening as a prototype of relaxation", *Phys. Rev.*, vol. 131, pp. 259–268, 1963.

[FAR 00] FAREBROTHER A.J., MEIJER A.J.H.M., CLARY D.C. *et al.*, "Formation of H_2 on a graphite surface via Eley-Rideal mechanism", *Chem. Phys. Lett.*, vol. 319, pp. 303–308, 2000.

[FEL 49] FELLGETT P.I., "On the ultimate sensitivity and practical performance of radiation detectors", *J. Opt. Soc. Am.*, vol. 39, pp. 970–976, 1949.

[FEL 58] FELLGETT P.I., "Les principes généraux des méthodes nouvelles en spectroscopie interférentielle – A propos de la théorie du spectromètre interférentiel multiplex", *Journal de physique et radium*, vol. 19, pp. 187–191, 1958.

[FER 02a] FERGUSON B., ZHANG X.C., "Materials for terahertz science and technology", *Nat. Mater.*, vol. 1, pp. 26–33, 2002.

[FER 02b] FERLAUTO A.S., FERREIRA G.M., PEARCE J.M. *et al.*, "Analytical model for the optical functions of amorphous semiconductors from the near-infrared to ultraviolet: Applications in thin film photovoltaics", *J. App. Phys.*, vol. 92, p. 2424, 2002.

[FEY 57] FEYNMAN R., VERNON F., HELLWARTH R., "Geometrical representation of the Schrödinger equation for solving maser problems", *J. App. Phys.*, vol. 28, pp. 49–52, 1957.

[FEY 65] FEYNMAN R.P., LEIGHTON R.B., SANDS M., *The Feynman Lectures on Physics, Vol. II: The Electromagnetic Field*, Addison-Wesley Longman, Reading, 1965.

[FEY 85] FEYNMAN R.P., QED: *The Strange Theory of Light and Matter*, Princeton University Press, Princeton, 1985.

[FEY 98] FEYNMAN R., *Quantum Electrodynamics*, Advanced Books Classics, Perseus Publishing, 1998.

[FIU 62] FIUTAK J., VAN KRANENDONCK J., "Impact theory of Raman line broadening", *Can. J. Phys.*, vol. 40, pp. 1085–1100, 1962.

[FLA 81] FLAUD J.M., CAMY-PEYRET C., TOTH R.A., "Water vapour line parameters from microwave to medium infrared", *An Atlas of* $H_2{}^{16}O$, $H_2{}^{17}O$ *and* $H_2{}^{18}O$ *Line Positions and Intensities between 0 and 4350* cm^{-1}, Pergamon Press, Oxford, p. 259, 1981.

[FLA 90] FLAUD J.M., CAMY-PEYRET C., RINSLAND C.P. *et al.*, *Atlas of Ozone Spectral Parameters from Microwave to Medium Infrared*, Academic Press, Boston, p. 600, 1990.

[FLE 63] FLETCHER R., POWELL M.J.D., "A rapidly convergent descent method for minimization", *Comput. J.*, vol. 6, pp. 163–168, 1963.

[FLE 70] FLETCHER R., "A new approach to variable metric algorithms", *Comput. J.*, vol. 13, p. 371, 1970.

[FLI 62] FLINN P.A., MARADUDIN A.A., "Distortion of crystals by point defects", *Ann. Phys.*, vol. 18, pp. 81–109, 1962.

[FLU 67] FLUGGE S., GENTZEL L., *Encyclopedia of Physics: Light and Matter*, Springer-Verlag, Berlin, 1967.

[FLY 63] FLYGARE W.H., "Molecular rotation in the solid state. Theory of rotation of trapped molecules in rare-gas lattices", *J. Chem. Phys.*, vol. 39, pp. 2263–2273, 1963.

[FOR 48] FORSTER T., "Intermolecular energy and fluorescence/Zwischenmolekulare Energiewanderung und Fluoreszenz," *Annalen der Physik*, vol. 437, nos 1–2, pp. 55–75, 1948.

[FOR 49] FORSTER T., "Experimental and theoretical study of the intermolecular transition of electron excitation energy", *Z Naturforsch.*, vol. A4, pp. 321–327, 1949.

[FOR 72] FORMANN R.A., PIERMARINI G.J., BARNETT J.D. *et al.*, "Pressure measurement made by the utilization of ruby sharp-line luminescence", *Science*, vol. 176, pp. 284–285, 1972.

[FOR 86] FOROUHI A.R., BLOOMER I., "Optical dispersion relations for amorphous semiconductors and amorphous dielectrics", *Phys. Rev. B.*, vol. 34, p. 7018, 1986.

[FOR 20] FORTIER T., BAUMANN E., "20 years of developments in optical frequency comb technology and applications", *Commun. Phys.*, vol. 6, pp. 1–16, 2020.

[FOX 10] FOX M., *Optical Properties of Solids*, Oxford University Press, Oxford, 2010.

[FRA 05] FRANTA D., NEGULESCU B., THOMAS L. *et al.*, "Optical properties of NiO thin films prepared by pulsed laser deposition technique", *Appl. Surf. Sci.*, vol. 244, pp. 426–430, 2005.

[FRI 65] FRIEDMANN H., KIMEL S., "Theory of shifts of vibration-rotation lines of diatomic molecules in noble-gas matrices. Intermolecular forces in crystals", *J. Chem. Phys.*, vol. 43, pp. 3925–3938, 1965.

[FRI 67] FRIEDMANN H., KIMEL S., "Rotation-translation-coupling effect in noble-gas crystals containing molecular impurities", *J. Chem. Phys.*, vol. 47, pp. 3589–3605, 1967.

[FUJ 07] FUJIWARA H., *Spectroscopic Ellipsometry Principles and Applications*, Wiley, London, 2007.

[GAB 96] GABART T., Etude des effets collisionnels dans les molécules tétraédriques. Applications au méthane perturbé par l'argon, PhD thesis, Université de Bourgogne, Dijon, 1996.

[GAM 98] GAMACHE R., LYNCH R., NESHYBA S.P., "New developments in the theory of pressure-broadening and pressure-shifting of spectral lines of H_2O: The complex Robert-Bonamy formalism", *J. Quant. Spectr. Rad. Transfer*, vol. 59, pp. 319–335, 1998.

[GAM 14] GAMACHE ROBERT R., LAMOUROUX J., BLOT-LAFON V., "An intercomparison of measured pressure-broadening, pressure shifting parameters of carbon dioxide and their temperature dependence", *J. Quant. Spectr. Rad. Transfer*, vol. 135, pp. 30–43, 2014.

[GAN 09] GANJI M.D., MOHSENI M., GOLI O., "Modeling complexes of NH_3 molecules confined in C_{60} fullerene", *J. Mol. Struct.*, vol. 913, pp. 54–57, 2009.

[GAU 80] GAUTHIER ROY B., Relaxation de l'énergie vibrationnelle du fluorure de méthyle en matrice inerte à basse temperature, PhD thesis, Université Paris SUD, Orsay, 1980.

[GAU 12] GAUTHIER T., CARRASCO N., MAHJOUB A. *et al.*, "Mid- and far-infrared absorption spectroscopy of Titan's aerosols analogues", *Icarus*, vol. 221, pp. 320–327, 2012.

[GER 77] GERBER R.B., BERKOWITZ M., "Role of rotational and translational local modes in vibrational relaxation in solids: A study of NH and ND in solid Ar", *Phys. Rev. Lett.*, vol. 39, p. 1000, 1977.

[GHA 91] GHANEM R.G., SPANGOS P.D., *Stochastic Finite Elements: A Spectral Approach*, Revised Ed., Springer Verlag, Berlin, 1991.

[GHA 12] GHADERI S.H., HAJIESMAILI E., "Molecular structural mechanics applied to coiled carbon nanotubes", *Comput. Mater. Sci.*, vol. 55, pp. 344–349, 2012.

[GHE 79] GHERISSI S., Etude à haute résolution des bandes de la molécule $^{12}CH_4$. Calcul de l'intensité des raies de vibration-rotation et des constantes moléculaires, PhD thesis, Université Pierre et Marie Curie, Paris, 1979.

[GHE 02] GHERMAN T., ROMANINI D., "Mode locked cavity-enhanced absorption spectroscopy", *Opt. Express*, vol. 19, pp. 1033–1042, 2002.

[GIA 08] GIANNOPOULOS G., KAKAVAS P., ANIFANTIS N., "Evaluation of the effective mechanical properties of single walled carbon nanotubes using a spring based finite element approach", *Comput. Mater. Sci.*, vol. 41, pp. 561–569, 2008.

[GIB 01] GIBB E.L., WHITTET D.C.B., CHIAR J.E., "Searching for NH_3 in grain mantles toward massive young stellar objects", *Astrophys. J.*, vol. 558, pp. 702–716, 2001.

[GIL 06] GILMORE R., *Lie Groups, Lie Algebras, and Some of their Applications*, Dover Books on Mathematics, Dover Publications Inc., New York, 2006.

[GIN 78] GINZBURG V., *Physique théorique et astrophysique*, Editions Mir, Moscow, 1978.

[GIR 85] GIRARDET C., LAKHLIFI A., "Infrared profile of NH_3 trapped in argon matrix", *J. Chem. Phys.*, vol. 83, pp. 5506–5519, 1985.

[GIR 86] GIRARDET C., LAKHLIFI A., "The orientation-inversion motions of ammonia trapped in matrix an inertial model", *Chem. Phys.*, vol. 110, pp. 447–464, 1986.

[GLA 63] GLAUBER R.J., "The quantum theory of optical coherence", *Phys. Rev. A*, vol. 130, p. 2529, 1963.

[GLA 67] GLAUBER R.J., "Photon fields and classical fields", *Proceedings of the Symposium on Modern Optics*, vol. 47, pp. 1–18, 1967.

[GLE 68] GLENN W.H., BRIENZA M.J., DE MARIA A.J., "Mode locking of an organic dye laser", *Appl. Phys. Lett.*, vol. 12, pp. 54–56, 1968.

[GOL 70] GOLDFARB D., "A family of variable metric algorithms", *Mathematical Computations*, vol. 24, pp. 24–26, 1970.

[GOL 03] GOLDSTEIN D., *Polarized Light*, Marcel Dekker, New York, 2003.

[GOO 76] GOODMAN L., BRUS L.E., "Mechanism of vibrational relaxation in molecular solids", *J. Chem. Phys.*, vol. 65, p. 3146, 1976.

[GOR 13] GORROTXATEGI-GARBAJO P., FASCI E., VENTRILLARD I., *et al.*, "Optical-feedback cavity-enhanced absorption spectroscopy with a quantum-cascade laser yields the lowest formaldehyde detection nlimit", *Applied Physics B - Laser and Optics*, vol. 110, pp. 309–314, 2013.

[GOU 77] GOULD G., Optically pumped laser amplifiers; light amplifiers employing collisions to produce a population inversion, U.S. patents 4,053,845 and 4,704,583, 1977.

[GRI 75] GRIFFITHS P.R., HASETH J.A., *Chemical Infrared Fourier Transform Spectroscopy*, John Wiley & Sons, New York, 1975.

[GRI 86] GRIFFITHS P.R., HASETH J.A., *Fourier Transform Infrared Spectroscopy*, John Wiley & Sons, New York, 1986.

[GRY 97] GRYNBERG G., ASPECT A., FABRE C., *Introduction aux lasers et à l'optique quantique*, Ellipses, Paris, 1997.

[GUB 09] GUBIN M., KIREEV A., KONYASHCHENKO A. *et al.*, "Femtosecond fiber laser based methane optical clock", *Appl. Phys. B*, vol. 95, pp. 661–666, 2009.

[GUE 15a] GUERINE A., EL HAMI A., WALHA L. *et al.*, "A perturbation approach for the dynamic analysis of one stage gear system with uncertain parameters", *Mech. Machine Theory*, vol. 92, pp. 113–126, 2015.

[GUE 15b] GUERINE A., EL HAMI A., FAKHFAKH T. *et al.*, "A polynomial chaos method to the analysis of the dynamic behavior of spur gear system", *Struct. Engi. Mech. Int. J.*, vol. 53, pp. 819–831, 2015.

[HAL 98] HALLOU A., Etude par spectroscopie Ir de la diffusion et de la photodissociation du chlorure de nitrosyle isole en matrice. Identification et photochimie de son complexe avec la molécule d'eau, PhD thesis, UPMC, Paris, 1998.

[HAL 06] HALL J. "Nobel lecture: Defining and measuring optical frequencies", *Rev. Modern Phys.*, vol. 78, pp. 1279–1295, 2006.

[HAM 62] HAMERMESH. M., *Group Theory and its Application to Physical Problems*, Addison-Wesley Publishing Company Inc., New York, 1962.

[HÄN 06] HÄNSCH T., "Nobel lecture: Passion for precision", *Rev. Modern Phys.*, vol. 78, pp. 1297–1309, 2006.

[HAR 64] HARGROVE L.E., FORK R.L., POLLACK M.A., "Locking of He-Ne laser modes induced by synchronous intracavity modulation", *Appl. Phys. Letts.*, vol. 5, pp. 4–5, 1964.

[HAR 17] HARLE C., ALLAL S.E., SOHIER D. *et al.*, "Numerical investigations of the thermal, pressure and size effects on 2D spin crossover nanoparticles", *J. Phys. Conf. Ser.*, vol. 936, pp. 012061-1–012061-5, 2017.

[HAU 75] HAUS H.A., "Theory of mode locking with a slow saturable absorber", *IEEE J. Quant. Elect.*, vol. 11, pp. 736–746, 1975.

[HAY 64] HAYFIELD P.C.S., WHITE G.W.T., *Ellipsometry in the Measurement of Surfaces and Thin Films: Symposium Proceedings Washington 1963*, PASSAGLIA E., STROMBERG R.R., KRUGER J. (eds), United States Department of Commerce, Washington, 1964.

[HEC 05] HECHT E., *Optics*, 4th ed., Pearson Education Inc., Berlin, 2005.

[HED 91] HEDBERG, K., HEDBERG, L., BETHUNE, D., *et al.*, "Bond lengths in free molecules of buckminsterfullerene, c60, from gas-phase electron diffraction", *Science*, vol. 254, issue 5030, pp. 410–410, 1991.

[HER 59] HERZFELD K.F., LITOVITZ T.A., *Absorption and Dispersion of Ultrasonic Waves*, Academic Press, New York, 1959.

[HIR 64] HIRSCHFELDER J.O., CURTISS C.F., BIRD R.B., *Molecular Theory of Gases and Liquids*, Wiley, New York, 1964.

[HOC 66] HOCKER O., KOVACS M.A., RHODES C.K. *et al.*, "Vibrational relaxation measurements in CO2 using an induced-fluorescence technique", *Phys. Rev. Lett.*, vol. 17, p. 233, 1966.

[HUA 97] HUARD S., *Polarization of Light*, Masson, Paris, 1997.

[HUA 05] HUANG-CHANG L., "Using N-D method to solve multi-response problem in Taguchi", *J. Intell. Manuf.*, vol. 16, pp. 331–347, 2005.

[HUM 05] HUMBERT C., PEREMANS A., SILIEN C., "Le laser", *Revue des questions scientifiques*, vol. 176, pp. 97–162, 2005.

[HUN 68] HUNTEN D.M., "Fourier transform spectroscopy of the planets", *Science*, vol. 162, pp. 313–318, 1968.

[JAC 48] JACQUINOT P., DUFOUR C., "Conditions optique d'emploi de cellules photo électriques dans les spectrographes et les interféromètres", *J. de Rech. du Centre National de Recherche Scientifique Laboratoire Bellevue, Paris. Am.*, vol. 6, pp. 91–103, 1948.

[JAC 54a] JACQUINOT P., "The etendue advantage", *XVII Congrès du GAMS, groupement pour l'avancement des méthodes spectrographiques*, Paris, 1954.

[JAC 54b] JACQUINOT P., "The luminosity of spectrometers with prisms, gratings, or Fabry-Perot etalons", *J. Opt. Soc. Am.*, vol. 44, pp. 761–765, 1954.

[JAC 60] JACQUINOT P., "New developments in interference spectroscopy", *Rep. Prog. Phys.*, vol. 23, pp. 267–312, 1960.

[JAC 98] JACKSON J.D., *Classical Electrodynamics*, 3rd ed., John Wiley & Sons, New York, 1998.

[JAM 02] JAMES J.F., *A Students Guide to Fourier Transforms*, Cambridge University Press, New York, 2002.

[JAS 94] JASMIN D., DAHOO P.R., BROSSET P. *et al.*, "Laser-induced infrared fluorescence of O_3 trapped in xenon matrices", *J. Chem. Phys.*, vol. 101, p. 7337, 1994.

[JAS 95] JASMIN D., Relaxation vibrationnelle de l'ozone en matrice inerte : émission stimulée et transferts non-radiatifs, PhD thesis, UPMC, Paris, 1995.

[JAS 98] JASMIN D., DAHOO P.R., BROSSET P. *et al.*, "Stimulated emission at 10 μm of O_3 trapped in low temperature matrices: Threshold measurements and quantum yield evaluation", *J. Chem. Phys.*, vol. 108, p. 2302, 1998.

[JAU 01] JAULIN L., KIEFFER M., DITRIT O. *et al.*, *Applied Interval Analysis: With Examples in Parameter and State*, Springer-Verlag, London, 2001.

[JAV 61] JAVAN A., BENNETT W.R., HENRIOTT D.R., "Population inversion and continuous optical maser oscillation in a gas discharge containing a he-ne mixture", *Phys. Rev. Lett.*, vol. 63, pp. 106–110, 1961.

[JAY 63] JAYNES E.T., CUMMINGS F.W., "Comparison of quantum and semiclassical radiation theories with application to the beam maser", *Proc. IEEE*, vol. 51, p. 81, 1963.

[JAY 05] JAYASANKAR C.K., VENKATRAMU V., BABU P. *et al.*, "High-pressure fluorescence study of Sm^{3+}-doped borate and fluoroborate glasses", *J. Appl. Phys.*, vol. 97, p. 093523, 2005.

[JEL 66a] JELLISON JR. G.E., "The calculation of thin film parameters from spectroscopic ellipsometry data", *Thin Solid Films*, vol. 290, pp. 40–45, 1966.

[JEL 66b] JELLISON JR. G.E., MODINE F.A., "Parameterization of the optical functions of amorphous materials in the interband region", *App. Phys. Lett.*, vol. 69, pp. 371–373, 1966.

[JON 41] JONES R., "A new calculus for the treatment of optical systems I. Description and discussion of the calculus", *J. Opt. Soc. Am.*, vol. 31, p. 488, 1941.

[JUR 16a] JURESCHI C.M., LINARES J., BOULMAALI A. *et al.*, "Pressure and temperature sensors using two spin crossover materials", *Sensors*, vol. 16, pp. 187-1–187-9, 2016.

[JUR 16b] JURESCHI C.M., POTTIER B.L., LINARES J. *et al.*, "Simulation of multi-steps thermal transition in 2D spin-crossover nanoparticles", *Phys. B Condens. Matter*, vol. 486, pp. 160–163, 2016.

[JUR 16c] JURESCHI C.M., LINARES J., DAHOO P.R. *et al.*, "Monte carlo entropic sampling applied to Ising-like model for 2D and 3D systems", *J. Phys. Conf. Ser.*, vol. 738, pp. 012051-1–012051-5, 2016.

[KAC 94] KACHANOV A., CHARVAT A., STOECKEL F., "Solid state intracavity laser absorption spectroscopy spectro temporal transient behavior of a Ti:Al2O3 laser", *Journal de physique*, vol. 4, (C4), pp. 619–624, 1994.

[KAS 50] KASTLER A., "Quelques suggestions concernant la production optique et la détection optique d'une inégalité de population des niveaux de quantification spatiale des atomes. Application à l'expérience de Stern et Gerlach et à la résonance magnétique", *J. Phys. Rad.*, vol. 11, p. 155, 1950.

[KAY 04] KAYMAZ I., MCMAHON C.A., "A probabilistic design system for reliability-based design optimization", *Struct. Multidisc. Optimi.*, vol. 28, pp. 416–426, 2004.

[KEI 12] KEILMANN F., AMARIE S., "Mid-infrared frequency comb spanning an octave based on an Er fiber laser and difference-frequency generation", *J. Infrared Milli TeraHz Waves*, vol. 33, pp. 479–484, 2012.

[KEL 72] KELLER R.A., ZALEWSKI F., PETERSON N.C., "Enhancement of absorption spectra by dye-laser quenching II", *J.O.S.A.*, vol. 62, pp. 319–326, 1972.

[KHA 04] KHARMANDA G., EL HAMI A., OLHOFF N., "Global reliability based design optimization", *Front. Glob. Optimi.*, vol. 255, 2004.

[KHA 14] KHARMANDA G., IBRAHIM M.-H., ABO AL-KHEER A. *et al.*, "Reliability-based design optimization of shank chisel plough using optimum safety factor strategy", *Comput. Electron. Agricul.*, vol. 109, pp. 162–171, 2014.

[KHE 14] KHETTAB M., Etude de l'influence du résinage au niveau de l'IML (Insulated Metal Leadframe), dans le packaging de module commutateur de courant mécatronique, PhD thesis, UVSQ, Versailles, 2014.

[KIT 96] KITTEL C., *Introduction to Solid State Physics.*, John Wiley & Sons, New York, 1996.

[KLE 92] KLEIBER M., HIEN T.D., *The Stochastic Finite Element Method*, John Wiley & Sons, New York, 1992.

[KOH 65] KOHN W.S., SHAM L.J., "Self-consistent equations including exchange and correlation effects", *Phys. Rev. A.*, vol. 140, pp. 1133–1138, 1965.

[KOL 97] KOLSKI C., *Interfaces Homme Machine*, Hermes, Paris, 1997.

[KON 83] KONO H., LIN S.H., "Spectroscopic properties and relaxation processes of impurity molecules in solids. I. Rotational spectra", *J. Chem. Phys.*, vol. 78, pp. 2607–2620, 1983.

[KOR 06] KOROBALEV O., BERTAUX FEDOROVA A., FONTEYN D. *et al.*, "SPICAM IR acousto-optic spectrometer experiment on Mars Express", *J. Geophys. Res.*, vol. 111, pp. 09S03, 1–17, 2006.

[KRO 85] KROTO H.W., HEATH J.R., O'BRIEN S.C. *et al.*, "C_{60}: Buckminsterfullerene", *Nature*, vol. 318, pp. 162–163, 1985.

[LAB 86] LABANI B., Elargissement collisionnel des raies de vibration-rotation de molécules toupies asymétriques : application à H_2O, Thesis, UFC, Besançon, 1986.

[LAK 84] LAKHLIFI A., GIRARDET C., "Interaction potential for symmetric top-diatom or atom pairs", *J. Mol. Struct.*, vol. 110, pp. 73–96, 1984.

[LAK 87a] LAKHLIFI A., Etude théorique de la molécule d'ammoniac piégée en matrice : interactions, mouvements et phénomènes relaxationnels, PhD thesis, Université de Franche-Comté, Besançon, 1987.

[LAK 87b] LAKHLIFI A., GIRARDET C., "Vibrational relaxation of ammonia trapped in rare gas matrices", *J. Chem. Phys.*, vol. 87, p. 4559, 1987.

[LAK 93] LAKHLIFI A., GIRARDET C., DAHOO R. *et al.*, "Interpretation of the infrared spectrum of ozone trapped in inert matrices", *Chem. Phys.*, vol. 177, p. 31, 1993.

[LAK 00] LAKHLIFI A., CHABBI H., DAHOO P.R. *et al.*, "A theoretical study of CO_2 and N_2O molecules trapped in an argon matrix: Vibrational energies, and transition moments for low-lying levels and IR bar spectra", *Eur. Phys. J. D.*, vol. 12, p. 435, 2000.

[LAK 11] LAKHLIFI A., DAHOO P.R., "Enhanced localization of tunneling of the NH_3 umbrella mode in nano-cage fullerene", *Chemi. Phys.*, vol. 386, pp. 73–80, 2011.

[LAK 12] LAKHLIFI A., DAHOO P.R., DARTOIS E. *et al.*, "Modeling IR spectra of CO_2 isotopologues and CH_4 trapped In type I clathrate", *Colloquium of the CNRS Interdisciplinary Initiative "Planetary Environments and Origins of Life"*, Paris, France, November 2012.

[LAK 15] LAKHLIFI A., DAHOO P.R., PICAUD S. *et al.*, "A simple van't Hoff law for calculating Langmuir constants in clathrate hydrates", *Chem. Phys.*, vol. 448, pp. 53–60, 2015.

[LAL 97] LALLOUET A., "DP-LOG : un langage logique data-parallèle", *Actes des 6 journées francophones de programmation logique et programmation par contraintes JFPLC'97*, Orléans, Hermes-Lavoisier, Paris, pp. 53–68, May 26–28, 1997.

[LAN 66] LANDAU L., LIFCHITZ E., *Théorie des Champs*, Editions Mir, Moscow, 1966.

[LAN 89] LANDAU L., LIFCHITZ E., PITAEVSKI L. *et al.*, *Electrodynamique Quantique*, Editions Mir, Moscow, 1989.

[LEG 77] LEGAY F., *Chemical and Biochemical Applications of Lasers*, vol. 2, MOORE C.B. (ed.), New York, 1977.

[LEH 96] LEHMANN K.K., Ring-down cavity spectroscopy cell using continuous wave excitation for trace species detection, US Patent Number 5, 28,040. June 18, 1996.

[LEI 11] LEINDECKER N., MARANDI A., BYER VODOPYANOV K.L., "Broadband degenerate OPO for mid-infrared frequency comb generation", *Opt. Express.*, vol. 19, pp. 6296–6302, 2011.

[LEV 44] LEVENBERG K., "A method for the solution of certain non-linear problems in least squares", *Quart. Appl. Mathe.*, vol. 2, pp. 164–168, 1944.

[LEW 41] LEWIS G.N., LIPKIN D., MAGEL T.T., "Reversible photochemical processes in rigid media. A study of the phosphorescent state", *J. Am. Chem. Soc.*, vol. 63, pp. 3005–3018, 1941.

[LIN 80] LIN S.H., *Radiationless Transitions*, Academic Press, New York, 1980.

[LIN 16] LINARES J., CHIRUTA D., JURESCHI C.M. *et al.*, "Simulations of edge effect in 1D spin crossover compounds by atom-phonon coupling model", *J. Phys. Conf. Ser.*, vol. 738, pp. 012052-1–012052-5, 2016.

[LIN 17] LINARES J., ALLAL S.E., DAHOO P.R. *et al.*, "Numerical simulation of a device with two spin crossover complexes: Application for temperature and pressure sensors", *J. Phys. Conf. Ser.*, vol. 936, pp. 012048-1–012048-5, 2017.

[LIN 18] LINARES J., SAHBANI I., DESVOIX F. *et al.*, "Ising-like model applied to a triangular lattice of 2D spin-crossover nanoparticles: Evidence of a re-entrant phase transition", *J. Phys. Conf. Ser.*, vol. 1141, pp. 012073-1–012073-5, 2018.

[LIN 20a] LINARES J., CAZELLES C., DAHOO P.R. *et al.*, "Shape, size, pressure and matrix effects on 2D spin crossover nanomaterials studied using density of states obtained by dynamic programming", *Comput. Mater. Sci.*, vol. 187, pp. 110061-1,110061-9, 2020.

[LIN 20b] LIND A.J., KOWLIGY A.S., TIMMERS H. *et al.*, "Mid-infrared frequency comb generation and spectroscopy with few-cycle pulses and χ (2) Nonlinear optics", *Phys. Rev. Lett.*, vol. 124, p. 133904, 2020.

[LIT 67] LITOVITZ T.A., "Theory of ultrasonic thermal relaxation times in liquids", *J. Chem. Phys.*, vol. 26, p. 469, 1967.

[LOE 66] LOEWENSTEIN E.V., "The history and current status of Fourier transform spectroscopy", *Appl. Opt.*, vol. 5, pp. 845–854, 1966.

[LOË 84] LOËTE M., Théorie des intensités absolues des transitions de vibration-rotation des molécules XY_4 application au méthane, PhD thesis, Université de Bourgogne, Dijon, France, 1984.

[LOU 64] LOUISELL W.H., *Radiation and Noise in Quantum Electronics*, 4th ed. McGraw-Hill, USA, 1964.

[LOU 07] LOUTETE-DANGUI E.D., VARRET F., CODJOVI E. *et al.*, "Thermal spin transition in $[Fe(NH_2\text{-}trz)_3]Br_2$ investigated by spectroscopic ellipsometry", *Phys. Rev. B*, vol. 75, pp. 184425-1–184425-7, 2007.

[LOU 08] LOUTETE-DANGUI E.D., CODJOVI E., TOKORO H. *et al.*, "Spectroscopic ellipsometry investigations of the thermally induced first order transition of RbMn $[Fe(CN)_6]$n derivative", *Phys. Rev.*, vol. 78, pp. 014303-1–014303-9, 2008.

[LU 12] LU X., HU Z., "Mechanical property evaluation of single-walled carbon nanotubes by finite element modeling", *Compos. B: Eng.*, vol. 43, pp. 1902–1913, 2012.

[LUG 94] LUGEZ C., Etude en matrice d'argon des réactions élémentaires de l'oxygène atomique avec le méthane, le dichloro(dibromo)méthane et le méthanol. Etude théorique de la dynamique de l'insertion de l'oxygène atomique singulet d dans une liaison OH et dans une liaison CO, PhD thesis, UPMC, Paris, 1994.

[LYN 96] LYNCH R., GAMACHE R., NESHYBA S.P., "Fully complex implementation of the Robert–Bonamy formalism: Half widths and line shifts of H_2O broadened by N_2", *J. Chem. Phys.*, vol. 105, pp. 5711–5721, 1996.

[MAC 95] MACCHI A., MARADUDIN A.A., TOGNETTI V., "Frequency moments of the spectral density of an atomic-displacement time-correlation function of a three-dimensional Lennard-Jones crystal in the classical and quantum regimes", *Phys. Rev. B*, vol. 52, pp. 241–252, 1995.

[MAD 06] MADDALONI P., MALARA P., GAGLIARDI G. *et al.*, "Mid-infrared fiber-based optical comb", *New J. Phys.*, vol. 8, p. 262, 2006.

[MAH 08] MAHIEUX A., BERKENBOSCH S., CLAIRQUIN R. *et al.*, "In-flight performance and calibration of SPICAV SOIR onboard Venus Express", *Appl. Opt.*, vol. 47, pp. 2252–2265, 2008.

[MAH 11] MAHJOUB A., SCIAMMA-O'BRIEN E., DAHOO P.R. *et al.*, "Optical constants of Titan's tholins produced in a RF plasma discharge in different N_2-CH_4 concentrations", *EPSC-DPS Joint Meeting*, October 2011, Nantes, France, 2011.

[MAH 12a] MAHJOUB A., CARRASCO N., DAHOO P.R. *et al.*, "Influence of methane concentration on the optical indices of Titan's aerosols analogues", *Icarus*, vol. 221, pp. 670–677, 2012.

[MAH 12b] MAHJOUB A., CARRASCO N., DAHOO P.R. *et al.*, "Optical properties of organic material produced in a dusty plasma", *ESCAMPIG 2012* (Europhysics Conference on the Atomic and Molecular Physics of Ionized Gases) Jul 2012, Viana do Castelo, Portugal. p. 2, 2012.

[MAH 12c] MAHJOUB A., DAHOO P.R., GAUTHIER T. *et al.*, "Influence of methane concentration on the optical indices of Titan's aerosols analogues", *Second Workshop on Titan Through Time 2012: Unlocking Titan's Past, Present and Future*, Greenbelt, MD,. NASA, p. 95, April 2012.

[MAH 12d] MAHMOUDINEZHAD E., ANSARI R., BASTI A. *et al.*, "An accurate spring-mass model for predicting mechanical properties of single-walled carbon nanotubes", *Comput. Mater. Sci.*, vol. 62, pp. 6–11, 2012.

[MAH 14] MAHJOUB A., CARRASCO N., DAHOO P.R. *et al.*, "Effect of the synthesis temperature on the optical indices of organic materials produced by N_2–CH_4 RF Plasma", *Plasma Process. Polym.*, vol. 11, pp. 409–417, 2014.

[MAI 60] MAIMAN T., "Stimulated optical radiation in ruby", *Nature*, vol. 187, pp. 493–494, 1960.

[MAL 08] MALARA P., MADDALONI P., GAGLIARDI G. *et al.*, "Absolute frequency measurement of molecular transitions by a direct link to a comb generated around 3 µm", *Opt. Express*, vol. 16, pp. 8242–8249, 2008.

[MAN 68] MANNHEIM P.D., "Influence of force-constant changes on the lattice dynamics of cubic crystals with point defects", *Phys. Rev.*, vol. 165, pp. 1011–1018, 1968.

[MAN 71] MANNHEIM P.D., COHEN S.S., "Force-constant changes and the crystal impurity problem", *Phys. Rev. B*, vol. 4, pp. 3748–3756, 1971.

[MAN 72] MANNHEIM P.D., "Localized modes and cell-model limit in the crystal impurity problem", *Phys. Rev. B*, vol. 5, pp. 745–748, 1972.

[MAR 63] MARQUARDT D.W., "An algorithm for least squares estimation of nonlinear parameters", *SIAM J. Appl. Math*, vol. 11, pp. 431–441, 1963.

[MAR 65] MARADUDIN A.A., "The effects of point defects on the vibrations of crystal lattices", *Rep. Progr. Phys.*, vol. 28, pp. 331–380, 1965.

[MAT 89] MATHIS J.S., WHIFFEN G., "Composite interstellar grains", *Astrophys. J.*, vol. 341, pp. 808–822, 1989.

[MAX 54] MAXWELL J.C., *A Treatise on Electricity and Magnetism*, 3rd ed., Dover Publications Inc., New York, 1954.

[MEI 94] MEIJER G., BOOGAARTS G.H.M., JONGMA R.T. *et al.*, "Coherent cavity ring-down spectroscopy", *Chem. Phys. Lett.*, vol. 217, pp. 112–116, 1994.

[MEI 15] MEIS C., *Light and Vacuum*, World Scientific Publishing Co. Pte. Ltd., Singapore, 2015.

[MEJ 08] MEJEAN G., KASSI S., ROMANINI D., "Measurement of reactive atmospheric species by ultraviolet cavity-enhanced spectroscopy with mode-locked femtosecond laser", *Opt. Lett.*, vol. 33, pp. 1231–1233, 2008.

[MEN 20] MENG B., SINGLETON M., SHAHMOHAMMADI M. *et al.*, "Mid-infrared frequency comb from a ring quantum cascade laser", *Optica*, vol. 7, pp. 162–167, 2020.

[MES 64] MESSIAH A., *Mécanique Quantique*, vols 1 and 2, Dunod, Paris, 1964.

[MES 04] MESCHEDE D., *Optics, Light and Lasers*, Wiley-VCH Verlag GmbH & Co., Weinheim, 2004.

[MIC 81] MICHELSON A.A., "The relative motion of the Earth and the luminiferous ether", *Ame. J. Sci.*, vol. 22, pp. 120–129, 1881.

[MIC 82] MICHELSON A.A., "Interference phenomena in a new form of refractometer", *Philos. Mag.*, vol. 13, pp. 236–242, 1882.

[MIL 67] MILLIGAN D.E., JACOX M.E., ABOUAF-MARGUIN L., "Vacuum-ultraviolet photolysis of acetylene in inert matrices: Spectroscopic study of the species C_2", *J. Chem. Phys.*, vol. 46, pp. 4562–4570, 1967.

[MIR 13] MIR F.A., BANDAY J.A., CHONG C. *et al.*, "Optical and electrical characterization of Ni-doped orthoferrites thin films prepared by sol-gel process", *Euro. Phys. J. App. Phys.*, vol. 61, pp. 10302–10305, 2013.

[MIZ 72] MIZUSHIMA M., *Theoretical Physics: From Classical Mechanics to Group Theory of Microparticles*, John Wiley & Sons, New York, 1972.

[MOH 10] MOHSINE A., EL HAMI A., "A robust study of reliability-based optimisation methods under eigen-frequency", *Comput. Methods Appli. Mech. Eng.*, vol. 199, pp. 1006–1018, 2010.

[MON 15] MONDELAIN D., SALA T., KASSI S. *et al.*, "Broadband and highly sensitive comb-assisted cavity ring-down spectroscopy of CO near 1.57 µm with sub-MHz frequency accuracy", *J.Q.S.R.T.*, vol. 154, pp. 35–43, 2015.

[MOO 66] MOORE R., BAKER R., *Introduction to Interval Analysis*, SIAM, Philadelphia, PA, 1966.

[MOR 61] MORET-BAILLY J., Sur l'interprétation des spectres de vibration-rotation des molécules à symétrie tétraédrique ou octaédrique, PhD thesis, Faculté de Paris, Université de Paris, Paris, 1961.

[MOR 68] MOROZOV V.N., NIKITIN V.V., SHERONOV A.A., "Self-synchronization of modes in a GaAs semiconducting injection laser", *Lett. to J.E.T.P.*, vol. 7, pp. 256–258, 1968.

[MOR 77] MORE J., "The levenberg-marquardt algorithm: Implementation and theory, lecture notes in mathematics", *Nume. Anal.*, vol. 630, pp. 104–116, 1977.

[MOR 03] MORISSET S., AGUILLON F., SIZUM M. *et al.*, "Quantum wavepacket investigation of Eley-Rideal formation of H_2 on a relaxing graphite surface", *Chem. Phys. Lett.*, vol. 378, pp. 615–621, 2003.

[MOR 04a] MORISSET S., AGUILLON F., SIZUM M. *et al.*, "Quantum dynamics of H_2 formation on a graphite surface through the Langmuir Hinshelwood mechanism", *J. Chem. Phys.*, vol. 121, pp. 6493–6501, 2004.

[MOR 04b] MORVILLE J., ROMANINI D., KACHANOV A.A. *et al.*, "Two schemes for trace detection using cavity ring-down spectroscopy", *Appl. Phys. B*, vol. 78, pp. 465–476, 2004.

[MOR 05] MORVILLE J., KASSI S., CHENEVIER M. *et al.*, "Fast, low-noise, mode-by-mode cavity-enhanced absorption spectroscopy by diode-laser self-locking", *Appl. Phys. B*, vol. 80, pp. 1027–1038, 2005.

[MOU 98] MOURBAT A., ABOUMAJD A., LOËTE M., "Simultaneous determination of force constants and dipole moment derivatives of methane", *J. Mol. Spectrosc.*, vol. 190, pp. 198–212, 1998.

[MUK 75] MUKAMEL S., JORTNER J., "Lifetimes for resonance fluorescence and near resonance Raman scattering", *J. Chem. Phys.*, vol. 62, pp. 3609–3615, 1975.

[MUL 48] MULLER H., "The foundations of optics", *J. Opt. Soc. Am.*, vol. 38, p. 661, 1948.

[MUR 02] MURAKI N., MATOBA N., HIRANO T. *et al.*, "Determination of thermal stress distribution in a model microelectronic device encapsulated with alumina filled epoxy resin using fluorescence spectroscopy", *Polymer*, vol. 43, pp. 1277–1285, 2002.

[MUS 99] MUSCOLINO G., RICCIARDI N., IMPOLLONIA N., "Improved dynamic analysis of structures with mechanical uncertainties under deterministic input", *Probab. Eng. Mech.*, vol. 15, pp. 199–212, 1999.

[NEG 91] NEGUS D.K., SPINELLI L., GOLDBLATT N. *et al.*, "Sub-100 femtosecond pulse generation by Kerr lens mode-locking in Ti:Al_2O_3", *Adv. Solid-State Lasers*, vol. 10, paper SPL, 1991.

[NEL 65] NELDER J.A., MEAD R., "A simplex method for function minimization", *Comput. J.*, vol. 7, pp. 308–313, 1965.

[NEV 06] NEVEJANS D., NEEFS E., VAN RANSBEEK E. *et al.*, "Compact high-resolution spaceborne echelle grating spectrometer with acousto-optical tunable filter based order sorting for the infrared domain from 2.2 to 4.3 micrometer", *Appl. Opt.*, vol. 45, pp. 5191–5206, 2006.

[NEW 74] NEW G.H.C., "Pulse evolution in mode-locked quasicontinuous lasers", *IEEE J. Quant. Electron.*, vol. 10, pp. 115–124, 1974.

[NGO 12] NGO N.H., TRAN H., GAMACHE R.R. *et al.*, "Pressure effects on water vapour lines: Beyond the Voigt profile", *J. Phil. Trans. R. Soc. A*, vol. 370, pp. 2495–2508, 2012.

[NIT 73] NITZAN A., JORTNER J., "Line shape of a molecular resonance", *Mol. Phys.*, vol. 24, pp. 109–131, 1973.

[NIT 74a] NITZAN A., MUKAMEL S., JORTNER J., "Some features of vibrational relaxation of a diatomic molecule in a dense medium", *J. Chem. Phys.*, vol. 60, p. 3929, 1974.

[NIT 74b] NITZAN A., SILBEY R.J., "Relaxation in simple quantum systems", *J. Chem. Phys.*, vol. 60, p. 4070, 1974.

[NOM 02] NOMURA Y., OCHI S., TOMITA N. *et al.*, "Modelocking in Fabry-Perot semiconductor lasers", *Phys. Rev. A.*, vol. 65, p. 043807, 2002.

[NOU 07] NOUN W., BERINI B., DUMONT Y. *et al.*, "Correlation between electrical and ellipsometric properties on high-quality epitaxial thin films of the conductive oxide LaNiO3 on STO (001)", *J. Appl. Phys.*, vol. 102, pp. 063709/1–063709/7, 2007.

[OHL 00] OHLIDAL I., FRANTA D., "Ellipsometry of thin film systems", *Progr. Opt.*, vol. 41, p. 181, 2000.

[OKE 88] O'KEEFE A., DEACON D.A.G., "Cavity ring-down optical spectrometer for absorption measurements using pulsed laser sources", *Rev. Sci. Instrum.*, vol. 59, pp. 2544–2554, 1988.

[OSA 70] OSAWA E., "Superaromaticity", *Rep. Progr. Phys.*, vol. 25, pp. 854–863, 1970.

[OUR 15] OURIR A., GALLAS B., BECERRA L. *et al.*, "Electromagnetically induced transparency in symmetric planar metamaterial at THz wavelengths", *Photonics*, vol. 2, pp. 308–316, 2015.

[PAK 70] PAKHOMYCHEVA L.A., SVIRIDENKOV E.A., SUCHKOV A.F. *et al.*, "Line structure of generation spectra of lasers with inhomogeneous broadening of the amplification line", *JETP Lett.*, vol. 12, pp. 43–45, 1970.

[PAL 98] PALDUS B.A., HARB C.C., SPENCE T.G. *et al.*, "Cavity-locked ring-down spectroscopy", *JAP*, vol. 83, pp. 3991–3997, 1998.

[PAP 97] PAPOUŠEK D., "Vibrational-rotational spectroscopy and molecular dynamics", *Adv. Ser. Phys. Chem.*, vol. 9, 1997.

[PAP 14] PAPP S.B., BEHA K., DEL'HAYE P. *et al.*, "Microresonator frequency comb optical clock", *Optica*, vol. 1, pp. 10–14, 2014.

[PAR 07] PARTAY L.B., JEDLOVSZKY P., HOANG P.N.M. *et al.*, "Free-energy profile of small solute molecules at the free surfaces of water and ice, as determined by cavity insertion widom calculations", *J. Phys. Chem. C*, vol. 111, pp. 9407–9416, 2007.

[PAU 30] PAULING L., "The rotational motion of molecules in crystals", *Phys. Rev.*, vol. 36, pp. 430–443, 1930.

[PEA 96] PEARSON J., ORR-EWING A.J., ASHFOLD M.N.R. *et al.*, "Dependent linewidths for the (110)-(000) band of the A˜1A″– ˜X 1A′ transition of HNO studied by cavity ring-down spectroscopy", *J. Chem. Soc. Faraday Trans.*, vol. 92, pp. 1283–1285, 1996.

[PEN 99] PENDRY J.B., HOLDEN A.J., ROBINS J. *et al.*, "Magnetism from conductors and enhanced nonlinear phenomena", *IEEE Trans. Trans. Microw. Theory Tech.*, vol. 47, p. 2075, 1999.

[PER 96] PERMOGOROV D., Spectroscopie intracavité laser des transistions du méthane, silane, germane et de leurs dérivés deutérés dans les domaines visible et proche infrarouge, PhD thesis, Université Joseph Fourier, Grenoble, 1996.

[PET 71] PETERSON N.C., KURYLO M.J., BRAUN W. *et al.*, "Enhancement of absorption spectra by dye-laser quenching", *J.O.S.A.*, vol. 61, pp. 746–750, 1971.

[PIC 95] PICKERING C., *Photonic Probes of Surfaces*, Elsevier Science B.V., North Holland, 1995.

[PIC 19] PICQUE N., HÄNSCH T.W., "Frequency comb spectroscopy", *Nat. Photon.*, vol. 13, pp. 146–157, 2019.

[PIM 60] PIMENTEL G.C., "Chapter 4" in BASS A.M., BROIDA H.P. (eds), *Formation and Trapping of Free Radicals*, Academic Press, New York, 1960.

[PIM 62] PIMENTEL G.C., "Matrix technique and its application in the field of chemical physics", *Pure Appl. Chem.*, vol. 4, pp. 61–70, 1962.

[PIM 63] PIMENTEL G.C., CHARLES W., "Ir spectral perturbations in matrix experiments", *Pure Appl. Chem.*, vol. 7, pp. 111–123, 1963.

[PIN 67] PINARD J., "Spectromètre de Fourier à très haute résolution", *J. de physique colloques*, vol. 28, pp. 136–143, 1967.

[PLA 86] PLATEAUX J.J., COURTOIS D., DELAHAIGUE A. *et al.*, "Spectromètre haute résolvance par transformée de Fourier géré par microordinateur", *Revue d'optique appliquée*, vol. 21, pp. 239–244, 1986.

[POI 92] POINCARÉ H., *Théorie mathématique de la lumière*, vol 2, Georges Carré, Paris, 1892.

[POR 50] PORTER G., "Flash photolysis and spectroscopy. A new method for the study of free radical reactions", *Proc. R. Soc. Lond. Ser. A. Math. Phys. Sci.*, vol. 200, pp. 284–300, 1950.

[POR 68] PORTER G., TOPP M.R., "Nanosecond flash photolysis and the absorption spectra of excited singlet states", *Nature*, vol. 220, pp. 1228–1229, 1968.

[POU 15] POUGNET P., DAHOO P.R., ALVAREZ J.L., "Highly accelerated testing" in EL HAMI, A., POUGNET, P. (eds), *Embedded Mechatronic Systems 2: Analyses of Failures, Modeling, Simulation and Optimization*, ISTE Press, London, and Elsevier, Oxford, 2015.

[POW 64] POWELL M.J.D., "An efficient method for finding the minimum of a function of several variables without calculating derivative", *Comput. J.*, vol. 7, pp. 155–162, 1964.

[PRE 86] PRESS W.H., FLANNERY B.P., TEUKOLSKY S.A. *et al.*, *Numerical Recipes in C*, Cambridge University Press, Cambridge, 1986.

[RAD 95] RADUCU V., Etude de complexes faiblement lies en matrice inerte : photodissociation de $CO:O_3$ et changement de conformation de CO : CO_2, PhD thesis, UPMC, Paris, 1995.

[RAG 80] RAGES K., POLLACK J.B., "Titan aerosols: Optical proprieties and vertical distribution", *Icarus*, vol. 41, pp. 119–130, 1980.

[RAH 05] RAHINOV I., GOLDMAN A., CHESKIS S., "Intracavity laser absorption spectroscopy and cavity ring-down spectroscopy", *Appl. Phys.*, vol. 81, pp. 143–149, 2005.

[RAM 02] RAMIREZ S.I., COLL P., DA SILVA A. *et al.*, "Complex refractive index of Titan's aerosol analogues in the 200–900 nm domain", *Icarus*, vol. 156, pp. 515–529, 2002.

[RAN 10] RANNOU P., COURS T., LE MOUELIC S. *et al.*, "Titan aerosols: Optical proprieties and vertical distribution", *Icarus*, vol. 208, pp. 850–867, 2010.

[REE 02] REE J., KIM Y.H., SHIN H.K., "Dynamics of H_2 formation on a graphite surface", *Chem. Phys. Lett.*, vol. 353, pp. 368–378, 2002.

[RHO 68] RHODES C.K., KELLY M.J., JAVAN A., "Collisional relaxation of the $10^0 0$ state in pure CO_2", *J. Chem. Phys.*, vol. 48, p. 5730, 1968.

[ROB 67] ROBERT D., Contribution à la théorie du spectre infrarouge des solutions liquides, UFC Thesis, Besançon, 1967.

[ROB 76] ROBERT D., GALATRY L., "Influence of molecular nonrigidity on the infrared absorption and Raman scattering line shape in dense media", *J. Chem. Phys.*, vol. 64, pp. 2721–2734, 1976.

[ROB 79] ROBERT D., BONAMY J., "Short range force effects in semiclassical molecular line broadening calculations", *J. Phys. Paris*, vol. 40, pp. 923–943, 1979.

[ROM 95] ROMANINI D., LEHMANN K.K., "Cavity ring-down overtone spectroscopy of HCN, $H^{13}CN$ and $HC^{15}N$ overtone bands", *J. Chem. Phys.*, vol. 102, pp. 633–642, 1995.

[ROM 97a] ROMANINI D., KACHANOV A.A., STOECKEL F., "Cavity ring-down spectroscopy", *Chem. Phys. Lett.*, vol. 264, pp. 316–322, 1997.

[ROM 97b] ROMANINI D., KACHANOV A.A., STOECKEL F., "Diode laser cavity ring-down spectroscopy", *Chem. Phys. Lett.*, vol. 270, pp. 538–545, 1997.

[ROS 67] ROSE M.E., *Elementary Theory of Angular Momentum*, Wiley, New York, 1967.

[ROS 88] ROSENMANN L., Etudes théorique et expérimentale de l'élargissement par collisions des raies de CO_2 perturbé par CO_2, H_2O, N_2 et O_2 : constitution d'une base de données infrarouge et Raman appliquée aux transferts thermiques et à la combustion, Thesis, Ecole Centrale, Paris, 1988.

[ROS 98] ROSENBERG E., VINTER B., *Optoélectronique*, Masson, Paris, 1998.

[ROT 45] ROTHEN A., "Reflectometry", *Rev. Sci. Instrum.*, vol. 16, p. 26, 1945.

[SAK 67] SAKHAROV A., "Violation of Cp invariance, C asymmetry, and baryon asymmetry of the universe", *JETP Lett.*, vol. 5, pp. 24–27, 1967.

[SAK 11] SAKURAI J.J., NAPOLITANO J., *Modern Quantum Mechanics*, 2nd ed., Addison-Wesley, Pearson, 2011.

[SAM 68] SAMUELSON H., LEMPICKI A., "Q-switching and mode locking of Nd^{3+}: $SeOCl_2$ liquid laser", *J. Appl. Phys.*, vol. 39, pp. 6115–6116, 1968.

[SAU 66] SAUER P., "On hindered rotation of linear molecules", *Zeitschrift für Physik*, vol. 194, pp. 360–372, 1966.

[SCH 52] SCHWARTZ R.N., SLAWSKY Z.I., HERZFELD K.F., "Calculation of vibrational relaxation times in gases", *J. Chem. Phys.*, vol. 20, p. 1591, 1952.

[SCH 58] SCHAWLOW A.L., TOWNES C.H., "Infrared and optical masers", *Phys. Rev.*, vol. 112, p. 1940, 1958.

[SCH 68] SCHMIDT W., SCHAFER F.P., "Self-mode-locking of dye-lasers with saturable absorbers", *Phys. Lett.*, vol. 26A, pp. 558–559, 1968.

[SCH 95] SCHERER J.J., VOELKEL D., RAKESTRAW D.J. *et al.*, "Infrared cavity ring-down laser absorption spectroscopy (IR-CRLAS)", *Chem. Phys. Lett.*, vol. 245, pp. 273–280, 1995.

[SCH 97] SCHERER J.J., PAUL J.B., O'KEEFE A. *et al.*, "Cavity ring-down laser absorption spectroscopy: History, development, and application to pulsed molecular beams", *Chem. Rev.*, vol. 97, pp. 25–51, 1997.

[SCH 12] SCHLIESSER A., PICQUE N., HÄNSCH T.W., "Mid-infrared frequency combs", *Nat. Photon.*, vol. 6, pp. 440–449, 2012.

[SCI 10] SCIAMMA-O'BRIEN E., DAHOO P.R., HADAMCICK E. *et al.*, "Optical constants determination of complex organic films by spectroscopic ellipsometry from the UV to the mid-IR", *Vth International Conference on Spectroscopic Ellipsometry, (ICSE-V)*, Albany, United States, May 2010.

[SCI 12] SCIAMMA-O'BRIEN E., DAHOO P.R., HADAMCIK E. *et al.*, "Optical constant from 370 nm to 900 nm of Titan tholinsproducted in a low pressure RF plasam discharge", *Icarus*, vol. 218, pp. 356–363, 2012.

[SEC 81] SECROUN C., BARBE A., JOUVE P. *et al.*, "Intensity of the v_3 band of ozone from a study of its anomalous dispersion at 10 µm", *J. Mol. Spec.*, vol. 85, pp. 8–15, 1981.

[SHA 70] SHANNO D.F.A., "Conditioning of quasi-Newton methods for function minimization", *Math. Comp.*, vol. 111, pp. 647–656, 1970.

[SHA 96] SHAH J., *The Ultrafast Spectroscopy of Semiconductors and Semiconductor Nanostructure*, Springer-Verlag, New York, 1996.

[SHU 09] SHUN LIEN C., *Physics of Photonic Devices*, John Wiley & Sons, New York, 2009.

[SIE 86] SIEGMAN A.E., *Lasers*, University Science Books, Sausalito, 1986.

[SIM 00] SIMMONS J.H., POTTER K.S., *Optical Materials*, Academic Press, New York, 2000.

[SLA 05] SLANINA Z., UHLIK F., ADAMOWICZ L. *et al.*, "Computing fullerene encapsulation of non-metallic molecules: $N_2@C_{60}$ and $NH_3@C_{60}$", *Mol. Simul.*, vol. 31, pp. 801–806, 2005.

[SLO 08] SLOAN E.D., KOH C.A., *Clathrate Hydrates of Natural Gases*, 3rd ed., CRC/Taylor & Francis, Boca Raton, 2008.

[SPE 91] SPENCE D.E., KEAN P.N., SIBBETT W., "60-fsec pulse generation from a self-mode-locked Ti-sapphire laser", *Opt. Lett.*, vol. 16, pp. 42–44, 1991.

[STA 67] STATZ H., DE MARS G.A., TANG C.L., "Self-locking of modes in lasers", *J. Appl. Phys.*, vol. 38, pp. 22127–2222, 1967.

[STE 78] STENFIELD J.I., *Laser and Coherence Spectroscopy*, Plenum Press, New York, 1978.

[STO 52] STOKES G.G., "The illumination and polarization of the sunlight sky on Rayleigh scattering", *Trans. Cambridge. Phil. Soc.*, vol. 9, p. 399, 1852.

[STO 66] STOECKEL F. MELIERES M-A., CHEVEVIER M., "Quantitative measurements of very weak H2O absorption lines by time resolved intracavity laser spectroscopy", *J. Chem. Phys.*, vol. 56, pp. 896–910, 1966.

[STO 08] STOWE M.C., THORPE M.J., THORPE M.J. *et al.*, "Frequency comb spectroscopy", in ARIMONDO, E., BERMAN, P.R., LIN, C.C. (eds)., *Advances in Atomic Molecular and Optical Physics*, vol. 55, Kluwer Academic Publishers, Springer, Norwell, MA, 2008.

[SUD 63] SUDARSHAN E.C.G., "Equivalence of semiclassical and quantum mechanical descriptions of statistical light beams", *Phys. Rev. Lett.*, vol. 10, pp. 277–279, 1963.

[SUG 65] SUGDEN T.M., KENNEY C.N., *Microwave Spectroscopy of Gases*, D. Van Nostrand Company Ltd, New York, 1965.

[SUN 68] SUN H.Y., RICE S.A., "Conjecture on the rate of vibrational relaxation of a diatomic molecule in a monatomic lattice", *J. Chem. Phys.*, vol. 42, p. 3826, 1968.

[SWA 62] SWALEN J.D., IBERS J.A., "Potential function for the inversion of ammonia", *J. Chem. Phys.*, vol. 36, pp. 1914–1918, 1962.

[TAG 86] TAGUCHI G., *Introduction to Quality Engineering: Designing Quality into Products and Processes*, ARRB Group Edition, Melbourne, 1986.

[TAN 68] TANGO W.J., LINK J.K., ZARE R.N., "Spectroscopy of K_2 using laser-induced fluorescence", *J. Chem. Phys.*, vol. 49, pp. 4264–4268, 1968.

[TAN 15] TANG X., Contribution à la simulation et l'expérimentation des nanotubes de carbones avec prise en compte des incertitudes, Mechanics PhD thesis, Ecole doctorale Sciences Physiques Mathématiques et de l'Information pour l'ingénieur, Saint-Etienne-du-Rouvray, 2015.

[TAR 65] TARRAGO G., Fréquences des transitions de vibration-rotation des molécules possédant un axe de symétrie ternaire. Calcul des corrections jusqu'au quatrième ordre, PhD thesis, Faculté de Paris, Université de Paris, Paris, 1965.

[TAU 66a] TAUC J., GRIGOROVICI R., VANCU A., "Optical properties and electronic structure of amorphous germanium", *Phys. Stat. Sol.*, vol. 15, pp. 627–637, 1966.

[TAU 66b] TAUC J., *The Optical Properties of Solids*, Academic Press, New York, 1966.

[TIF 67] TIFFANY W.B., MOOS H.W., SCHAWLOW A.L., "Selective laser photocatalysis of bromine reactions", *Science*, vol. 157, pp. 40–43, 1967.

[TOM 99] TOMPKINS H.G., MCGAHAN W.A., *Spectroscopic Ellipsometry and Reflectometry*, Wiley, New York, 1999.

[TOM 05] TOMPKINS H.G., IRENE E.A., *Handbook of Ellipsometry*, Springer, New York, 2005.

[TOW 75] TOWNES C.H., SCHAWLOW A.L., *Microwave Spectroscopy*, Dover Publications, New York, 1975.

[TRA 03] TRAN B.N., JOSEPH J.C., FERRIS J.P. *et al.*, "Simulation of haze formation using a photochemical flow reactor: The optical constants of the polymers", *Icarus*, vol. 165, pp. 379–390, 2003.

[TRO 13] TROTS D.M., KURNOSOV A., BALLARAN T.B. *et al.*, "The Sm:YAG primary fluorescence pressure scale", *Solid Earth*, vol. 118, pp. 5805–5813, 2013.

[TSA 62] TSAO P.J., CURNUTTE B., "Line-widths of pressure-broadened spectral lines", *J. Quant. Spectr. Rad. Transfer*, vol. 2, pp. 41–91, 1962.

[ULV 14] ULVILA V., PHILLIPS C.R., HALONEN L. *et al.*, "High-power mid-infrared frequency comb from a continuous-wave-pumped bulk optical parametric oscillator", *Opt. Express.*, vol. 22, pp. 10535–10543, 2014.

[URB 53] URBACH F., "The long-wavelength edge of photographic sensitivity and of the electronic absorption of solids", *Phys. Rev.*, vol. 92, p. 1324, 1953.

[VAN 29] VAN VLECK J.H., "On sigma-type doubling and electron spin in the spectra of diatomic molecules", *Phys. Rev.*, vol. 33, pp. 467–506, 1929.

[VAS 03] VASSEROT A.M., GAUTHIER ROY B., CHABBI H. *et al.*, "Spectroscopic studies of $^{13}CO_2$ trapped in rare gas matrices", *J. Mol. Spec.*, vol. 220, p. 201, 2003.

[VED 98] VEDAM K., "Spectroscopic ellipsometry: A historical overview", *Thin Solid Films*, vol. 313, p. 1, 1998.

[VER 92] VERNOV A., STEELE W.A., "The electrostatic field at a graphite surface and its effect on molecule-solid interactions", *Langmuir*, vol. 8, pp. 155–159, 1992.

[VES 68] VESELAGO V.G., "Electrodynamics of substances with simultaneously negative values of ε and μ", *Sov. Phy. Uspekhi.*, vol. 10, pp. 509–518, 1968.

[VIA 07] VIAL A., LAROCHE T., "Description of dispersion properties of metals by means of the critical points model and application to the study of resonant structures using the FDTD method", *J. Phys. D: Appl. Phys.*, vol. 40, p. 7152, 2007.

[VIE 73] VIEZEE W., COLLIS R.T.H., LAWRENCE J.J.D., "An investigation of mountain waves with lidar observations", *J. Appl. Meteor.*, vol. 12, pp. 140–148, 1973.

[VIL 08a] VILLARD E., L'instrument SPICAV d'étude de l'atmosphère de Vénus sur la mission venus express : caractérisation instrumentale et observations en vol, PhD thesis, UVSQ, Versailles, 2008.

[VIL 08b] VILLANUEVA G.L., MUMMA M.J., NOVAK R.E. *et al.*, "Identification of a new band system of isotopic CO2 near 3.3 µm: Implications for remote sensing of biomarker gases on Mars", *Icarus*, vol. 195, pp. 34–44, 2008.

[WAL 94] WALTER E., JAULIN L., "Guaranteed characterization domains via set inversion", *Automat. Control Trans.*, vol. 39, pp. 886–889, 1994.

[WAT 68] WATSON J.K.G., "Simplification of the molecular vibration-rotation Hamiltonian", *Mol. Phys.*, vol. 15, pp. 479–490, 1968.

[WEB 71] WEBER M.J., "Luminescence decay by energy migration and transfer: Observation of diffusion-limited relaxation", *Phys. Rev. B*, vol. 4, p. 2932, 1971.

[WHI 96] WHITTET D.C.B., SMITH R.G., ADAMSON A.J. *et al.*, "Interstellar dust absorption features in the infrared spectrum of HH 100-IR: Searching for the nitrogen component of the ices", *Astrophys. J.*, vol. 458, pp. 363–370, 1996.

[WHI 08] WHITENER K.E., FRUNZI M., IWAMATSU S. *et al.*, "Putting ammonia into a chemically opened fullerene", *J. Am. Chem. Soc.*, vol. 130, pp. 13996–13999, 2008.

[WHI 09] WHITENER K.E., CROSS R.J., SAUNDERS M. *et al.*, "Methane in an open-cage [60]fullerene", *J. Am. Chem. Soc.*, vol. 131, pp. 6338–6339, 2009.

[WHI 10] WHITENER K.E., "Theoretical studies of CH_4 inside an open-cage fullerene: Translation-rotation coupling and thermodynamic effects", *J. Phys. Chem. A*, vol. 114, pp. 12075–12082, 2010.

[WIL 36] WILSON E.B., HOWARD J.B., "The vibration-rotation energy levels of polyatomic molecules. I. Mathematical theory of semirigid asymmetrical top molecules", *J. Chem. Phys.*, vol. 4, pp. 262–267, 1936.

[WOL 06] WOLFRAM T., ELLIALTIOGLU S., *Electronic and Optical Properties of D-Band Perovskites*, Cambridge University Press, Cambridge, 2006.

[WON 10] WONG S.T., VODOPYANOV K.L., BYER R.L., "Self-phase-locked divide-by-2optical parametric oscillator as a broadband frequency comb source", *J.O.S.A. B*, vol. 27, pp. 876–882, 2010.

[WOO 68] WOOD R., SCHWARZ S.E., "Passive mode locking of a CO2 laser", *Applied Phys. Lett.*, vol. 12, pp. 263–265, 1968.

[WOO 72] WOOTEN F., *Optical Properties of Solids*, Academic Press, New York, 1972.

[WOO 00] WOOLLAM J.A., "Ellipsometry, variable angle spectroscopic", in WEBSTER J.G. (ed.), *Wiley Encyclopedia of Electrical and Electronics Engineering*, John Wiley & Sons, New York, 2000.

[YAR 65] YARIV A., "Internal modulation in multimode laser oscillators", *Appl. Phys.*, vol. 36, pp. 388–391, 1965.

[YAR 84] YARIV A., YEH P., *Optical Waves in Crystals*, John Wiley & Sons, New York, 1984.

[YE 10] YE J., CUNDIFF T., *Femtosecond Optical Frequency Comb: Principle, Operation, and Applications*, Kluwer Academic Publishers, Springer, Norwell, MA, 2010.

[YOS 71] YOSHIDA Z., OSAWA E., "Aromaticity", *Rep. Progr. Phys.*, vol. 28, pp. 174–178, 1971.

[ZAD 65] ZADEH L., "Fuzzy sets and systems", *Proc. of the Symposium on System Theory*, Polytechnic Press of Polytechnic Institute of Brooklyn, New York, 1965.

[ZAR 12] ZARE R., "My life with LIF: A personal account of developing laser-induced fluorescence", *Annu. Rev. Analy. Chem.*, vol. 5, pp. 1–14, 2012.

[ZEW 00] ZEWAIL A., "Femtochemistry: Atomic-scale dynamics of the chemical bond", *J. Phys. Chem. A*, vol. 104, pp. 5660–5694, 2000.

[ZON 85] ZONDY J.J., GALAUP J.P., DUBOST H., "Coherent pulsations in the vibrational fluorescence of CO and NO in solid N_2", *J. Luminescence.*, vol. 38, p. 255, 1985.

[ZON 86] ZONDY J.J., Fluorescence IR de $^{13}C^{18}O$ en matrice induite par un laser à centres colorés : émission stimulée vibrationnelle et pompage anharmonique V-V des très hauts niveaux ($v \leq 41$), Science PhD thesis, UPSUD, Orsay, 1986.

[ZUM 78] ZUMOFEN G., "*A priori* rates of phonon-assisted *v–v* transfers among CO and N_2 molecules in N_2 matrices", *J. Chem. Phys.*, vol. 69, p. 4264, 1978.

[ZWA 61] ZWANZIG R.W., "Theory of vibrational relaxation in liquids", *J. Chem. Phys.*, vol. 34, p. 1931, 1961.

[ZWE 98] ZWEIFEL H., *Stabilisation of Polymeric Materials*, Berlin, Springer, 1998.

Index

F, G

H, I

L, M

N, O

P, R

S, T, V

Other titles from

iSTE

in

Waves

2021

DAHOO Pierre-Richard, LAKHLIFI Azzedine
Infrared Spectroscopy of Symmetric and Spherical Top Molecules for Space Observation 1 (Infared Spectroscopy Set – Volume 3)

2020

DANIELE Vito G., LOMBARDI Guido
Scattering and Diffraction by Wedges 1: The Wiener-Hopf Solution - Advanced Applications (Waves and Scattering Set – Volume 1)

Scattering and Diffraction by Wedges 2: The Wiener-Hopf Solution - Advanced Applications (Waves and Scattering Set – Volume 2)

SAKHO Ibrahima
Introduction to Quantum Mechanics 2: Wave-Corpuscle, Quantization & Schrödinger's Equation

2019

BERTRAND Pierre, DEL SARTO Daniele, GHIZZO Alain
The Vlasov Equation 1: History and General Properties

DAHOO Pierre-Richard, LAKHLIFI Azzedine
Infrared Spectroscopy of Triatomics for Space Observation
(Infrared Spectroscopy Set – Volume 2)

RÉVEILLAC Jean-Michel
Electronic Music Machines: The New Musical Instruments

ROMERO-GARCIA Vicente, HLADKY-HENNION Anne-Christine
Fundamentals and Applications of Acoustic Metamaterials: From Seismic to Radio Frequency
(Metamaterials Applied to Waves Set – Volume 1)

SAKHO Ibrahima
Introduction to Quantum Mechanics 1: Thermal Radiation and Experimental Facts Regarding the Quantization of Matter

2018

SAKHO Ibrahima
Screening Constant by Unit Nuclear Charge Method: Description and Application to the Photoionization of Atomic Systems

2017

DAHOO Pierre-Richard, LAKHLIFI Azzedine
Infrared Spectroscopy of Diatomics for Space Observation
(Infrared Spectroscopy Set – Volume 1)

PARET Dominique, HUON Jean-Paul
Secure Connected Objects

PARET Dominque, SIBONY Serge
Musical Techniques: Frequencies and Harmony

RÉVEILLAC Jean-Michel
Analog and Digital Sound Processing

STAEBLER Patrick
Human Exposure to Electromagnetic Fields

2016

ANSELMET Fabien, MATTEI Pierre-Olivier
Acoustics, Aeroacoustics and Vibrations

BAUDRAND Henri, TITAOUINE Mohammed, RAVEU Nathalie
The Wave Concept in Electromagnetism and Circuits: Theory and Applications

PARET Dominique
Antennas Designs for NFC Devices

PARET Dominique
Design Constraints for NFC Devices

WIART Joe
Radio-Frequency Human Exposure Assessment

2015

PICART Pascal
New Techniques in Digital Holography

2014

APPRIOU Alain
Uncertainty Theories and Multisensor Data Fusion

JARRY Pierre, BENEAT Jacques N.
RF and Microwave Electromagnetism

LAHEURTE Jean-Marc
UHF RFID Technologies for Identification and Traceability

SAVAUX Vincent, LOUËT Yves
MMSE-based Algorithm for Joint Signal Detection, Channel and Noise Variance Estimation for OFDM Systems

THOMAS Jean-Hugh, YAAKOUBI Nourdin
New Sensors and Processing Chain

TING Michael
Molecular Imaging in Nano MRI

VALIÈRE Jean-Christophe
Acoustic Particle Velocity Measurements using Laser: Principles, Signal Processing and Applications

VANBÉSIEN Olivier, CENTENO Emmanuel
Dispersion Engineering for Integrated Nanophotonics

2013

BENMAMMAR Badr, AMRAOUI Asma
Radio Resource Allocation and Dynamic Spectrum Access

BOURLIER Christophe, PINEL Nicolas, KUBICKÉ Gildas
Method of Moments for 2D Scattering Problems: Basic Concepts and Applications

GOURE Jean-Pierre
Optics in Instruments: Applications in Biology and Medicine

LAZAROV Andon, KOSTADINOV Todor Pavlov
Bistatic SAR/GISAR/FISAR Theory Algorithms and Program Implementation

LHEURETTE Eric
Metamaterials and Wave Control

PINEL Nicolas, BOURLIER Christophe
Electromagnetic Wave Scattering from Random Rough Surfaces: Asymptotic Models

SHINOHARA Naoki
Wireless Power Transfer via Radiowaves

TERRE Michel, PISCHELLA Mylène, VIVIER Emmanuelle
Wireless Telecommunication Systems

2012

LALAUZE René
Chemical Sensors and Biosensors

LE MENN Marc
Instrumentation and Metrology in Oceanography

LI Jun-chang, PICART Pascal
Digital Holography

2011

BECHERRAWY Tamer
Mechanical and Electromagnetic Vibrations and Waves

BESNIER Philippe, DÉMOULIN Bernard
Electromagnetic Reverberation Chambers

GOURE Jean-Pierre
Optics in Instruments

GROUS Ammar
Applied Metrology for Manufacturing Engineering

LE CHEVALIER François, LESSELIER Dominique, STARAJ Robert
Non-standard Antennas

2010

BEGAUD Xavier
Ultra Wide Band Antennas

MARAGE Jean-Paul, MORI Yvon
Sonar and Underwater Acoustics

2009

BOUDRIOUA Azzedine
Photonic Waveguides

BRUNEAU Michel, POTEL Catherine
Materials and Acoustics Handbook

DE FORNEL Frédérique, FAVENNEC Pierre-Noël
Measurements using Optic and RF Waves

FRENCH COLLEGE OF METROLOGY
Transverse Disciplines in Metrology

2008

FILIPPI Paul J.T.
Vibrations and Acoustic Radiation of Thin Structures

LALAUZE René
Physical Chemistry of Solid-Gas Interfaces

2007

KUNDU Tribikram
Advanced Ultrasonic Methods for Material and Structure Inspection

PLACKO Dominique
Fundamentals of Instrumentation and Measurement

RIPKA Pavel, TIPEK Alois
Modern Sensors Handbook

2006

BALAGEAS Daniel *et al.*
Structural Health Monitoring

BOUCHET Olivier *et al.*
Free-Space Optics

BRUNEAU Michel, SCELO Thomas
Fundamentals of Acoustics

FRENCH COLLEGE OF METROLOGY
Metrology in Industry

GUILLAUME Philippe
Music and Acoustics

GUYADER Jean-Louis
Vibration in Continuous Media

Printed and bound by CPI Group (UK) Ltd, Croydon, CR0 4YY

13/09/2026

14961036-0008